Telecommunications Deregulation and the Information Economy

Second Edition

For a listing of recent titles in the *Artech House Telecommunications Library,*
turn to the back of this book.

Telecommunications Deregulation and the Information Economy

Second Edition

James K. Shaw

Artech House
Boston • London
www.artechhouse.com

Library of Congress Cataloging-in-Publication Data
Shaw, James K., 1952–
    Telecommunications deregulation and the information economy / James K. Shaw.—
2nd ed.
      p.   cm.—(Artech House telecommunications library)
      1st ed. published under title: Telecommunications deregulation. 1998.
      Includes bibliographical references and index.
      ISBN 1-58053-276-4 (alk. paper)
      1. Telecommunication—Deregulation—United States.   2. Telecommunication
policy—United States.   3. Telecommunication—Information services—Economic
aspects—United States.   I. Shaw, James, 1952–   Telecommunications deregulation.
II. Title.   III. Series.
HE7781 .S53   2001
384′.041—dc21

                                 2001022206

British Library Cataloguing in Publication Data
Shaw, James K.
    Telecommunications deregulation and the information economy.—2nd ed.—
(Artech House telecommunications library)
    1. Telecommunication—United States   2. Telecommunication—
Deregulation—United States.   3. Telecommunication policy—United States
I. Title
384′.043′0973

    ISBN 1-58053-276-4

Cover design by Gary Ragaglia

© 2001 ARTECH HOUSE, INC.
685 Canton Street
Norwood, MA 02062

International Standard Book Number: 1-58053-276-4
Library of Congress Catalog Card Number: 2001022206

10 9 8 7 6 5 4 3 2 1

For Anne

Contents

Preface

Shortly after the passage of the Telecommunications Act in 1996, I was working as a corporate consultant in telecommunications price forecasting. The day after President Clinton signed this legislation into law, I was having lunch with a number of professionals drawn from marketing, sales, engineering, regulatory affairs, and strategic planning. Not surprisingly, everyone had something to say about this landmark legislation. I made a point of listening carefully to my colleagues, studying their expressions, and displaying, I thought, the requisite empathy they sought in their consultant. It was at that moment that I decided to write *Telecommunications Deregulation,* published in 1997.

This book, *Telecommunications Deregulation and the Information Economy,* is the second edition of the initial publication. In this volume I have attempted to broadly address the suggestions and criticisms associated with the first edition. *Telecommunications Deregulation* has been adopted at 30 universities since 1998, and I have received a number of comments from academics in engineering, business, and economics since the book's publication. Other critiques have been gleaned from industry professionals and consultants.

Two overriding suggestions have been made with respect to the first edition of this research. One request has focused on an expansion of economic and strategic analysis associated with the 1996 statute. I have complied by adding an extensive array of economic data, domestic and global, highlighting the growing interaction between telecommunications innovation and economic growth. A second suggestion has stressed the need to reevaluate the competitive environment in light of deregulatory forces. The first edition addressed, however nominally, this dimension of deregulation in its earliest phase. We now have empirical evidence dating back four years and are thus in a better position to examine this vital area of inquiry.

Reflecting upon these legitimate requests, I have concluded that additional commentary on the emergence of hypercompetitive markets—euphemistically described by some as the "New Economy"—would be especially valuable for industry strategists. It is impossible to address the reconfiguration of the telecommunications industry in the wake of deregulation without assessing the direct impact of networked commerce on the competitive landscape. This feature, too, has been added to this revision. Because the communications industry now acts as the physical infrastructure for our economic system, it might be more appropriate to term this recent metamorphosis as the "Information Economy." The linkage between regulatory reform and our redefined economic environment constitutes the rationale for the change in title.

Proponents of telecommunications deregulation, the underlying spur to this legislation, immediately hailed the revision of the Communications Act of 1934 as the single most important "economic" statute ever passed by Congress. For critics, the law represented, at best, a marginal step in the right direction. Like most things in life, the truth probably rests somewhere between the extremes. I will concede that I am a bit more enthusiastic than many critics in projecting what I believe to be the prospective rewards for regulatory relaxation. My optimism is a function of what I believe to be the indirect benefits associated with the convergence of the telephony, broadcasting, and computer sectors; this convergence has been given significant impetus via deregulation.

I maintain great respect for my colleagues, both advocates and critics of communications reform. I counsel patience to all: telecommunications deregulation is a work in progress, and its effect cannot be accurately calibrated for some years to come. Presently, there is little consensus about the fate of the global communications industry, but substantive inferences will be drawn in the years immediately ahead. We are poised on the cutting edge of industry rivalry, as key features of the 1996 Act approach their blast effect.

It is hoped that this modest effort clarifies the principal strategic options open to professionals as this decade unfolds.

Acknowledgments

I am indebted to countless students who have enrolled in advanced graduate coursework in which the first edition of this book, *Telecommunications Deregulation*, has been employed. Apart from the reactions of students assigned this text at the University of San Francisco, Saint Mary's College, and the University of California, Berkeley Extension, I have received valuable commentary from students and faculty at other institutions in both the United States and Western Europe. A serious effort has been made to incorporate suggestions forwarded by these observers, particularly with regard to the linkage between deregulation and strategy in corporate enterprises. I also extend my appreciation to Dr. Edward Yardeni, who provided access to an extensive array of economic databases; this research is reflected in the Appendixes of this book.

I am grateful to Anne, my wife, for her unwavering support and cajoling presence. I also wish to thank the administration at the College of Professional Studies, University of San Francisco, and particularly Dean Larry Brewster, for his flexibility and farsightedness in advancing projects of this kind. I also wish to thank Tom Rauch and Quentin Burrows, colleagues at the University of California, Berkeley, for their encouragement and assistance.

James K. Shaw
Lafayette, California
April 2001

1

Telecommunications Deregulation: An Era Begins

The passage of the Telecommunications Act of 1996 has spawned a new era in the communications industry, a period that may well transform both the domestic and international economies. The enactment of this deregulatory legislation represents one of the most important economic decisions ever produced by Congress. In removing or diminishing regulatory constraints in the telecommunications industry, government has simultaneously introduced profound structural change to all sectors of American business. A new era has unfolded in the global economy—one promoting new businesses, receding passive enterprises, and restructuring veteran public, nonprofit, and private organizations.

A number of technological, economic, and political forces have impelled deregulation of the communications industry. The interaction of these forces has altered the mind-set of both American and global regulators and, in doing so, energized engineers, scientists, corporate strategists, and entrepreneurs on an international scale. In the immediate period following the enactment of this critical legislation, we are witnessing spectacular growth in new business formations, strategic alliances and partnerships, as well as mergers and acquisitions. No one can estimate with precision the long-term social, cultural, political, and economic consequences associated with deregulation; one can only be assured that the structure and delivery of public and private services are hereafter permanently realigned. It is the transformation of competitive behavior in the telecommunications industry that this book examines.

The initial chapter of the text examines the historical dynamics dictating telecommunications reform, identifies the macroeconomic significance of communications for the domestic and global economies, elaborates the rise of intellectual capital resulting from regulatory relaxation, and lists those industries expanding, receding, or restructuring under terms of revitalized competition. Two caveats should be revealed as we proceed to a serious discussion of tele-communications deregulation: (1) American and global regulatory reform will unfold on a *gradual*, systematic basis, and therefore the full impact of telecommunications competition will not be felt until after 2002; and (2) we must eschew the instinct to measure business adaptation to new communication technologies merely in terms of acquisition and application of these tools. Rather, we must broadly evaluate the tertiary implications of such change to project future business opportunities, profitability, and consumer adoption rates. With these admonishments in mind, we move to the specifics of the how, why, when, and where of telecommunications reform.

1.1 Historical Overview of Telecommunications Deregulation

The building blocks of communications regulation and the direct intervention of the Federal Communications Commission (FCC) in applying such policy evolved in the years following 1907. This evolution is described in detail in Chapter 2. The historical forces prompting telecommunications deregulation unfolded during the two decades following World War II [1]. In an effort to deliberately provoke AT&T, the nation's communications monopoly, several firms innovated new equipment intended to network with the larger infrastructure. In a key ruling in 1968, the FCC decided that one of these innovations— the Carterfone—could be networked with the Bell system without unilaterally being disconnected by the monopoly. As a result, this decision superseded interstate tariffs applied to customer-owned equipment connected to AT&T. Although viewed as a relatively inconsequential ruling at the time, this regula-tory decision had two significant repercussions: (1) small firms now had an incentive to innovate new telephone equipment of several kinds, and (2) a precedent had been established for relaxed government regulation of the com-munications sector [2].

A number of technological breakthroughs led the FCC to continue its policy of permitting new business entrants to promote hardware innovations. Shortly after the Carterfone decision, Microwave Communications, Inc. was allowed to construct a microwave system that facilitated private telephone com-munications in business. This system was allowed interconnection with the Bell system [2]. The precedent had been established for specialized private carriers to offer quasi-telephone communications. With this ruling, the rever-berations felt because of the Carterfone decision were reinforced. More importantly, although the impact of these decisions could not be fully foreseen at the time, the trickling momentum toward eventual deregulation was encouraged.

Similar decisions in 1974, 1977, and 1980 paved the way for continued technical innovation and enhanced competition in the telecommunications industry. The single most vital development in the movement toward eventual deregulation was a lawsuit filed by MCI in 1974. This action, filed on antitrust grounds, asserted that the Bell system had illegally restrained marketplace competition. The historic rationale for the Bell monopoly had always pivoted on the economic principle that the consumer was best served by a single networked system that applied universal service throughout the country. Yet, by the 1970s and 1980s it had become unmistakable that many new innovative technologies that had immediate consumer and business applications were no longer purely the product of Bell ingenuity. New firms were now seeking access to vast consumer markets.

AT&T had sought to delay a final decision by the Justice Department and circuit courts, but by 1981 it became evident that the monopoly could not prevent an eventual breakup of its system. Judge Harold Greene orchestrated an agreement by which AT&T would eventually divest itself of key elements of its system; it was permitted to retain its research labs, long-distance business, and Western Electric facilities. This first step toward complete divestiture also held a carrot for the giant; if regulation had historically insulated AT&T from competitive threat, the Modified Final Judgment (MFJ) also provided entry into new markets, particularly with respect to the computer and ancillary industries. In time, as outlined in the next chapter, AT&T and other large communications enterprises were permitted to enter essentially any and every market, fueling competition throughout the entire economy.

From the consumer's viewpoint, unfolding deregulation since 1984 has generally meant falling prices without significant diminution of quality [2]. Quality and service have actually improved in several areas of telecommunications delivery. In long-distance delivery, 500 companies compete aggressively for new customers. In telecommunications hardware deployment, a half-dozen providers now compete in earnest with Lucent Technologies. In data transmission and Internet access, dozens of providers are available to consumers and organizations regardless of region or location. Satellite delivery of voice, image, and data has been measurably enhanced during the past several years. The newly reconfigured Regional Bell Operating Companies (RBOCs), separated from AT&T after 1984, are now confronting intense competition from both large and small firms. One might refer to the dozen years immediately following divestiture as the cutting edge of deregulation; we are now only beginning to experience the longer-term effects of this strategy. Full deregulation and resultant competition will only emerge shortly after 2002, as the final federal and state regulatory restraints are relaxed or eliminated [3]. As one contemplates the future, what emerges in this uncertain era is a series of challenges, opportunities, and pitfalls. Strategic planners face the competitive dynamics of primary, secondary, and tertiary markets concurrent with a multiplicity of expanding services.

1.2   Social Dimensions of Telecommunications Deregulation

1.2.1   Telecommunications Deregulation and Organizational Response

Unparalleled opportunities and challenges, with attendant risks, now define the communications environment. We begin first by isolating the key players transforming the industry and whose efforts now instigate organizational redesign. These companies can be categorized as follows:

- *Mega-enterprises:* billion-dollar corporations that leverage capital to build, integrate, and sustain infrastructure;
- *Entrepreneurial enterprises:* formed by individuals, partnerships, or alliances, such companies aim to create or penetrate a market base that previously was the sole domain of established corporations;
- *Intrapreneurial firms:* new business entrants formed within corporate structures that seek to innovate new products and services by drawing upon the capital and expertise of their hosts;
- *Regional corporations:* small and large, these business entities seek entry to both national and international markets through a variety of tactics, including acquisitions, mergers, partnerships, and alliances;
- *Convergent firms:* can only deliver a product through the complementary expertise of a partner ancillary to the firm's originating mission;
- *Spin-offs:* former elements of parent corporations, managed as independent enterprises to enhance their productivity and creative initiative.

For purposes of this classification, AT&T, MCI, Sprint, and other long-distance firms would fit comfortably in the first category, though it is essential to emphasize the evolving role of corporations: as each year unfolds in the deregulatory environment, these companies seek access to any and all markets, and thus their image, reputations, and missions remain in a state of flux. With respect to the second category, emerging enterprises such as Pixar, 3Com,

Next, Oracle, and others—whose origins may or may not fall within the domain of telecommunications—are now looking upon vast communications markets with great appetite.

Intrapreneurial firms typically start within the framework of a corporate umbrella, drawing capital and in-house expertise to promulgate new products to market. Often, these firms are eventually spun off from the corporate parent and seek an independent role in defining a destiny. Because original corporate backing resulted in a substantial equity interest at the point of "spin-off," these entities still maintain obligations to the parent. Regional telecommunications corporations, presently and chiefly the domain of the RBOCs (SBC, PacTel, Ameritech, US WEST, and Bell South), are devoted in the short run to the maintenance and upgrading of infrastructure but in the long run to the development of emerging technologies.

Convergent firms, now given impetus by the passage of telecommunications reform, are formed through partnerships or alliances. The convergent enterprise reflects the growing reality of the deregulatory environment: no single firm can simultaneously innovate new services while dominating veteran product lines. Figure 1.1 depicts the mechanics of telecommunications convergence. The result is that established and new firms integrate their expertise to promote rapid and competitive entry to market. Often, converging companies combine the talents of a firm operating within the telecommunications industry with those of a firm committed to an ancillary objective. Pacific Bell, for example, launched such an alliance with CAA, a Hollywood talent agency, to fuse content with service delivery [4]. Other communications firms have sought alliances with computer companies, broadcasting affiliates, and cable television companies. In each case, strategic positioning, competitive advantage, and nimble response to market changes have been the catalysts for business convergence.

When large corporations determine that their structures stifle innovation or otherwise limit opportunities to seize new markets, spin-offs can result. A spin-off is a firm whose origins lie within a corporate parent; the parent identifies a factor, such as bureaucratic inertia, that prevents the smaller entity from otherwise thriving. The emerging enterprise is given its independence under a corporate holding umbrella but operates as an independent firm. Such companies, if successful, often are prime candidates for absorption by competitors. Lucent, an AT&T spin-off, assigned stock to its independent status and established an immediate and important communications presence in 1996. As corporations attempt to exploit their own strategic positioning, that is, seize upon new market opportunities, we may project continued decentralization and devolution of organizational structures. In this way, established firms can more flexibly meet the challenges of new business entrants [5].

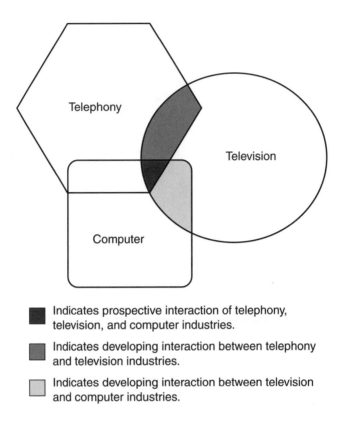

Indicates prospective interaction of telephony, television, and computer industries.

Indicates developing interaction between telephony and television industries.

Indicates developing interaction between television and computer industries.

Figure 1.1 Telecommunications convergence. Area size conforms to market penetration of technologies. Market penetration for the telephony and television industries is 98%. Market penetration for the computer industry approximates 33% in households and exceeds 80% for all businesses.

1.2.2   An Economic Paradigm for Telecommunications Deregulation

We may project three discrete economic stages for regulatory reform based on similar historical experiments in other industries [6]. These stages, or phases, of reform can be divided into short-term (initial five years), mid-term (5–10 years), and long-term (beyond 10 years) time frames and permit us to anticipate the impact of deregulation for the industry. Given deregulatory strategy adopted in the late 1970s and early 1980s—most notably in aviation, trucking, banking, and other commerce—it appears likely that the industry will witness the following.

- *Stage One (short term)*: Rapid expansion in new business creations combined with dramatic efforts to consolidate market share. Stage

One is characterized by a significant increase in mergers, acquisitions, alliances, and partnerships. New business entrants and *initial public offerings* (IPOs) position themselves to innovate new products and create new markets; often, their own long-term objective is to create an enterprise sufficiently attractive to procure a buyout from a veteran company.

- *Stage Two (mid-term)*: The wave of mergers and acquisitions begins to slow but remains a prominent strategic feature embraced by corporations in telephony, computer, and cable services. The principal emphasis now is on securing market share and preventing competitors from "seizing" traditional customers. The wave of IPOs also slows, and new business entrants do not typically constitute an immediate strategic threat. The rate of absorption of new firms by older, established corporations accordingly diminishes.

- *Stage Three (long term)*: New business entrants slow to a trickle, and competition declines as larger firms extend consolidation—the practice of absorbing competitors to expand market share—to primary and secondary markets. This stage, extending from two to three decades into the future, often provokes consumer groups to demand re-regulation of the industry so as to mitigate the perceived harmful effects of industry consolidation. The initial intent of deregulation—to maximize competitive benefits—is defeated by the systematic formation of an oligopolistic market environment; a handful of firms now dominates the market with respect to price, service, and maintenance of quality.

Collectively, these three stages of industry restructuring form a paradigm or model of future economic activity. If one accepts the notion of "historical analogy" as a useful device to forecast industry trends and extrapolates the experiences of other sectors that have undergone similar deregulation, these sequential outcomes seem probable. However, it should be noted that the Telecommunications Act of 1996, as outlined in Chapter 2, is so sweeping in its focus and so dramatic in its strategic effect that such a scenario may represent a trivialization of this experiment. Those who view with suspicion the analogy of other industrial deregulation for the communications sector have noted several critical factors that may differentiate telecommunications.

1.2.2.1 Consumer Customization

Unlike the aviation, energy, and transportation industries, the communications industry facilitates the freedom and independence of the customer. In short, in an industry whose rapid innovations provide enhanced choice and encourage creativity, the consumer can customize a menu of services unique to his professional and personal needs. Other industries continue to standardize their product lines; communication providers paradoxically elaborate value-added services to secure market share but, in doing so, build platforms that urge consumers to drop providers that cannot sustain or improve value and service. A cycle of continued value-added improvement, based on technological innovation, may compel consumers to seek only the most reliable firms. Such accelerated innovation is not characteristic of other industries that have recently experienced deregulation.

1.2.2.2   Globalization

Unlike other recently deregulated industries, the telecommunications sector is presently experiencing a relaxation of government restrictions throughout the world (see Chapter 9 for an elaboration of these developments). New cross-country alliances, partnerships, mergers, and acquisitions are unusual in their scope and influence. The long-term global regulatory structure pivots on deregulatory experiments principally engineered in the United States and the United Kingdom, as hereafter evaluated. If this reduction in government intervention is deemed successful by other countries, we may anticipate a wave of new ventures joining entrepreneurs in many nations. This would be an unprecedented development in the history of the global economy and would simultaneously influence the restructuring of other industries dependent on communications infrastructure.

1.2.2.3   The Rise and Advance of Intellectual Capital

Intellectual capital is the economic value that labor contributes to an organization. Communications in its most tangible form (voice, video, or data) is a crucial ingredient in the management and dissemination of information. Information, in turn, underpins intellectual capital—the projection of organizational productivity is directly and proportionately related to the expertise of labor. The three factors of economic production (i.e., finance, labor, and technology) are now interwoven in a way unimaginable a decade earlier. The rise and advance of intellectual capital is a necessary ingredient and precondition for a firm's success; the loss or stagnation of such capital, through job mobility, attrition, or incompetent management, signals a firm's decline. The value of labor is thus inextricably linked to the exploitation of telecommunications infrastructure and the resultant generation, interpretation, and transmission of information. The rise and advance of intellectual capital is unique to the service economy in which we now find ourselves [7]. This factor thus differentiates the communications industry from other sectors already deregulated.

1.2.2.4 FCC Deregulatory Charting

A final factor may differentiate telecommunications from prior deregulatory experiments. Unlike examples heretofore identified, the Federal Communications Commission, charged with the responsibility of deregulating this sector, has promised to monitor and accentuate regulatory relaxation according to market response. In other words, the FCC retains the authority to adjust policy if anticompetitive practices surface. Chairman Reed Hundt stated in 1997 that the FCC would not permit tacit cooperation among telecommunications giants as a substitute for competition [8]:

> There should be no mistake. We want competition, not coopetition. We want a competitive war, not a standoff in which incumbent companies warily eye each other but never really enter each other's markets. . . . We may have the forming of détente between the cable companies and telephone companies. . . . Détente is not what the Telecommunications Act was supposed to be about. . . .

Given the consensus that the White House, Congress, and Judiciary have achieved—a consensus not apparent in prior regulatory reforms—some observers believe that it would be an error to presume that the industry will repeat the experience of other sectors. Telecommunications deregulation, argue these analysts, is a unique experience in the history of business regulation.

1.3 Economic Dimensions of Telecommunications Deregulation

1.3.1 The Macroeconomic Significance of Telecommunications

For reasons elaborated upon in Chapters 3 to 6, unfolding telecommunications deregulation will have a profound future impact on microeconomic analysis—the process by which business sets prices, computes the cost of introducing new product lines, and estimates future profitability. The methodology of applied microeconomics is discussed in the following chapters, but several critical macroeconomic tools—those methods aimed at forecasting the general impact of telecommunications technology on the national and global economies—must be discussed. Ordinarily, macroeconomic analysis, as practiced by economists, is devoted to the evaluation of the nation's money supply, resultant interest rates, employment rates, inflation, and government budgeting. Viewed from the perspective of telecommunications professionals, this traditional definition is myopic; increasingly, business people are concerned about the underlying macroeconomic trends vital to their own industry. In this context, telecommunications development may be construed as both an independent and dependent variable impinging upon future macroeconomic performance. Telecommunications innovation transforms the nation's economy, while macroeconomic trends influence future consumption of telecommunications products and services.

Key macroeconomic developments currently and prospectively influencing demand for telecommunications services include *demographic trends, productivity shifts, consumer spending and savings, personalized customization,* and the *industrial replacement of niche marketing in favor of strategic particle marketing.* These six variables, combined with the traditional macroeconomic concerns of economists, define a scenario gaining credence among strategic planners. The nation appears to be headed for a telecommunications appetite that can broadly be described as requiring communication tools to contend with the emergence of a watershed change in the history of economic development [9].

1.3.2 The Demographic Determinants of Telecommunications Consumption

The demand for telecommunications products and services is the product of both macroeconomic and microeconomic factors. We refer to the factors defining demand for communication services as *determinants.* The demographic determinants of telecommunications consumption, viewed from a macroeconomic perspective, include income, savings, investment, and debt accumulation. These combined determinants influence consumer consumption; and the determinants of business consumption include income, investment, and capital acquisition.

The demographic composition of society constitutes the single most valuable information available to telecommunications planners. As we project American demographics ahead 15 years, we note the following [10]:

1. The Baby Boom generation, born between 1946 and 1964 and nearly 80 million strong, is now entering a period of peak consumption. Two-thirds of the labor force consists of Baby Boomers, and we can anticipate repeated waves of substantial spending extending to at least 2010.

2. Inflation and interest rates may achieve a stability not known since the mid-1960s. Typically, inflation and high interest rates decline as

a disproportionate fraction of the labor force approaches its highest level of productivity.

3. The Baby Boomers are beginning to inherit, and will continue to inherit, vast sums of wealth extending past 2010. Spending capacity will therefore be enhanced significantly over the next 15 years.

These factors auger well for rising consumer and business consumption for telecommunications products. Products aimed at professional advancement, business productivity, education, entertainment, and management of the home form the foundation of future telecommunications demand [11]. Additional applications will be derived from these core industries, thereby stimulating the creation of new companies in tributary markets hardly imaginable today. From a demographic standpoint, it is unmistakable that telecommunications development serves multiple needs in multiple markets at a time when consumers can afford to augment their consumption. The importance of demographic determinants for telecommunications development is elaborated further in Chapter 10, with models introduced to explore alternative business scenarios based on these data.

1.3.3 Productivity and Telecommunications Development

The innovation and development of telecommunications services and products are directly linked to the nation's productivity and competitive position. We may equate productivity (P) with efficiency and assign the following equation to its computation [12]:

$$P = O/R$$

where O = output, and R = resources. Thus, productivity is determined by dividing a firm's output (goods or services) by the resources required to produce that output, which include the costs of labor, capital, and technology. This is an oversimplified view of the productivity process, but it represents a straightforward method of estimating organizational efficiency. One should bear in mind that a firm's relative productivity can rise, fall, or stagnate. For the past 30 years, the rate of improvement in American productivity has generally lagged behind a number of its competitors in the global market.

The macroeconomic significance of productivity lies in the fact that it is directly correlated to a nation's material standard of living: higher rates of productivity are tied to enhanced profitability and resultant higher wages [13]. Equally significant, higher productivity can mitigate inflation while enhancing competitiveness abroad. Should productivity lag, prices may rise and prevent the private sector from elevating wages. Productivity, therefore, defines future consumption, savings, investment, and debt accumulation. Viewed quantitatively, we note the following equation for the estimation of national wealth:

$$GDP = C + I + G + Ex - Im$$

where C = consumer spending, I = business investment, G = government spending, Ex = exports, Im = imports, and GDP (Gross Domestic Product) represents annual national wealth. Productivity is the pivotal element in stimulating higher rates of consumer spending and business investment, both of which determine the long-term economic health of the private sector. Because two-thirds of GDP output is directly related to consumer spending (C), business investment (I) and government spending (G) hinge on shifts in productivity. Business does not invest until the consumer makes economic decisions; government spending is chiefly a function of its ability to tax consumer and business transactions. We note that American productivity was rising by an average annual factor of twice our current level in the 25 years immediately following World War II [14]. Hence, it is not surprising that many Americans have expressed their frustration over a stagnant standard of living [15].

In sum, a rising level of productivity provides the following benefits to the nation's economy:

- Increases standard-of-living via higher wages and benefits;
- Enlarges profitability;
- Improves the nation's competitive position by restraining costs of production;
- Improves exports through competitive pricing;
- Enhances governmental fiscal policy by elevating taxable revenues;
- Restrains inflation and mitigates otherwise higher interest rates.

There is a vested interest by all components of the American economy in identifying new methods of raising productivity. Telecommunications, as earlier noted, plays a critical role in the nation's economy because it prospectively improves a firm's sales position, efficiency, and competitive strategy. Therefore, many anticipate that telecommunications will emerge as the key catalyst in raising future productivity.

It should also be noted, however, that despite advances in information and communications technology, it is a fact that average annual productivity in the United States has not risen appreciably in recent years. The incongruity of advanced technology with stagnant productivity has baffled many forecasters. This phenomenon is referred to as the "productivity paradox." At the same time, one must consider that advanced telecommunications services have yet to penetrate a critical mass of both consumers and business users. Data transmission, Internet usage, advanced wireless applications, and interactive technologies of various kinds remain very much in their infancy but offer a promise of increased efficiency for many occupations. When a market penetration of 90% or more of the population is attained, we will have introduced a new dynamic in the management and transmission of information. The solution to this paradox may well lie in the successful integration of the television, telephony, and computer industries. When both consumers and business people are able to fully exploit the information generation, transmission, retrieval, and storage powers of these technologies, levels of efficiency should rise. It takes time to learn these technologies, integrate them into personal and professional life, identify their appropriate applications, and synthesize information to speed decision making. There is, in short, an ambiance to productivity in the age of telecommunications deregulation: we must experience a period of assimilation with these emerging technologies to understand and exploit their greatest value.

1.3.4 Consumer Spending for Telecommunications Products and Services

Consumer spending for telecommunications products and services is growing exponentially throughout the world. The appetite for communications services is fueled by a growing need in developed nations to generate, monitor, distribute, and interpret information efficiently. Such consumption is positively correlated with the complexity of work and need for enhanced productivity, as earlier cited. However, it is also fueled by an insatiable demand for state-of-the-art information in key industries: education, medicine, the sciences, entertainment, health care, and business management. Eventually, the whole of personal, governmental, and industrial activity and all their tributaries will sustain a demand described by marketers as "habitual" and "dependent." In effect, advanced communications services that today are regarded as more exotic than necessary will be permanently woven into the fabric of everyday life in the near future [16]. The specifics and determinants of consumer spending for telecommunications are described in detail in Chapter 5. As we approach this discussion, it is important to bear in mind a unique phenomenon associated with information as a driving force for communications innovation: the more information that is generated and used, the greater the need for advanced telecommunications infrastructure [17]. In other words, unlike other resources that sustain life in modern economies (capital, labor, physical assets, land, minerals, and so forth), an attribute of information is that its creation and availability are inversely related to its consumption. The more information people consume, the more they require it to perform work and facilitate leisure. Thus, unlike other resources, information cannot be depleted and, indeed, expands at a pace that exceeds an individual's capacity to assimilate.

The economic significance of this paradoxical phenomenon is that the communications industry must now prepare for accelerated demand for all means of distributing information—voice, video, and data through both wireline and wireless transmission. This development cannot define winners or losers in the industry; it merely projects a future in which demand unfolds at an exponential rate. It also suggests that exponential demand for information in post-industrial economies like the United States, Canada, Western Europe, and Japan, will be followed by comparable consumption in those nations whose enterprises must interact with these economies. The advent of the "Internet age" has defined a future in which organizations can only thrive by exchanging and assimilating information. The need for information will drive the innovation of telecommunications infrastructure; to understand the economics of communications, we must first comprehend this peculiar characteristic of information management.

1.3.5 The Economics of Customized Information

The value of information to both individuals and firms will increasingly be a function of its customization [18]. We may define "customized information" as the organization, management, and distribution of data, voice, and image for each individual according to his or her needs. Tailoring information to the unique needs and tastes of each customer thus becomes an imperative for the telecommunications enterprise. The higher the degree of customization, the greater the value obtained. We refer to progressively elevated levels of value as "value-added" economics, a key to securing and sustaining a communication firm's future profitability.

Customized information and the value added by telecommunications firms in organizing, managing, and distributing that information will define the industry's future winners and losers. A discussion of value-added economics and its computation are elaborated in Chapter 3. For the moment, we pause to observe the inevitability of traditional communications providers (AT&T, MCI, the RBOCs, broadcasters, and cable television corporations) to infiltrate each other's markets. If a firm's profitability is directly tied to market demand for the refinement of information content and transmission, we may infer a new priority in the age of deregulation, that is, the need to translate the economic value of customized information into strategic planning. To the extent that providers can define future value-added pricing in converging markets, they can also establish the most effective competitive strategies. By extension, this consideration now heightens the contribution of labor in all organizations and certainly the future wealth generated by a provider's employees [19]. It will not be adequate for a communications firm to integrate state-of-the-art technology into its operations; it must forecast the customized needs of its clients and project new methods of enhancing the value of that customization.

1.3.6 Particle Marketing and Telecommunications

In the post–World War II period, corporate marketing (pricing, promotion, and distribution of goods and services) has undergone several metamorphoses. Through the mid-1970s "market segmentation" strategy was the preferred method of introducing products to market. Basic segmentation strategy essentially divided an entire market according to demographic data, including income, occupation, and age, as well as many other determinants of demand. This primary emphasis on demographic analysis was translated into product development that was tied to high-income, middle-income, and low-income segments. Producers originated new products and then sought customers via defined market segments to sustain demand before product life cycles were exhausted.

During the 1970s and 1980s American industry progressively sought further differentiation of these segments in the form of "niche marketing." Niche marketing consisted of narrowed definitions of taste as opposed to essential need; in short, product lines were expanded by differentiating tastes—style, form, color, and whatever factors circumscribed narrowed consumer preferences. Where an entire market had formerly consisted of three or four defined segments, perhaps nine, 12, or 15 segments had now become the linchpin of market research. The transition from market segmentation to niche marketing, accompanied by growing consumer demand for personalized delivery of services, has altered the contemporary marketing paradigm. Basic niche marketing is gradually giving way to what some have now termed "particle marketing" [20]. Particle marketing may be defined as the pricing, promotion, and distribution of any good or service intended to appeal to each consumer. Expressed differently, each consumer now becomes a market unto him- or herself. Consumers that now watch 30 or 40 cable television stations express disappointment at not having a greater array of channels available; customers that use multiple services are frustrated at the inability to obtain all of them from a single provider; a substantial fraction of the population claims that it requires wireless services but cannot afford them. These were precisely the factors cited by Congress in negotiating telecommunications deregulation. With regulatory reform now installed, providers must move rapidly to particle marketing as a means to develop customized services and secure market share. A model projecting the future impact of particle marketing is formulated in Chapter 10. It should be noted that the transition to particle marketing may require providers to divert substantial capital from basic to applied research, from infrastructure to sales and promotion, and from standardized personnel training to service innovation. A lingering concern remains that the traditional managers of infrastructure—the major corporate providers—will depart from that responsibility in favor of building a new clientele based on particle marketing strategy. Whatever these concerns, relaxed regulatory constraints will encourage major telecommunications providers to move in the direction of highly inventive market research methodologies. Communications firms now move in the cumbersome direction of improving infrastructure while competing for consumers. Given the financial resources now committed, the outcome of these conflicting twin objectives cannot be predicted with precision. However, individualized customization of information via content and service delivery has been asserted and will likely intensify.

1.3.7   Systems Analysis and the Economics of Telecommunications

In assessing the current status of the telecommunications market vis-à-vis regulatory reform and projecting resultant market opportunities, we will make reference to the application of economic systems analysis. For purposes of this discussion, we are not considering the technical dimensions of systems theory, that is, the engineering of telecommunications systems. Instead, we are concerned with applying systems analysis so as to ascertain probable market supply and demand for telecommunications; additionally, we wish to focus on the long-term consequences of new product and service introduction. We choose to apply systems theory, in short, because it permits us to identify future tertiary demand for goods and services based on competitive interaction in today's marketplace. Figure 1.2 isolates the components of economic systems analysis as applied to today's deregulated marketplace.

Four essential elements combine to facilitate an economic systems evaluation of the communications marketplace: *inputs*, *process*, *outputs*, and *feedback*. In this regard, we note that inputs consist of technological innovation, business and individualized demand for information, regulatory constraints, legal encumbrances, political machinations, and cultural and philosophical changes. The market collectively processes this information to define new products intended to meet the demand generated by the interaction of these variables. These new products, and the strategies that support them, are referred to as outputs. The products, once launched, can be gauged, monitored, and evalu-

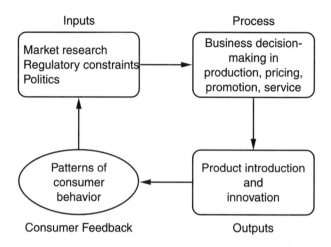

Figure 1.2 Economic systems analysis and telecommunications deregulation.

ated through the feedback loop; feedback thus defines and adjusts future inputs to satisfy the market process that, in turn, generates new products in response. These emerging products can then be evaluated via use of the feedback loop. Viewed from this perspective, a firm is treated as a living organism susceptible to the application of forecasting methodologies later discussed in this book. We must treat each communications enterprise, operating in a deregulated environment, as a firm that will constantly adjust its strategic plans in relation to market change. Only the fusion of micro- and macroeconomics with systems analysis can produce credible forecasts of emerging telecommunications supply and demand.

1.3.8 The Economics of Externalities and Opportunity Costs

Two final considerations underscore the macroeconomic significance of tele-communications: economic *externalities* and strategic trade-offs associated with *opportunity costs*. Externalities are third-party consequences precipitated by transactions between producers and consumers [21]. In other words, externali-ties, or external effects, influence the course of society in a way that was not taken into account or anticipated at the time a transaction was consummated. Externalities may be positive or negative in nature and are now a source of government concern as America embarks on the "information age" promul-gated by telecommunications progress. Examined from a macroeconomic per-spective, positive externalities associated with telecommunications innovation include new methods of delivering educational and training instruction to cus-tomers in remote regions. From the microeconomic viewpoint of a firm, such innovation can divert personnel from their primary managerial tasks, given the rapid proliferation of information. Will the generation and dissemination of information via advanced infrastructure raise or lower individual and organizational productivity? Will telecommunications elevate society to new levels of material gain or merely divert us from important tasks at hand?

Externalities, no matter the perspective, have economic implications. Because externalities are not borne by the primary parties involved in transactions, it is extremely difficult to estimate their short-term and long-term ramifications. Nevertheless, communications firms will inevitably be impelled to forecast the business and consumer trends associated with technological innovation, and the analysis of externalities will play a critical role in this regard. Chapter 3 elaborates in greater detail the identification and prediction of externalities and their implication for strategic planning in telecommunications corporations. Two vital characteristics denote externalities: (1) typically, external factors are not calculated by sellers and buyers in transactions, and (2) the benefits and costs intrinsic to such transactions are not borne by either party involved in an agreement. Thus, in the absence of government intervention, positive and negative third-party effects are inevitable in the interconnected domestic and global economies.

Network externalities are unique to the telecommunications market [22]. A network externality occurs when customer use of a communications service is solely a function of its availability to others. Simply, business and consumer consumption for telecommunications—that is, the price a provider can charge for its services—is positively correlated to the number of people who use that service. Fluctuations in price are not merely tied to the demand exerted in relation to supply introduced; higher prices result when the value of a service rises proportionate to the number of users who find information utility in that product. Significantly, the "type" of consumer attracted to a service also defines a network externality; higher income and better educated users, for example, will produce important externalities in the industry. When alerted to changes in consumer preference, advertisers will adjust their strategy to exploit such externalities, as will all others in business.

Opportunity costs are ordinarily defined as economic benefits forgone by pursuing one goal as opposed to others [23]. In short, to the extent that one activity is assumed and another dismissed, the benefit of a pursuit rejected constitutes its opportunity cost. An individual or firm cannot simultaneously pursue multiple objectives with equal zeal, hence the path it chooses among that multiplicity defines its future success. If a telecommunications company chooses to invest its revenue in the development of physical infrastructure, for example, it cannot devote those same resources to better support services or the retraining of staff. The price paid for these strategic trade-offs is represented by opportunity costs. By projecting opportunity costs among such trade-offs, a firm may clarify its short-, mid-, and long-term strategic priorities. We will note in Chapter 8 that the consolidation—mergers, acquisitions, alliances, and partnerships—now unfolding in the communications industry will carry important long-term financial consequences for consumers and governments throughout the world. The valuation of consolidation rests significantly on the computation of opportunity costs, as hereafter discussed.

1.4 Telecommunications and National Transformation

1.4.1 Telecommunications and the Advance of Post-Industrial Society

Structural changes in a nation's economy pinpoint its future requirements with respect to telecommunications usage. Economists refer to three essential national economies now comprising the global economy: pre-industrial (or agrarian) economies, industrial (manufacturing) economies, and post-industrial (service) economies [24]. Pre-industrial economies include most African nations as well as those whose economic status is typically defined as "underdeveloped." Wages are low in pre-industrial economies because productivity is constrained. Thus, the government tax base and revenue are limited, the educational and infrastructure base of such economies is restricted, and material progress is painfully slow.

Industrial economies—including Russia, the Czech Republic, Hungary, and Brazil—enjoy a higher standard of living because productivity is higher. Manufacturing produces jobs providing greater material benefit and escalating levels of consumption; mounting consumption, in turn, sustains these expanding economies. Economies evolving from either the pre-industrial or industrial stages are termed "transitional." Transitional economies include China, which is gradually shifting from pre-industrial to industrial positioning; and Singapore and South Korea, which are edging to post-industrial status.

The post-industrial or "service" economies include the United States, Canada, Germany, France, Great Britain, Italy, most of Western Europe, and Japan. The world's three largest economies—the United States, Japan, and Germany—represent the foundation of the post-industrial system and, by implication, define the future of the global economy. Wages, benefits, profits, and the material standard of living are highest in the post-industrial economies. Because service-based businesses depend most heavily on information-based technologies to facilitate their commerce, the greatest need for advanced telecommunications descends from post-industrial to industrial to pre-industrialized phases. As demonstrated in this volume, however, expansive empirical evidence moves us to infer that all economies will require advanced communications services to market their products. Indeed, labor in agriculture, manufacturing, and all service economies is destined for "knowledge-based" decision making based on state-of-the-art infrastructure.

In connecting the growing importance of telecommunications with the advance of the world's economy, we may integrate the following characteristics of business development in the post-industrialized world. Table 1.1 identifies key decision makers in agrarian, industrialized, and service economies while isolating the determinants of power and wealth in each stage of progress. We note that the world is gradually shifting from an era in which the influence of landholders and financiers is transitioning to technocrats who have the technical and managerial experience to direct organizations.

Technocrats understand and manipulate the arcane language (accounting, finance, marketing, computing, and law) unique to their professions— languages heavily reliant on the communications infrastructure that efficiently distributes information. As we contemplate the global economic future, it becomes clear that the systematic progress of a nation's economy is tied to its capacity to compete; competitive outcomes are chiefly a function of the value added by a firm's labor. As noted earlier, the intellectual capital of an organization constitutes its competitive advantage. Hence, the selection, application, and integration of telecommunications services play a pivotal role in defining the future value contributed by a firm's personnel.

1.4.2 Comparative Advantage and Leveraging in the Post-Industrial Society

Because the global economy essentially consists of three discrete stages of development, as outlined above, omnipresent competition will result in conflicting national priorities. In a dynamic competitive environment, it is certain that some organizations will recede in importance while others will emerge as vi-

Table 1.1
Business Development in the Post-Industrial Era

Economic Era	Critical Resource	Determinants of Wealth	Power
Pre-industrial	Land	Agriculture	Landholders
Industrial	Capital	Manufacturing	Financiers
Post-industrial	Information	Services	Knowledge workers

brant enterprises. A number of factors will decide the fate of the contemporary firm, but effective use of telecommunications to establish "comparative advantage" will dichotomize such competitors.

The firm that adopts telecommunications technology to enlarge its competitive advantage—superior products and services at competitive prices—will succeed. A firm unable to train staff to fully exploit such technologies is vulnerable to the insight of the competitor that foresees the advantage. The key to competitive advantage, therefore, lies in leveraging telecommunications: identifying and applying those technologies that maximize the productivity of employees at minimal cost. We are just entering the cutting edge of strategic planning designed to efficiently exploit communications technology. Formulas have been developed to compute the economic value of leveraging, and they are discussed in subsequent chapters.

1.4.3 Receding Versus Emerging Post-Industrial Enterprises

Major innovations in technology generally induce higher levels of demand, provided personal or professional objectives are achieved by their adoption. In this regard, we can differentiate those American industries in various stages of development based on the integration of communications [25]. We can label these industries as *receding, emerging,* or *restructuring*. In other words, increasingly telecommunications integration outlines competitive change in a firm's development. In this context, newly emerging industries that are exclusively the product of advances in communications include distance education, Internet commerce, virtual libraries, virtual college campuses, electronic publishing, telecommuting, groupware, shareware, electronic messaging, interactive television, robotic medicine, on-line medical support, and outsourcing [26]. Industries now receding, or on the verge of serious decline or capitulation, include traditional broadcasting and telephony, the postal service, and video rental. Industries coping with such technological change, and thus restructuring or reengineering their firms in response, include virtually all traditional manufacturing and service sectors. The communications tools they will employ to meet such volatility include videotext, videoconferencing, neural networking, home shopping and multimedia networking, and wireless services.

In each instance, we note that the organizational restructuring characteristic of the post-industrial era is closely linked to changes in communications technology. It is telecommunications that is fueling this transformation and its resultant benefits; telecommunications is also indirectly responsible for the anxiety and concern that many employers and employees claim to now experience [27]. The unpredictability implicit in such a transition has undermined traditional assumptions about security and stability in the American economy. The task at hand for a firm is to energize its staff to accept such change as a necessary ingredient to unleashing creativity and productivity. Telecommunications is thus redefining marketing, sales, management, finance, and all other activities required to sustain an organization [28].

1.5 The Limitations of Economic Analysis in the Post-Industrial Era

We explore in Chapters 3 to 5 the application of traditional economic tools in measuring and predicting the organizational impact of telecommunications integration. Traditional techniques designed to monitor and forecast the efficient use of communications include the basic laws of supply and demand; estimation of equilibrium, elasticities, and value-added benefits; identification of optimal pricing strategies; application of productivity analysis; and computation of costs and profits.

While it is clear that traditional economic analysis will and should play a prominent role in meeting these changes, there is much about telecommunications progress that should humble the veteran economist. We may now be witnessing the development of a new paradigm for the economic valuation of telecommunications services. Consider, for example, that some firms presently exploiting such technologies can maintain profit margins on a volume of production a fraction of former output. With the use of communications systems acting in concert with other innovations, for instance, the law of economies of scale (i.e., maximizing profit by increasing production at lower unit prices) may now be rewritten. Other extraordinary examples are now surfacing in the post-industrial era. The author concedes that incisive economic analysis should be tempered by humility in appraising the economics of deregulation. One cannot be dogmatic in the adherence to economic tools that were initially written in the industrial age. This period represents a new era in the progress of organizations and the personnel that manage them.

References

[1] National Telecommunications and Information Administration, *Telecommunications in the Age of Information*, Washington, D.C.: Department of Commerce, Oct. 1991, pp. 201–284.

[2] Stranahan, Linda, "Telephones: History and Development," *Jones Telecommunications Encyclopedia*, New York: Jones Digital Century Inc., 1995.

[3] Title I: Telecommunications Services, *Telecommunications Act of 1996*, Washington, D.C.: United States Congress, Feb. 1996.

[4] Wood, Wally, "Can Telecos Survive," *Telephony,* March 4, 1996, pp. 22–29.

[5] Pinchot, Gifford, and Elizabeth Pinchot, *The End of Bureaucracy and the Rise of the Intelligent Organization,* San Francisco: Berrett-Koehler Publishers, 1994, pp. 61–76.

[6] Shaw, James K., "Future Scenarios for the Telecommunications Industry," *New Telecom Quarterly,* Vol. 4, No. 4, 1996.

[7] Stewart, Thomas A., "Your Company's Most Valuable Asset: Intellectual Capital," *Fortune,* Oct. 3, 1994, pp. 68–74. See also Leif Ericsson and Michael S. Malone, *Intellectual Capital,* New York: Harper-Collins, 1997, for a discussion of the measurement and evaluation of effective intellectual capital contributions to organizational performance.

[8] Rockwell, Mark, "FCC Charts Course for '97," *Comm. Week,* Jan. 6, l997, p. 23.

[9] Gilbert, Steven, "How to Think and Learn," *AGB Trusteeship Magazine,* Special Issue 1996.

[10] Dent, Harry S., *Job Shock,* New York: St. Martin's Press, 1996, pp. 3–64.

[11] Dent, Harry S., *The Great Boom Ahead,* New York: Hyperion Press, 1993, pp. 192–219.

[12] Case, K. E., and R. C. Fair, *Principles of Economics,* New York: Prentice-Hall, 1995, pp. 905–1007.

[13] Dent, *op. cit.,* 1993, pp. 97–128.

[14] Colander, David, *Economics,* Homewood, IL: Irwin Press, 1995, pp. 159–167.

[15] Blanchard, Olivier, *Macroeconomics,* New York: Prentice-Hall, 1997, pp. 449–450.

[16] Cleveland, Harlan, "Information Is the Critical Resource of the Future," *The Futurist,* Bethesda, MD: World Future Society, Jan.–Feb. 1997, p. 13.

[17] Brand, Stewart, "Better to Part 'Phile' and Part 'Phobe'," *The Futurist,* Bethesda, MD: World Future Society, Jan.–Feb. 1997, pp. 12–13.

[18] KPMG Consulting, "Views of Information, Communications, and Entertainment," *Business Wire,* Dec. 17, 1996.

[19] Brand, *op. cit.*

[20] Pettis, Chuck, *TechnoBrands,* New York: AMACOM Press, 1995, pp. 136–164.

[21] Leftwich, R., and A. Sharp, *The Economics of Social Issues,* Plano, TX: Business Publications, Inc., 1994, pp. 63–92 and 187–204.

[22] Brennan, Tom, "ISDN Finds Its Niche," *Telephony,* Dec. 11, 1995, pp. 26–32.

[23] Young, P., and J. McAuley, *The Portable MBA in Economics,* New York: John Wiley and Sons, 1994, p. 118.

[24] Cetron, M., and O. Davies, *American Renaissance: Life at the Turn of the 21st Century,* New York: St. Martin's Press, 1989, pp. 133–148. Also see Toffler, Alvin, *The Next Wave,* New York: Random House, 1991. The seminal book produced in the field of the social and economic implications of the post-industrial society is: Bell, Daniel, *The Coming of Post-Industrial Society,* New York: Random House, 1975.

[25] Centron and Davies, *op. cit.,* pp. 273–284.

[26] Varian, Hal, and Carl Shapiro, *Information Rules,* Boston: Harvard Business School Press, 1999, pp. 173–220.

[27] Marien, Michael, "Top Ten Reasons Information Revolution Is Bad for Us," *The Futurist,* Bethesda, MD: World Future Society, Jan.–Feb. 1997, pp. 11–12.

[28] Eisenach, Jeffrey, Thomas Lenard, and Stephen McGonegal, *The Digital Economy,* Washington, D.C.: Progress and Freedom Foundation, 2000, pp. 74–87.

2

Analysis of the Telecommunications Act of 1996

Political and economic self-interest intersected in 1996 to compel the enactment of the Telecommunications Act of 1996 (hereafter denoted "Act"). Initially labeled the Telecommunications Reform Act (TRA) by its legislative sponsors, this legislation represents the linchpin of future telecommunications competition. As is the case in all legislative matters involving diverse business and consumer interests, the adoption of this statute reflected political compromise; as a result, no interest group was fully placated by the Act. However, every competing interest—in addition to politicians who reflected those interests—gained something in the resultant legislation.

Several factors should be noted as a backdrop to the passage of the Act. First and foremost, technological innovation incited and sustained congressional debate that precipitated the call for telecommunications deregulation. As suggested in Chapter 1, the fact is that accelerated advances in communications technology, particularly those inspired by firms other than AT&T, obviated traditional regulatory enforcement. By the 1980s, it had become increasingly difficult to supply an economic rationale that justified a national monopoly. To seriously contend that the American public was best served by FCC enforcement of the 1934 Communications Act was to argue that the public should be deprived of the benefits of technologies otherwise promulgated by the marketplace. Even if one were philosophically persuaded that such a monopoly was in the consumer's best interest, obvious legal grounds were mounting that the monopoly could not be maintained as originally constituted and regulated. Concerns ranging from restraint of trade to antitrust violations now impelled the government to rewrite communications law [1].

Secondly, an ambiance of deregulatory fervor swept the country beginning in the late 1970s. Many economists, including traditional liberals and conservatives, increasingly called for relaxation or reduction of regulatory sanctions applied to AT&T [2]. Although opposition to government intrusion had long been associated with Republican Party rhetoric, a significant number of Democrats joined in criticism of contemporary regulatory enforcement [3]. The first major experiments in the deregulation of the aviation and trucking industries in 1978, for example, provided hard empirical evidence that the introduction of competitive forces in place of government decision making could mitigate price increases. In the atmosphere of this era, marked by rising inflation throughout the economy, reductions in consumer prices in these two key sectors demonstrated the power of competition. In such an environment, a bipartisan chorus called for deregulatory experiments in many sectors of the economy; the Reagan administration moved to deregulate the petroleum, natural gas, and financial services industries. Deregulatory strategies continued during the Bush and Clinton administrations. Viewed in a broader context, and against the background of global privatization and deregulation, the United States had not merely encouraged but responded to an international shift in regulatory philosophy. Inevitably, as the 1990s unfolded, the telecommunications industry had become a target of consumer groups seeking the perceived benefits of deregulation already associated with aviation and other industries.

Thirdly, producers in a number of sectors had projected declining communication costs under conditions of deregulation. Organizations in fields as diverse as energy, education, and manufacturing foresaw significant reductions in the cost of doing business; this vision, complemented with the prospect of having a greater array of services from which to choose, spurred support for statutory revision of the 1934 law. Labor-intensive firms, particularly those depending heavily on communications to conduct and monitor their relations with clients, suddenly became proponents of some form of regulatory relaxation. In short, producers in firms outside the communications sector had adopted a political posture not unlike that of consumer groups [4].

Finally, "grass-roots" activity inspired congressional representatives to examine the effectiveness of regulatory enforcement in an era characterized by rising market competition. A key to spurring congressional review of the effectiveness of such enforcement, particularly that originating at the Federal Communications Commission, was rising local opposition to cable television rates [5]. Disaffected cable television subscribers instigated regulatory reform through interest-group pressure. The impetus toward deregulation was given important momentum through the collective action of consumers.

These factors—technological innovation, the shift in political ambiance, and interest-group pressure directed from both producers and consumers—led Congress to reevaluate traditional regulatory policy in telecommunications.

Given comparable experiments in aviation, trucking, and energy, it was perhaps inevitable that deregulation would be applied as an alternative to government direction in pricing and related matters. Congress seriously debated a restructuring of the Communications Act of 1934 beginning in 1993 and, in doing so, reexamined the historical evolution of regulatory guidance in telecommunications.

2.1　The Meaning and Dispute of Telecommunications Deregulation

Deregulation was defined in Chapter 1 as the relaxation, modification, or elimination of government rules and regulations intended to influence business performance. There exists a spectrum of deregulatory application: Government regulations can be massaged, reduced, or removed from the agenda of the guiding regulatory body. Given this continuum of regulatory adjustment, precisely what constitutes deregulation? Why do some observers believe that the Telecommunications Act of 1996 represents genuine deregulation, while others contend the opposite?

The meaning and underlying dispute of deregulation is both literal and philosophical. For some, authentic deregulation is only promulgated when public agencies and their regulatory apparatuses are fully dismantled. For others, a modest shift away from government intrusiveness constitutes deregulation. Because the term "deregulation" is used liberally in different contexts, it is perhaps inevitable that philosophical (as well as economic) disputes would arise as to the efficacy of any law seeking the benefits of deregulation. As noted, the Telecommunications Act of 1996 reflects an attempt to satisfy, however nominally, multiple constituencies. The author will concede the following bias in contending that the Act does indeed represent a form of deregulation: although government regulatory power via the FCC remains unabated, a number of the law's passages attempt to redirect the pattern of such enforcement. Given congressional intent, this surely was a key motive in enacting the legislation.

It should also be indicated that empirical evidence discussed in Chapters 8 and 9 discloses that shifts in capital investment among the telecommunications sectors—telephony, broadcasting, computer, and their affiliates—imply an emerging confidence by the private sector that a fuller blossoming of deregulation may emerge over the long run. In this regard, it is doubtful that the wave of mergers, acquisitions, partnerships, affiliations, and alliances that have been completed during the year following the Act's passage would have surfaced in the same way. A literalist might be relatively dissatisfied by FCC enforcement of the Act's deregulatory features, but others regard the law as providing a necessary ambiance to further deregulatory undertakings. Deregulation is a slow, lumbering process in telecommunications. This may not be the favored approach to deregulation, but in a vast industry of competing interests such a contingency may be a necessary precondition for the exploitation of deregulation's benefits. In other words, the current application of deregulation is but a first step in the direction of something potentially larger and more significant for the future.

2.2 The Prehistory of Telecommunications Regulation: 1876–1934

The essence of telecommunications regulation in the United States lies, in many ways, in a historical recapitulation of sanctions applied to AT&T. After the invention of the telephone in 1876, Bell Telephone, later syndicated as AT&T, was formed to exploit the new technology through a policy underscoring the economics of universal service [6]. As envisioned by a partner of Alexander Graham Bell, Thomas Vail, the telephone had the potential for wide consumer acceptance because every new user represented a potential marketer of the product; in short, wide—universal—acceptance was inevitable as each user instigated exponential growth on the part of associates, colleagues, family, and friends [7]. Although viewed as a luxury at the time of its inception, telephone usage carried the seeds of its own expansion. Universal service, later to be redefined in social, cultural, and political terms, was to become a cornerstone of AT&T's strategic planning in the century to come [7].

Just as universal service became the linchpin of AT&T policy, so too, did emphasis on aggregation of market power with resultant cash flow benefits. In building one of the world's great corporations, AT&T implicitly followed a strategy applied by other entrepreneurs, such as J. D. Rockefeller [8]. In pursuing the objective of accumulating competing telephone companies, AT&T simultaneously reduced competition while expanding market share. In doing so, the company acquired vast numbers of telephone subscribers who provided a continuing source of monthly cash flow. With few competing alternatives available and the value of the telephone reinforced by a growing need for communication, AT&T customers created a formula for corporate success in the communications industry: identify a market base, minimize competitive threats, and build the organization through the reliability and predictability of monthly cash flow. This simple strategy has indeed been adopted in every sector of the telecommunications industry. It is cash flow, more so than profitability, that drives the creation and maintenance of new telecommunications products. In generating cash flow, a firm retains the ability to borrow upon need or issue stock—key mechanisms for driving a firm forward during the initial and intermediate stages of growth.

In linking universal service with acquisition strategy, AT&T set in motion powerful twin engines of market concentration, a sure invitation for antitrust enforcement by the Justice Department. In retrospect, one can appreciate the keen insight of the firm's founders: the history of the regulatory and antitrust administration, hereafter discussed, had revealed in the railroad and petroleum industries that government intervention was often a modest price to pay for such intrusiveness [9]. If regulation could be contained in such a way as to ensure the perpetuation of these aforementioned tactics, the corporation could be assured of continued opportunities for growth. Moreover, as a concession for submission to government sanctions, the early history of federal regulation had illustrated that, within parameters, regulated monopolies could operate under "cost-plus" economic structures (i.e., be assured of recapturing cost under prescribed profit). These circumstances would propel AT&T to accept a succession of rules and regulations that immediately preceded 1934.

During its early stages of critical and decisive growth, the first 20 years of the twentieth century, AT&T supported its strategic plan with secondary objectives including emphasis on research and development as well refusal to permit competitors to network with its infrastructure [10]. Rather than endure the risks of protracted litigation with the Department of Justice, AT&T agreed to a compromise in 1912 that facilitated divestiture of Western Union from the parent firm; in addition, AT&T would be compelled to submit to Interstate Commerce Commission approval relative to any future acquisition of independent enterprises. Most significantly, independent telephone companies would be permitted, as a result of this agreement, to network with the Bell system [10]. Bolstered by dominant market share and confident of continued growth in the world's fastest growing economy, AT&T proceeded on a path of government-sanctioned monopoly.

With the advent of commercial use of the electromagnetic radio spectrum in the 1920s, AT&T did not confine itself to standard telephony. The corporation asserted its research and development expertise in new technologies: emerging market opportunities in radio, motion-picture sound, and the teletype industry. It became clear that, to the extent that profitability could be projected in its telephony operations, AT&T was simultaneously poised financially to enter, and potentially dominate, new communication technologies. This factor, combined with corporate and government desire to set clarifying ground rules for this vital young industry, led to the adoption of the landmark Communications Act of 1934.

2.3　The Communications Act of 1934: Foundation of Telecommunications Law

The success of AT&T in developing new communications products and services, its assertiveness in entering new markets, and its adroit method of managing the vicissitudes of bureaucratic regulation had alarmed many observers by the early 1930s. In particular, competitors began to lobby the federal government to further strengthen restraints placed on the telephony giant. Such concerns were not limited to traditional objections of independent firms or political interests historically opposed to monopolies. Gradually as the value of communications became tangibly demonstrated, various industries also came to view with concern so heavy a reliance on a single provider.

The use of government to outline or influence private markets is often broadly defined as regulation. In fact, government intervention occurs on several fronts, each embedded with a multiplicity of public policy options. We may categorize government involvement in setting prices, wages, jurisdictional boundaries, working conditions, safety concerns, and other matters in the following fashion [11].

- *Bureaucratic Policy:* broad policy statements on the part of administrative agencies, such as the Federal Communications Commission, intended to influence current or future business development. Often such policy pronouncements are written with some vagueness or ambiguity to promote flexibility and adaptability in the private sector.
- *Bureaucratic Rule Making:* specific statements and procedures aimed at defining with some precision the permissibility of prospective business planning.
- *Administrative Law:* that body of public law which guides bureaucrats in their decision making.
- *Antitrust Law:* that body of statutory law (statutes are laws passed by legislative bodies) aimed at preventing market concentration (i.e., anticompetitive practices of firms seeking to aggregate their market influence through horizontal or vertical integration). The most notable antitrust laws include the Sherman, Clayton, Federal Trade Commission, and Robinson–Patman Acts. The Sherman Act, passed in 1890, is the principal statute of federal enforcement and prohibits contracts aimed at monopolistic practices and restraint of trade. The Clayton Act, passed in conjunction with the Federal Trade Commission Act, focused on exclusive contracts; that is, the Clayton statute prevented the negotiation of contracts resulting in transactions in which customers were compelled to deal exclusively with a single seller or provider. The Robinson–Patman Act, enacted in 1936, outlawed price discrimi-

nation—in short, a restraint designed to prevent large suppliers from uniformly destroying small firms through quantity discount pricing. These statutes, collectively, were designed to prevent anticompetitive market concentration, pricing, and distribution practices [12].

The federal application of bureaucratic and antitrust laws is reinforced by comparable legislation passed in most states. On the federal level, communications corporations are primarily sensitive to potential litigation or rule making originating from the following agencies.

- *Department of Justice/Anti-Trust Division:* that section of the Justice Department charged with the responsibility of filing suit in response to anticompetitive behavior relative to market concentration, pricing, or the like. Often the Federal Trade Commission generates initial data and other evidence in support of these activities.

- *Federal Trade Commission:* an agency of vast responsibilities, the FTC intervenes to ensure consumer protection while promoting business competition. The FTC does not instigate prosecution of antitrust and similar legal sanctions, discharging these responsibilities to the Department of Justice. Nevertheless, a key feature of the Federal Trade Commission Act contains this nebulous clause [13]: "Unfair methods of competition in commerce are hereby declared unlawful." It is this ambiguous phrase that has stimulated many lawsuits in commercial transactions vis-à-vis the public and private sectors.

- *Federal Communications Commission:* established by the 1934 statute, the agency is charged with the responsibility of implementing the intent of federal communications statutes. The FCC draws its bureaucratic authority from both the 1934 Act and 1996 communications statutes.

- *National Telecommunications and Information Administration (NTIA):* the NTIA is designated with the assignment of encouraging research and formulating policy positions for various government agencies. As is the case with all federal bureaucracy, the NTIA is directly accountable to the executive branch. White House policy pronouncements on telecommunications often stem from initial work completed at this agency.

If one fuses the work of these agencies—policy making set at NTIA, rule-making authority exercised by the FCC, FTC emphasis on consumer protection, and continuous antitrust monitoring at the Department of Justice—we have the regulatory menu with which telecommunications corporations must contend. Clearly, the expansion of the communications sector prompted inevitable adoption of the first Communications Act. This 1934 statute is reprinted in Appendix A2; this condensed business applications version (the entire 1934 law is available via the Internet at www.fcc.gov) reveals the original intent of regulatory enforcement:[1]

> For the purpose of regulating interstate and foreign commerce in commu-
> nication by wire and radio so as to make available to all the people of the
> United States . . . with adequate facilities at reasonable charges . . . for the
> purpose of the national defense . . . for purpose of promoting safety of life
> and property . . . a more effective execution of this policy by centralizing
> authority . . . the Federal Communications Commission.

In adopting this statute, Congress sought multiple ends. First and foremost, Congress hoped to clarify jurisdictional boundaries of wireline and radio communications. Simply, the federal government, using the FCC as its regulatory arm, would now define rules of wireline and wireless communications. Such rules included technical safety standards, broadcasting provisions, interstate commerce guidelines, conditions of proposed acquisitions, and most significantly, the methods by which universal service would be available to all sections of the country. To augment additional public policy objectives, Congress adjoined FCC policy-making activity to broader purposes of national defense. For some in communications, it was troubling that a new regulatory agency with vast new powers was formed to regulate all aspects of the industry. Nevertheless, a salient outcome of this statute was the stability that accompanies basic ground rules about which all participants were compelled to abide.[2]

In the years following its adoption, the Communications Act of 1934 was periodically modified by Congress. Advances in technology, trends in mergers and acquisitions, transformations of industry infrastructure, and competing uses of the electromagnetic spectrum would impel FCC authorities to revisit Congress and request additional legislative authority for proposed rewriting of this seminal law. The FCC's principal objective during World War II was to ensure the nation's defense, and thoughtful consideration of these business concerns was deferred until the end of the war. Immediately following World War II, early work on commercial television forced the FCC to weigh carefully future uses of wireline and wireless services. Accompanying the problems of emerging technologies was the extraordinary and rapid growth in basic

---

1. Interpretation of legal doctrine regarding the 1934 Act is available under Appendix C. For updated interpretations of both the 1934 and 1996 Acts, access the Web site specified under the Michigan URL.
2. It is a fact that some broadcasters actively lobbied Congress for regulatory intervention by the federal government, a largely unprecedented phenomenon in American business at that time.

telephone service. Additionally, research and development at Bell Labs during World War II now yielded prospective market applications [14].

While the communications industry began to grow rapidly in response to technical innovation and exponential market demand, there unexpectedly developed a parallel phenomenon: traditional AT&T business strategy—most especially its ownership of Western Electric and its policy of dealing exclusively with this subsidiary as its supplier—invited the Justice Department to file suit under the auspices of the Sherman Anti-Trust Act. After six years, in 1955, a U.S. district court determined that, while AT&T would not be required to divest itself of Western Electric, it would (1) manufacture through its subsidiary only telephone-related equipment (with the exception of other forms of manufacturing targeted exclusively for government operations), (2) engage in only common-carrier operations and their related activities, and (3) end the practice of exclusivity through its licenses and distribution of technical information [15]. When viewed in its totality, this landmark decree signals the first fundamental change in the post–World War II communications environment with respect to established firms. While not a deregulatory decision per se, the ensuing 1956 Consent Decree decision—in conjunction with those events specified in Chapter 1 as influencing the 1983 divestiture decree—effectively set Congress on a path that inevitably led to the fundamental reorganization of communications policy.

One might infer from the historical development of technology, government policy, and interest group participation, an important economic dichotomy in the history of telecommunications regulation. During the first half of the twentieth century, government was committed to the regulation of a single monopoly that controlled the destiny of the nation's communication infrastructure. The second half of the twentieth century began at first as a creeping, then an unfolding attempt to contain the market power of AT&T; as the 1950s gave way to the following two decades—a period of marked improvement in technologies often introduced by AT&T competitors—it became clear that basic elements of the 1934 Act were gradually approaching obsolescence as to their effective regulatory implementation. Technology was racing ahead of legislative oversight. By the 1980s, it became obvious that the court system was inadequate to the systematic task of meeting disputes among industry participants. For one thing, the financial and time constraints associated with litigation could undermine both domestic and global competitive opportunities. Moreover, legal (as opposed to market) remedies only rarely fully satisfied even the winner of such a dispute. The most immediate mechanism, for better or worse, in guiding telecommunications development is the test of the market: Can a firm produce a service or product such that consumers adopt it? One may contest the social value of market adoption in deciding the fate of such a critical industry to the nation's progress. In purely economic terms, however, it can hardly be challenged that market tests are the most efficient methods of determining the outline of future communications development.

In defining the current and prospective development of telecommunications services, we are therefore left with these vehicles by which to decide policy:

- *Legal:* combining both private and public legal challenges to perceived inequities in the design, production, or distribution of such products;
- *Regulatory:* relying exclusively on bureaucratic authority to lay ground rules for private transactions;
- *Economic:* the test of market acceptance.

It has been the historic presumption since 1934 that telecommunications policy in the United States is best effected by a strategy combining all three elements. As noted in Chapter 1, however, deregulation in other industries, accompanied by a general change in political philosophy since the late 1970s, formed an ambiance favorable to recent legislation. As discussed later in this book, however, the conditions precipitating adoption of deregulatory strategy are fully capable of creating its antithesis with the passage of time. It is this paradox that drives the continued reevaluation and monitoring of telecommunications policy.

2.4 Divestiture and Communications Policy: 1984–1996

Although cracks in AT&T's monopoly of local and long-distance service had become apparent as early as the 1950s, serious prospective competition in the form of legal challenges had mounted by the 1970s. In 1974 the Department of Justice, acting in concert with MCI, filed an antitrust suit against AT&T. MCI had implicitly challenged the underlying economic rationale behind one of the world's great monopolies: *the doctrine of natural monopoly* [16]. A natural monopoly is one that is afforded exclusivity of markets, owing to economies of scale [17]. In other words, a natural monopoly theoretically operates in the public interest because it provides its service at a price lower than would otherwise be possible if multiple competitors were permitted to enter that market.

As the 1974 case unfolded, it was evident that AT&T could not sustain its "natural monopoly" status. Significantly, it could be demonstrated that it was not necessary for MCI and other firms to construct parallel infrastructure in order to offer telephone service. The industry had evolved in such a way that the existing infrastructure could be used to support both veteran and emerging technologies. With these technological developments, the chaos that the FCC had feared—multiple companies creating multiple systems with incompatible equipment—had disappeared. If the infrastructure and airwaves ultimately belonged to the public and the notion of exclusivity of markets no longer made economic sense, AT&T's historic foothold had been shaken.

In a U.S. district court presided by Judge Harold Greene, a Modified Final Judgment was adopted on January 8, 1982 [18]. The essence of the MFJ provided that AT&T would be permitted to keep its holdings, including Western Electric, Bell Labs, and its extensive long-distance operations. In addition, the corporation was permitted to enter new areas of business previously prohibited. From AT&T's viewpoint, something had been given and something had been taken away: If they had lost their monopoly status, they had gained the right to be a competitor in emerging areas of communications. Just as important, they had gained access to the growing computer industry, a market not regulated by government. The agreement effectively consolidated 22 Bell operating companies into six independent regional Bell units; these became known as Ameritech, Bell Atlantic, BellSouth, Pacific Telesis, Southwestern Bell (later to be renamed SBC), and NYNEX. The divestiture, or breakup, of AT&T was finalized on January 1, 1984.

In retrospect, one can see with greater clarity the importance of the period extending from the mid-1980s through 1995. This was an era in which, through initial experimentation with deregulation, the nation would experience unprecedented services. It was an era of change that brought great confusion to many consumers long accustomed to the predictability of a single provider. For business users, the gains were immediate—declines in long-distance pricing were accompanied by a proliferation of emerging services. The exploitation of these new products and services would later lead to organizational reconfiguration elsewhere in the economy. Whether these same services would have been inaugurated and developed by AT&T at a rate dictated by competitive interaction is a matter of conjecture. What is clear is that, in areas where the MFJ permitted competition, new services were introduced to market by multiple competitors (hundreds of long-distance providers surfaced, for example), and prices generally diminished. In areas where regulation was extended, most notably in local jurisdictions, prices remained at comparable levels or increased. A debate would ensue as to whether deregulation was truly effective, or instead, whether local providers were simple monopolists and should thus be suppliers of infrastructure to other branches of the industry. Inevitably, some began to call for the extension of MFJ strategy to all areas of the telecommunications industry.

2.5   The Administrative Procedures Act and Telecommunications

The Federal Communications Commission's administrative process—the procedure by which its regulatory decisions are promulgated—is governed by the Administrative Procedures Act. The APA, passed by Congress in 1946, specifies two important judicial functions that oversee regulatory decisions: (1) to determine whether proper procedure was followed by regulatory authorities, if private sector complaints are filed before the courts; and (2) to evaluate the substance or content of those decisions, subject to judicial discretion, in instances where there exists dispute as to whether such bureaucratic actions are justified. The APA statute governs all regulatory practices in federal agencies; and many, but not all, state governments maintain statutes comparable to the APA.

In essence, FCC authority manifests itself in four ways:

1. *Rule-making power:* the authority to assign rules and regulations to fulfill statutory intent (the Telecommunications Act of 1996 is the legislative vehicle that now inspires FCC regulatory actions);

2. *Directing power:* the authority to issue orders by which private concerns are directed to take or not take certain actions;

3. *Investigatory power:* the authority to call witnesses to testify or otherwise secure records so as to make rule-making determinations;

4. *Licensing power:* the authority to grant, refuse, recall, renew, or modify licenses (typically assigned within the electromagnetic spectrum).

Private concerns who wish to challenge the process or content of regulatory decisions may petition either administrative law judges (ALJs), operating within regulatory agencies, or the court system in redressing grievances when they arise. Both avenues can be time-consuming but nevertheless constitute useful venues for the potential reversal of unfavorable regulatory decisions.

2.6   The Politics of Telecommunications

As the 1990s unfolded, several factors would encourage certain economic and political interests to propose full deregulation for the telecommunications industry. These interests were given impetus by several academic and other studies reviewing the economic impact of deregulation in similar capital-intensive industries. An atmosphere began to permeate the national political dialogue about the efficacy and long-term effect of full competition in this and other industries [19].

One intriguing aspect of emerging political support for deregulation concerned its growing bipartisan nature. Although the Democratic Party had historically been associated with encroaching government regulation since the 1930s, it was the Carter administration that launched the first cycle of contemporary deregulation. The Republican Party, traditionally and historically, had eschewed government intervention in favor of expansive competition. The sharp distinction between the two parties was gradually disappearing as alternatives to government regulation were reexamined.

As majorities in both political parties galvanized in support of deregulation, economic interests of various kinds, not all of which were confined to the telecommunications industry, foresaw the inevitable path to freer markets. It has been noted that the confluence of technology, law, and politics encouraged Congress to rewrite communications law. It is no less significant that the strategic and tactical value of telecommunications became apparent to the nation's lawmakers. Legislators were advised in the early 1990s through investigation and testimony[3] that:

- The convergence of the computer, broadcasting, and telephony industries had rendered obsolescent the distribution of communication services, as defined by the 1934 Act. There was simply no valid economic reason why telephony firms could not—and should not—offer television service, for instance; no substantive reason why cable television companies could not offer telephone service, and so on. To the extent that these new competitive forces were suppressed by the conventions of law, consumers were now deprived of the very protection of an implicit promise of regulation.

- The growing significance of trade to the American economy was arbitrarily restrained by highly regulated communication markets via the first three sections of the 1934 law (and its subsequent abridgments). The ability of American firms to export communication services and products would, in theory, be measurably enhanced by greater flexibility in moving in and out of emerging market structures. In this era, it was argued, flexibility, adaptability, and responsiveness to change would further enlarge this important segment of the American economy.

- The telecommunications sector was projected to grow in excess of $1 trillion shortly after 2000, and perhaps double again by 2010. The competitive undergirding of communications afforded to other contingents of the economy was thought imperative; state-of-the-art com-

---

3. Testimony was given in Congress beginning in 1993 and extending through early 1995 prior to passage of the Act.

munications was thus perceived to enhance the competitive vitality of all areas of the economy. In addition, sponsors of deregulation argued that such a statute was necessary for improving the productivity of the health and education sectors.

An additional factor fueled debate regarding deregulation. Americans had become disenchanted by 1992 with the quality and pricing of cable television services. In response, Congress adopted the 1992 Cable Television Act, an amendment to regulatory treatment found in a 1984 abridgment to the 1934 Communications Act. In discussions preceding the adoption of the 1996 Telecommunications Act, lines of division emerged: A general consensus had surfaced, cutting across party lines, that deregulation was the appropriate path to take in the years to come. However, some legislators on both sides of the political aisle expressed concern about the timing, fairness, and universal application of deregulation.

Congress embarked on deregulation by attempting to meet two concurrent goals in cable television service: maximizing competition while protecting consumers in those areas of the nation not served by multiple providers. In short, the sponsors of the Cable Act attempted to reassure consumers that deregulation would unleash necessary capital to invite new suppliers of video delivery. But in certain geographic areas, given their limited population, little incentive would exist for new market entrants. The Cable Act was passed in October 1992. In framing this unique language in communications law, the country gained a keener insight into the political compromises necessary to engineer a broader reworking of the governing statute.

2.7 Statutory Law and Telecommunications Transformation

2.7.1 The Telecommunications Act of 1996: A New Commercial Paradigm

After a period of three years of intense debate, testimony, and investigation, Congress enacted the Telecommunications Act of 1996 on February 1, 1996. The legislation was signed into law by President Clinton on February 8, 1996, and took immediate effect. Its sponsors argued that this comprehensive rewriting of communications law would fundamentally alter the nature of both the industry and American economy [20]. Although some observers found the bill lacking in some respects, most agreed—including those who contended that the breakup of the natural monopoly was not desirable—that a restructuring of the industry was inevitable.

Where the 1934 Act had focused on the development of the telephony and radio broadcasting industries and combined features of universal service with national defense and other policy considerations, the 1996 Act augmented these tenets with five priorities: (1) redefinition of and deregulation of telephone service, (2) development of Internet and related computer services, (3) revised procedures for radio and television broadcasting, (4) cable television services, and (5) the manufacturing of telecommunications equipment and related standards. These provisions were augmented with the reinstatement of key elements of the 1934 Act, including a renewed emphasis on universal service and the regulatory authority of the FCC.

From a purely economic perspective, most significantly, the Act preempted all state laws that prevented, inhibited, or otherwise restricted competition in telephone service, including local and long-distance markets. The law also obviated consent decrees formerly invoked with respect to AT&T and its rival, GTE. Anticompetitive behavior is expressly forbidden by the Act, and powers are given uniformly to the FCC to ensure sustained competition in all communications markets. The FCC thus became the arbiter of optimal competition in all areas of the industry.

While the doctrine of universal service was sustained in this rewriting, key phraseology provided that universal service would now be treated as an "evolving" phenomenon; in other words, additional technological and economic criteria would now be factored into its enforcement. This was a tacit acknowledgment that government could not forecast what veteran and emerging services would be decisive for consumer use. With state administrative bodies now diminished in regulatory powers, with the relaxation of constraints imposed on new market entry, and with a more flexible interpretation as to universal service requirements, the industry gained a remarkable degree of freedom in market interaction.

An important distinction must be made with respect to the concept and importance of "convergence" as described in Chapter 1. While the telephony and broadcasting industries—regulated under both the 1934 and 1996 Acts—remain guided, in part, by public policy, such is not the case with respect to the computer sector. Computing, software, and hardware have sustained their development in the absence of government regulation (except in the area of export restrictions). We thus underscore the interaction of three industries, one of which remains unregulated, but all of which influence national communications policy. A continuing source of legal and regulatory dispute is likely to be the freedom with which the computer industry formulates its business affiliations with its telecommunications counterparts. Those acting in the absence of regulatory constraints are more likely to be nimble in the competitive jostling predicted to come.

The Act effectively repeals prohibitions on communication manufacturing by the RBOCs. A new market is thus open to regional corporations seeking new market opportunities. Adjoined to this new freedom is the right of local telephone companies to engage in electronic publishing, projected for substantial growth over the long term.

In cable television, the Act repeals key features of the 1992 Cable Act. Among other things, in 1999 "rate capping" was eliminated, except for basic programming. The ability of cable firms to raise rates was thus bolstered, although the FCC retained the right to examine challenges to unfair or inadequate service provision. The freedom of communications firms to raise rates is therefore encouraged, but hardly an unrestrained privilege.

One of the more contentious areas of debate immediately preceding adoption of the 1996 law concerned provisions governing radio and television broadcasting. Two issues of long-term importance are etched in compromise: (1) major broadcasting corporations would be permitted to operate under conditions of relaxed market concentration (i.e., media giants would be afforded greater latitude to engage in vertical and horizontal strategies), and (2) spectrum "loan" provisions would be applied to broadcasters (facilitating adaptation to HDTV services) and their transition from analog to high-definition frequencies.

In the area of policy pertaining to computer and Internet usage, the Act specified provisions prohibiting "lewd, indecent, obscene . . . behavior." Additionally, the repeated harassment of any person or persons was now expressly forbidden by the Act (this feature became known as antiflaming). Implied questions of privacy and freedom are ambiguously intermingled in this phraseology and have future commercial ramifications. It is this delicate language that instigated much debate but little resolution. Commercial interests are involved in developing the Internet, and thus legal and regulatory disputes are very likely to follow in the years to come. Unless Congress expressly forbids the FCC from regulating the Internet, authorities clearly retain the right to do so. The extent of any such regulation, if assumed, remains nebulous.

2.7.2 The Telecommunications Act of 1996: An Economic Perspective

Appendix A1 elaborates the full text of the Act. Indicated here are the key elements, detailed section by section in the legislation. Note that the Act consists of seven titles, in excess of 100 pages, with dozens of sections imparting important prospective economic and market information about the delivery of telecommunications products and services.

Each section of the Act can be categorized, in terms of its initial business impact, as follows:

• Relaxes, gradually phases out, or eliminates FCC regulatory powers;
• Identifies areas of present and/or future regulatory contentiousness;

- Leads to consolidation and merger/acquisition activity;
- Creates and sustains intense competition, with probable price reduction in services and/or equipment;
- Spurs probable technological innovation or creation of new business formations;
- Promotes economic convergence through prospective integration of telephony, cable, or computer industries;
- Diminishes state regulatory power and influence through federal intervention.

The author acknowledges the contentiousness implicit in this methodology of categorization. There is much room for interpretative disagreement among observers as to the impact of the legislation's language. The author has attempted to isolate the unfolding *primary, long-term* effect of regulatory application. One may presume that two or more of these categorical descriptions will continue to interact to impel future change. The motive here is to provide a sense of business reconfiguration associated with the terms of this legislation. For purposes of this analysis, we assume a time horizon of five to ten years following passage of the Act. The vertical bars appearing in Appendix A should be used to gauge the economic impact and resultant business opportunities in telecommunications through 2006.

As one reviews the full text of the Act, it is apparent that some features have greater long-term import than others. Language pertaining to the first and last categories define business response to market opportunities, while the second and third categories more accurately reflect developments in the year following passage of the legislation. The benefits sought by Congress in enacting the legislation—categories four, five, and six—are slowly evolving.

2.7.3 Key Sections of the Act

In reviewing the Act, it becomes apparent that features are of concrete interest to providers responsible for the nation's communications infrastructure. In Title I (see Appendix A1), the following are pertinent.

Section 103.   Public utilities as defined under the Public Utility Holding Company Act are permitted to diversify into telecommunications and other services through single-purpose subsidiaries.

Section 104.   Nondiscrimination in telecommunications services is reaffirmed by this reworking of the Act.

Section 201.   Advanced television licenses are assigned to incumbents.

Section 202.   Relaxes restrictions on the aggregation of broadcast outlets as owned or controlled by single entities, corporations, or conglomerates.

Section 205.   The FCC maintains exclusive authority and regulatory jurisdiction over the development of direct broadcast services (DBS).

Section 251.   Elaborates a model for integrated interconnection. The RBOCs are required to provide equal access to interexchange carriers and other information service providers. This element of the law is arguably the most critical short-term issue with which the FCC will have to contend in orchestrating eventual full deregulation. In the absence of systematic interconnection of networks, full deregulation of the industry is impossible.

Section 252.   State agencies retain the right to approve interconnection agreements but yields to the FCC in policy-making authority if it fails to act expeditiously in approving such agreements.

Section 253.   State and local governments are prohibited from adopting legislation aimed at preventing interstate and intrastate services. Large corporations engaged in product and service delivery, as well as information content, are vitally concerned with this section of the law.

Section 254.   Universal service is reaffirmed but subject to changing interpretation based on advances in technology and other trends. The creation of a Federal-State Board to determine the efficacy of universal service policy is provided. Additional responsibilities are imposed regarding discounted services provided to educational and health care institutions.

Section 256.   Provides for infrastructure interoperability on a voluntary basis; that is, the FCC assumes the responsibility for encouraging diverse interests to develop interoperability standards and coordination.

Section 257.   The FCC is to isolate and eliminate market entry barriers to the formation or creation of entrepreneurial ventures in telecommunications. This is tacit legislative support intended to encourage multiple market entrants—to stimulate maximum competition in all communications markets.

Section 259.   Local exchange carriers are required to share their networks with competitors that have qualified for universal service subsidies.

Section 261. The FCC and state governments preserve their authority to maintain existing regulations not inconsistent with the Act. In other words, those elements of the 1934 Act that empower public agencies to regulate are maintained as long as they do not supersede the intent of the 1996 Act.

Section 271. A Bell Operating Company is permitted to offer interLATA service outside its defined region. Interconnection requirements through a "checklist" must be met and approved by federal and state authorities. The RBOCs must offer intraLATA toll dialing at the commencement of interLATA service.

Section 273. Equipment manufacturing prohibitions imposed by the MFJ are eliminated. Thus, RBOCs may now manufacture and market equipment, provided interLATA service authorization is secured via the checklist identified previously.

Section 274. Electronic publishing by the RBOCs is now permitted, provided such commerce is directed through separate subsidiaries or nonexclusive joint ventures.

Section 301. Rate regulation ended on March 31, 1999.

Section 302. Telephone companies may now penetrate the cable subscription market.

Section 303. Local franchising regulators are prevented from prohibiting cable television providers from delivering multiple services; local government power is thus diminished in regulating this industry.

Section 401. The Act mandates that the FCC forbear applying any elements or sections of the law in industries that do not require intervention in order to facilitate competition; expressed simply, the Act prevents regulators from arbitrarily intervening in the private sector. Because the phraseology of this section is ambiguous, we may project possible future litigation in private tests of FCC regulatory authority. Additionally, this section prohibits states from applying federal statutes and regulations that the FCC itself has chosen not to implement.

Section 402. Following 1998, and in two-year cycles that follow thereafter, the FCC is authorized to evaluate all regulations and eliminate those not deemed necessary to carry out the purposes of the Act.

Section 403. The FCC is given the authority to eliminate any provisions of the 1934 Act that are irrelevant, immaterial, or arbitrary in regulatory enforcement. Additional authority is given to the FCC to formulate advisory boards to facilitate rapid advancement of the Act's provisions.

Section 552.   All forms of telecommunications services—broadcasting, satellite, cable, telephony, and other distributors—are strongly encouraged to establish a fund to encourage television equipment manufacturers to develop new technologies aimed at parental control of video information.

Section 601.   Consent decrees are now superseded by application of the Act's provisions.

Section 602.   Local taxation of the emerging direct broadcast satellite market is prohibited.

Section 652.   Telephone companies are prevented from assuming greater than a 10% interest in a cable television system in the same market. This regulation shifts according to the number of video service providers incumbent in a region.

Section 702.   The RBOCs assume the responsibility for the confidentiality, thus privacy, of proprietary information they produce or transmit.

Section 705.   Commercial mobile radio service (CMRS) providers are not required to adhere to equal access standards associated with local telephone service; this element of the Act is crucial to the development of personal wireless communication services.

Section 706.   The FCC and its companion state regulatory commissions are required to expedite deployment of emerging telecommunications services for all Americans. The FCC retains the right to design new incentives for accelerated advancement of important technologies.

Section 707.   Establishes a Telecommunications Development Fund, intended to provide low-interest loans and other financial incentives for small businesses engaged in expanding and developing the telecommunications sector.

Section 706.   The FCC and its companion state regulatory commissions are sector.

Section 708.   Establishes a private, nonprofit corporation whose objective it is to transmit federal funding to educational projects; an additional feature provides incentives to encourage private investment in the educational technology infrastructure.

2.7.4 A Note on the Electromagnetic Spectrum

In omnibus legislation passed in 1993, Congress granted the FCC authority to launch auctions of electromagnetic spectrum (radio frequencies). The Act's terms reinforced FCC responsibility to continue the policy of auctioning off radio frequencies for purposes of launching new wireless services in interactive television, personal communication services, specialized mobile radio, paging, and related areas. While stimulating the development of new services remained the principal goal of this tactic, of high priority was the generation of new public revenues. Licenses had formerly been granted without payment to encourage development of the industry.

The passage of deregulatory legislation in 1996, combined with the transition in philosophy illuminated by this policy shift, provides a clear indication that promotion of the industry remains a paramount objective for the federal government; speed to market of new services remains, in theory, the overriding intent of the Act.

2.7.5 The Telecommunications Act of 1996: A Four-Year Retrospective

In the year following passage of the Act, observers began the process of assessing deregulation's impact. We had previously defined telecommunications as the transmission of voice, video, or data over wireline or wireless networks. We had also identified the emergence of convergence—the interaction of the telephony, broadcasting, and computer industries—as a tangible influence on the outline of both veteran and emerging product lines. A reasonable judgment regarding the efficacy of deregulation must therefore be defined by: (1) relative quality, service, and pricing of products; (2) speed to market of new businesses and their products; and (3) competitive interaction [21].

An additional note must be emphasized in making this assessment. Unlike other industries in which deregulation was engaged, substantial legal and regulatory remedies for those firms disaffected by FCC decisions were built into the initial stages of implementation. Additionally, experiments in deregulation dating back to the 1970s were accompanied by the dismantling of the agencies in charge of regulatory enforcement. In interstate commerce and aviation, for example, the agencies as well as regulations were simply ended or phased out. A salient feature of the Act is that the FCC maintains great discretionary authority in designing and carrying out communications policy.

By early 1997 it became clear that deregulation's benefits were falling short of the lofty rhetoric accompanying passage of the Act. Critics argued that the predicted increase in mergers and acquisitions, though a reality, had meant little in the wave of price reduction [22]. While prices declined modestly in certain areas of communications, notably cellular and Internet service, the fact was that local telephone rates remained persistently high. Moreover, some noted that extraordinary costs of market entry, including high marketing expenses, discouraged competition and its projected favorable impact on pricing and service.

Several noteworthy developments in pending telecommunications competition were evidenced in April and May of 1997 [23]. For one thing, local and long-distance exchange carriers went on a "public relations blitz" during the first quarter of the year. The principal objective was to persuade the public that their strategic plans were in the public interest. Long-distance carriers reminded the public through television and other advertising that many experts estimated that local calling rates were as much as 30% higher than was justified by underlying costs. Local carriers contended that they were serving the public interest by assuming the costs of infrastructure development, particularly with respect to subsidized wiring of educational institutions.

In April 1997 a number of states, including Massachusetts, Illinois, and Georgia, pursued alternative strategies to encourage greater competition in local telephony [24]. These states, buoyed by a circuit court of appeals decision to stay an FCC local exchange interconnection order, considered measures to punish anticompetitive practices by the RBOCs. The Baby Bells had sought the right to offer long-distance services, but had been opposed by state regulators who felt they had prevented competitors from entering the local arena. Illinois regulators secured from the legislature the right to fine firms for anticompetitive behavior, and such enforcement has not been applied to Ameritech. Significantly, these developments suggest that a spur to regional competition may come from state regulatory agents and that their actions could conceivably be more significant for the local environment than those actions taken by the FCC.

In early May 1997 the FCC unveiled a plan to overhaul the system of access charges imposed on long-distance firms by local exchange carriers. The plan called for a phased-reduction of $18.5 billion over a five-year period, commencing in July 1997. The FCC decision was intended to reduce the cost of long-distance calling by lowering this underlying cost of doing business. Most long-distance carriers, including AT&T, had promised publicly that they would pass on such access reductions in the form of cheaper rates. Most forecasters anticipated marginal reductions in long-distance rates; some foresaw original increases in the cost of local dialing. It may take as long as five years to fully gauge the impact of this decision on prospective competition in local telephony.

It should also be noted that implementation of the Act reinforced legislative intent to bring state-of-the-art telecommunications operations to the nation's schools. A subsidy in excess of $2 billion was imposed by the FCC on telecommunications providers (with additional subsidies designated for a telecommunications research and development fund). With new subsidies overlaying former subsidies, could one conclude that deregulation was a material outcome of the Telecommunications Act of 1996? Some concluded that only an overhaul of the 1996 Act—in the extreme, the elimination of most FCC regulatory powers—constituted genuine deregulation. Others deduced that a modest first step in the direction of deregulation represented a portent of deeper competition yet to be instigated. For these observers, a compromise between vigorous competing groups, balanced by a nominal consensus on the public interest, justified passage and implementation of the 1996 law.

It is important to recall, however, that the Act's complex, often ambiguous language defined variance in the timing of market entry and new product development. In short, competition was destined to develop more rapidly in emerging as opposed to veteran product lines, in wireless as opposed to wireline services, in arenas of multiple providers as opposed to newly deregulated monopolists. Deregulatory influence therefore could not be accurately gauged, positively or negatively, until time allowed all features of the Act to be fully implemented.

The FCC retains the power to modify its regulatory decisions to maximize competition and could therefore alter its policies at any point in the future. Some observers believed that nearly a decade might pass before deregulation took full effect in local telephony. Apart from exhausting its regulatory appeals during the early stages of deregulation, local exchange carriers would be unable to divert massive amounts of capital to new products until their competitive positioning in long-distance service had been asserted. Deregulation in telecommunications is not an instantaneous, uniform process; under terms of the Act, it is a deliberate, systematic procedure [25].

2.7.6 The Telecommunications Act of 1996: A Strategic Perspective

The regulatory changes embedded in telecommunications reform have naturally precipitated reconsideration of strategic initiatives planned in the industry. The process of strategic planning to maximize market penetration is discussed in Chapters 6 and 7. A note should be made here regarding the formulation of strategy in response to this reworking of communications law.

We may describe the collective economic impact of the Act, and its unfolding business implications, in the following fashion.

- Any company presently or prospectively involved in the communications industry may generally enter any segment of this sector.

- Liberalization of regulatory constraints is likely to accelerate capital formation.
- Waves of mergers and acquisitions in the short run and partnerships and alliances in the long-run are likely to surface.
- The United States may well gain a competitive edge in international communications commerce, given additional freedom in the private sector to exploit emerging technologies.

While historical evidence regarding deregulation in other industries tends to support these inferences, there is no assurance that they will be replicated in telecommunications. Given many passages of the Act that are necessarily ambiguous so as to promote regulatory flexibility, we may also anticipate prolonged legal challenges from time to time. Congress, however, was guided in its adoption of this legislation by these strategic and competitive imperatives. If a firm acts in concert with the convergence of the telephony, computer, and broadcasting industries, there is a greater probability of "strategic fit." A firm's fit—its economic synergy, compatibility, and complementary character—determines its prospects. With this criterion in mind, and armed with some insight into the economics of telecommunications deregulation, we can assert the following strategic findings relative to the past four years.

Some features of the Act have greater strategic import than others, of course; other components have their greatest impact in terms of time horizon (short- versus long-term consequences); still others hold the promise of future regulatory contentiousness that cannot be forecasted with precision. What is most revealing about the construction of the statute, however, is its generalized and sometimes vague phraseology. It is the ambiguities associated with these generalities which hold the promise of continued litigation between local and long-distance providers, between large and newly emerging firms, between the federal and state governments, between all government jurisdictions and large providers in telephony, broadcasting, and cable enterprises. Despite these discontinuities—problems not dissimilar to regulation encountered in other industries—the Act establishes a procedure for mediating disputes.

Other sections of the Act bear pertinence, directly or indirectly, for the strategic management process. The items indicated above, however, are those which directly affect pricing, promotion, distribution, or emerging product design. In short, firms cannot hope to compete assertively or aggressively in the absence of an assessment of the extent to which these features of the 1996 statute directly impact their response to market forces.

The impact of deregulation has been uneven with respect to the evolution of the communications industry. A contemporary definition of telecommunications had been asserted earlier in this volume: the transmission of video, voice, or data over wireline or wireless networks. These networks, in turn, are governed increasingly by the interaction of the telephony, broadcasting, and computer sectors. Thus, a communications firm can no longer afford to rely on a strategic management process exclusively focused on the parameters of its own sector: A broadcaster cannot define its strategic plan, as a result of deregulatory considerations, without considering the potential competition from a local exchange carrier; a telephone company cannot permit itself to ignore opportunities or threats emerging from cable television enterprises; a computer firm cannot dismiss an unfolding avenue in internet telephony or content development, and so forth. The shifting business landscape—the product of technological progress, relaxed government regulation, and intensified global competition—requires a broadening of strategic planning focus. The modern telecommunications firm is, in short, primarily a provider of communications products or services potentially distributed to market through:

- Multiple technical platforms;
- An expanding array of business formations, including mergers, acquisitions, intrapreneurial and entrepreneurial ventures, alliances, affiliations and partnerships;
- Coopetitive competitive structures, in which successful firms learn the process of adapting to transitional agreements with firms whose objectives temporarily conform to their own;
- Global penetration, particularly with respect to markets newly exposed to competitive market structures;
- Dynamic marketing mechanisms, culled from data warehousing and mining, which facilitate customized products for both businesses and household consumers.

Subsequent chapters explore cutting-edge methods of designing strategy to accommodate the dictates of deregulation for the decade ahead. It is now obvious that those firms which have experienced optimal market effectiveness, as adjudged by the past four years, have been the "first-movers" in exploiting strategic avenues precipitated by the Act.

References

[1] Colander, David, *Economics*, Chicago, IL: Irwin Press, 1995, pp. 645–648.

[2] Weidenbaum, Murray, *Business, Government, and the Public,* Englewood Cliffs, NJ: Prentice-Hall Inc., 1986, pp. 456–479.

[3] Eisner, Marc, *Anti-Trust: The Triumph of Economics*, Chapel Hill, NC: University of North Carolina Press, 1991, pp. 228–236.

[4] Egan, Bruce L., *Information Superhighways Revisited: The Economics of Multimedia*, Norwood, MA: Artech House, 1996, pp. 284–287.

[5] Bell, Trudy, John Adam, and Sue Lowe, "Communications Technology 1996," *IEEE Spectrum*, Jan. 1996, pp. 30–41. See also Steve Ditlea, "The PC Goes Ready to Wear," *IEEE Spectrum,* October 2000, pp. 34–39. Complementary analysis may be secured at: "Creating Convergence," *Scientific American*, Nov. 2000, pp. 50–56.

[6] Lipartito, Kenneth, *The Bell System and Regional Business*, Baltimore, MD: Johns Hopkins University Press, 1989, pp. 7–38.

[7] Calhoun, George, *Wireless Access and the Local Telephone Network*, Norwood, MA: Artech House, 1992, pp. 39–40.

[8] Peterson, H. Craig, *Business and Government*, New York: Harper-Collins Publishers, 1989, p. 99.

[9] Colander, *op. cit.*, p. 647.

[10] Garnet, Robert, *The Telephone Enterprise*, Baltimore, MD: Johns Hopkins University Press, 1985, pp. 155–159.

[11] See McCraw, Thomas K. (ed.), *Regulation in Perspective*, Boston, MA: Harvard Business School, 1981, for a review of categorical designations.

[12] Weidenbaum, *op. cit.*, pp. 13–33.

[13] Eisner, *op. cit.*, pp. 178–179.

[14] Head, S. W., and C. H. Sterling, *Broadcasting in America*, Boston, MA: Houghton-Mifflin Co., 1990, pp. 99–102.

[15] Kennedy, Charles, *Introduction to Telecommunications Law*, Norwood, MA: Artech House, 1994.

[16] Case, K. E., and R. C. Fair, *Principles of Economics*, Englewood Cliffs, NJ: Prentice-Hall, Inc., 1992, pp. 381–384.

[17] Colander, *op. cit.*, pp. 645–646.

[18] *Ibid.*, pp. 647–648.

[19] Hasin, B. R., *Consumers, Commissions and Congress*, New Brunswick, NJ: Transaction Books, 1987, pp. 197–218.

[20] Colander, *op. cit.*

[21] Alexander, Peter, "What Hath Telecom Reform Wrought," *Telephony*, June 24, 1996, pp. 52–56. See also Hamilton, E., and S. Virostek, "A Regulatory Burden or Hidden Opportunity," *WB&T Journal*, Dec. 1996, pp. 36–37; and Rose, John, "Trouble with the Telecom Act," *America's Network*, Dec. 15, 1996, p. 10 for different perspectives on the regulatory effect of the Act.

[22] "Business and Consumer Groups Announce Unique Reform Plan," *PRNewswire*, April 16, 1997.

[23] "FCC Releases Plan to Reduce Phone-Industry Access Fees," *Dow Jones,* May 7, 1997.

[24] "Commentary: Let the Dogfight Begin," *Business Week,* April 7, 1997, pp. 19–20.

[25] For an analysis of emerging regulatory developments directly affecting telephony firms, as spurred by the Telecommunications Act of 1996, refer to: Henk Brands, and Evan T. Leo, *The Law and Regulation of Telecommunications Carriers*, Norwood, MA: Artech House, 1999. For a broader analysis of political, social and legal ramifications of the Act, see Patricia Aufderheide, *Communications Policy and the Public Interest,* New York, Guilford Press, 1999.

3

Market Structure and Competitive Behavior

The most immediate impact of the Telecommunications Act of 1996 is the transformation of industry market structure. Before 1996 the telecommunications industry consisted of a set of subdivided markets in which consumers purchased products and services under conditions ranging from substantial competition to regulated monopoly. Passage of the Act reconfigures industry market structure by relaxing or eliminating constraints that formerly existed in telephony and broadcasting. The third tier of modern communications, the computer sector, remains unregulated and is now primed to aggressively interact with its two counterparts.

This chapter provides an overview of market structures, including *perfectly competitive, competitive, monopoly, oligopoly,* and *monopolistic competitive* environments. These five market structures continue to characterize elements of the economy, although regulated monopolies are systematically disappearing in the United States. By categorizing telecommunications goods and services according to the market structures in which they are distributed, we may deduce the short-term, mid-term, and long-term competitive influence anticipated by authors of the Act. Market structures set the environment for competition and, thus, ultimately provide a context for projecting capital investment, business opportunities, and profitability. The identification of transitional market structures, adjusted for time horizons, clarifies emerging competitive threats while refining organizational objectives.

3.1 Fundamentals of Market Competition

3.1.1 Market Structures: Definitions and Economic Significance

Market structure is a term used to describe the characteristics of interaction between competitors in the marketplace. Characterizing the physical interaction of firms necessarily involves a consideration of three main activities: (1) the transmission of information between companies, the key element of which is pricing; (2) physical barriers to entry; and (3) the number of firms and the market share they control or influence [1]. The significance of these three intersecting forces is that the speed at which new communications services are introduced to market will be a function of the restructured marketplace now defined by the Act. Deregulation has changed the rules of telecommunications competition primarily by eliminating cross-ownership restrictions between firms operating within the converging framework of the telephony, broadcasting, and computer industries. Deregulation has thus altered structural characteristics of the market and reshaped strategies by which products are distributed to consumers [2].

Rules guiding the sale of communications services are contingent upon the market structure in which products are distributed. A spectrum of rules running the gamut between perfectly competitive and monopoly structures define the market. Within these parameters, variations in market structures—competitive, oligopoly, and monopolistic competition—play a vital role in the immediate future of the telecommunications industry. The spectrum of market structures and adjoining rules are noted in Table 3.1.

At one end of the spectrum we note that public monopoly structures are fast disappearing in the global communications industry. Today, there exist no unregulated monopolies; a good many regulated monopolies, many of them publicly owned, exist throughout the world. This is particularly true in the utility industry, where it has long been held that simplicity in designing networks—communications, energy, water—minimized cost in distributing service [3]. Such perceived efficiencies have determined public policy guiding these sectors throughout the twentieth century. However, sophisticated networking has obviated the value of these economies of scale; as underscored in those sections of the Act encouraging deregulation, the price paid for suppressing competition may exceed any benefits associated with regulation in today's market environment. While the argument remains a contentious one, as a practical matter the Act now provides us with a forum to test the hypothesis: Does the Act instill market competition and, if so, will resultant benefits to consumers now exceed the advantages of regulation?

Table 3.1

Telecommunications Market Structures and Relationships

	Market Structure			
	Perfect Competition	**Oligopoly**	**Monopolistic Competition**	**Monopoly**
Number of firms	Unlimited	Few	Many	One
Production decisions	No restrictions set by firms	Output somewhat restricted	Output restricted by product differentiation	Output restricted subject to government sanctions and rules
Structural relationships	Firms act independently of each other	Strategic pricing decisions are interactive	Firms act independently	No competing firms
Dynamics of structural change	Rapidly changing relationships	Slowly evolving relationships	Changing relationships based on consumer tastes	Little or no change in structure of business

Concretely, we may define each category of market structure in the following fashion.

- *Perfect Competition:* optimal number of firms jostling in the market, characterized by dynamic changes in price, quality, and service options, with extremely limited market share held by any single firm;
- *Competition:* large number of firms moving in and out of markets, with no firm able to set market price;
- *Monopolistic Competition:* a market structure in which many firms sell segmented, differentiated product lines, often characterized by some producers who specialize in offering one product line but providing substantial variations of choice within that line;
- *Oligopoly:* few firms, many of them acting interdependently in the market, controlling large market share, often through vertical and horizontal strategies that invite antitrust investigation;
- *Monopoly:* a single producer dominating an entire market but submitting to the dictates of government regulation.

Several obvious conclusions can be drawn about the behavior of both domestic and international market structures following deregulation. First, the effect of regulatory relaxation is felt unevenly in the telephony, broadcasting, and computer sectors [4]; as a result, industry structure will respond in different ways, adjusted for different time horizons, to prospective competitive threats. While deregulation effectively ends the artificial boundaries that prevent these three industries from interacting, the fact remains that the computer industry remains an unregulated sector. As a result, the computer market is subject to wider price fluctuations and greater strategic change.

Second, in an industry characterized by turbulence in product development, it is reasonable to anticipate that larger firms will seek immediate access to markets heretofore unopened. Consequently, larger firms could well engage in vertical and horizontal market strategies for purposes of gaining strategic advantage [5]. By acquiring communications firms both within and allied to their own sectors, these entities secure access to new consumers and mitigate the dangers of prospective competition. There are, of course, great risks associated with this strategy: The decision to design an acquisitive plan reduces available capital that could otherwise be diverted to unexpected contingencies. In any case, the wave of mergers and acquisitions following passage of the Act clearly influences market structures in all three sectors.

Third, industry convergence in the short run will configure different market structures based on a combination of factors: product development and innovation, the multiplicity of new entrants to market; intensified globalization, ubiquitousness based on favorable pricing; the speed with which federal and state authorities approve or disapprove proposed mergers and acquisitions; and assertiveness of peripheral players (i.e., emerging competitors drawn from the utility, entertainment, publishing, and other industries) [6]. There is no way to anticipate fully the market structure descriptive of each sector a decade or two hence; such speculation is pure conjecture (note future scenarios described in Chapter 10), but these decisive variables outline a period of great instability. There is little certainty about product acceptance, consumer preferences, or the precise nature of demand amid proliferating choice. Nevertheless, in the short run strategic positioning will require a grasp of alternating market structures operating at variance in each of these three industries.

3.1.2 Market Structures and Concentration Ratios

Following deregulation, the three components of the communications industry can best be characterized as operating within three market structures—competitive, monopolistic competitive, and oligopoly. Monopoly structures have been disbanded, and perfect competition has yet to materialize. The degree of market concentration and market share denotes descriptive structure.

Market concentration is defined as that fraction of output controlled by the leading firms operating within that sector [7]. If minimal concentration and market share are controlled by industry leaders, a competitive structure is said to exist; in those industries in which 60% or more concentration is shared by four firms, we note the presence of an oligopolical structure [7].

Critics of telecommunications deregulation have contended that a departure from government guidelines will only lead to industry consolidation and concentration. Those less concerned about this potentiality note that consolidation can lead to efficiencies not otherwise attainable; in addition, they note, unleashing constraints may act as a catalyst for innovations generated by peripheral enterprises. In distinguishing short-run from long-run market developments, therefore, industry concentration will dictate accompanying market structures. If a dozen or more firms share less than 50% market share in their respective sectors, competitive or monopolistic competitive structures will prevail; if three to six corporations dominate the production and distribution of services, the industry will have transitioned to an oligopoly. An accepted tool used to calculate concentration ratios is the Herfindahl Index, based on the interaction of four-firm market structures [8]. This and derivative tools will be designed to track market consolidation and concentration in the years following adoption of deregulation and inspire Department of Justice merger and acquisition guidelines.

3.2 Market Environment and Organizational Behavior

3.2.1 Market Structure and Strategic Decision Making

Strategic decision making, in economic terms, is the process by which a firm plans for the anticipated competitive response to its pricing and other decisions [9]. In other words, strategic planners seek to account for competitors' behavior prior to launching their own plans. This procedure is a complicated and delicate one and is based substantially on the market structure governing sector behavior. In competitive arenas, strategic planners operate under the presumption that rivals can respond adroitly to tactics launched by one or more companies. In oligopolical structures, the speed with which change occurs is muted because market share (and thus pricing influence) is controlled by a handful of players.

Strategic decision making will become a watchword for the early years of telecommunications deregulation. With little experience in marketing, competitive research, and strategic decision making in competitive environments (see Chapter 6), large players may operate at a disadvantage relative to those with such experience [10]. In a competitive market, one may generally assume that market players respond rationally; therefore, their behavior is theoretically predictable and thus can be modeled.

3.2.2 Issues in Strategic Decision Making for Deregulated Markets

As deregulation transforms market structures, strategic planners will contend with the following issues [11]:

- Developing market plans that enhance customer relations;
- Recruiting new customers for diversified product lines;
- Retaining customers amid proliferating competitors and choices;
- Identifying lifetime projected values for customer bases;
- Computing break-even and roll-out response rates for each product line;
- Building customer loyalty.

The initial years of deregulation will bring limited competition to some areas of telecommunications, while others are likely to experience intense competitive interaction. Strategic decision making will follow the contour of market structure; intense competition will expedite strategic decision making while oligopical competition will diminish the necessity of rapid response. In either case, strategic planners in every sector of the industry will inevitably meet these challenges. It is reasonable to infer, given industry convergence, that to the extent the computer industry invades its companion sectors, it will stimulate creative strategic decision making that would otherwise take longer to unfold [12]. The essential lesson is clear: the necessity of engaging in creative, innovative strategic decision making has now been thrust upon all industry players, and the indicated agenda will be of immediate priority.

3.3 Market Environment and Product Introduction

3.3.1 Market Structure and Product Development: Five Sector Perspectives

One method of illustrating the impact of deregulation on various market structures is to review questions that strategists are themselves addressing relative to the Act's implementation [13]. If we isolate five sectors in the communications industry—local telephony, long-distance telephony, wireless voice and data, Internet applications, and video services—the following materializes.

- *Local Telephony:* How and when will local carriers open their networks? How competitive will local resale and loop unbundling be in the near future? How will local exchange carriers integrate new product lines in their bundled packages? Will local exchange carriers merge, acquire, or partner with prospective rivals to develop their product lines?

- *Long-Distance Telephony:* How will long-distance and international exchange carriers (IXCs) position themselves with respect to pending competition launched by RBOCs? How quickly will IXCs penetrate markets other than standard telephony? What is the threat of Internet telephony relative to standard long-distance voice transmission? Will competition emerge from global as well as domestic competitors?

- *Wireless Voice and Data:* How will incumbent cellular providers respond to emerging competitors [paging, specialized mobile radio (SMR), personal communications services (PCS)]? Should the wireless industry immediately diversify product lines (stressing multifunctionality) or hold back to gauge competitive developments? To what extent is portability of diversified services a need or desire of consumers?

- *Internet Applications:* To what extent will Internet usage define future commerce? Will multimedia applications supplant other technologies in voice and video transmission? Will interactive applications introduce new competitive threats to other areas of the communications industry? Does prospective Internet ubiquitousness form a new market structure likely to influence other segments of the industry? What long-term impact does flat-rate pricing have on local and long-distance telephony networks?

- *Video Services:* Does the interaction of direct broadcasting systems (DBS), *cable television* (CATV), multipoint microwave distribution systems (MMDS), and other video services fundamentally alter short-term pricing? What will be the impact of local exchange carriers' (LECs) entry into this field? Is there a market for interactive television? Does high-definition television (HDTV) have any short-term effect on the video market structure?

In synthesizing these questions, we can discern that each of these five sectors can be characterized by a tangible market structure [14]. Local telephony, the least competitive environment in telecommunications, sustains this mode in the first two years following deregulation. As described in Chapter 2, federal and state regulations are only gradually phased out; this factor, combined with RBOCs' resistance to a restructuring of their markets, has forestalled greater competition. Long-distance telephony, the first market to undergo the rigors of competition extending back to 1984, remains a competitive environment. Wireless voice and data communications is in the infancy of intense competition. Internet usage is growing exponentially worldwide and is presently the most dynamic and competitive communications market; one must dichotomize networking—with its reliance on traditional telephony structure—from emerging service provision. In this sense, the Internet remains a highly dynamic, largely unregulated phenomenon. Its potential power to supplant traditional wireline services make it the great market unknown for the future. Because of substantial capital construction costs, full competition in video services is not projected in many markets after 2005.

3.3.2   Market Structure and Pricing Strategy

As suggested, the initial phase of deregulation, embracing the period extending through 2002, will be characterized by a keener attempt to establish close customer relations and post-purchase satisfaction. Those who adhere to this view point to the strategies already employed in other industries where competition has accelerated due to both domestic and global forces. The assumption is that the communications industry will, in the end, be another component of the world's growing competitive marketplace. In economic terms, this perspective is also predicated on a faith in competitive pricing emerging from the interaction of dynamic market structures now evolving in the telephony, broadcasting, and computer industries.

From the viewpoint of the firm, the question remains: How do we price our product in the midst of a changing market structure? This question becomes an ever more difficult one to address when one considers the unpredictable extent to which these industries invade and withdraw from competitive territories [15]. A software innovation, for example, might suddenly redefine the method by which voice communication is delivered over the Internet. An advance in hardware design in telephony might alter future video delivery. An unanticipated improvement in wireless voice communications might reduce the need for wireline delivery. A partnership between the telephony and broadcasting industries might suddenly place greater demands on the software industry to generate new content and support services for such interaction. The multiplicity of mergers, acquisitions, affiliations, partnerships, and alliances is ceaseless. Thus, strategic decision making inevitably defines methods of *strate-*

gic pricing: setting prices so as to maximize market share while suppressing competitive threats and sustaining customer loyalty and retention [16]. This development is extraordinary for the telecommunications industry and forges a new pricing paradigm. Some firms may not be able to effectively or expeditiously adjust to this new model of pricing.

Because the telephony, broadcasting, and computer industries all price their products and services within different market structures, we must differentiate three key variables in price determination. These include (1) current market structure, based on deregulatory influence; (2) identification of variables to be included in price determination; and (3) the impact of a firm's pricing on the behavior of competitors acting in response. In other words, we must be able to *describe*, *prescribe*, and *project* the collective pricing influence of all competitors in the market. Short-term instabilities associated with deregulation make these procedures exceptionally difficult; the economic stability many anticipate for a later "industry shakeout" will ease the burden and forecasting of strategic pricing.

The fusion of description, prescription, and projection relative to pricing allows a firm to simultaneously take into account both its own actions as well as the competitive response of others. In competitive market structures, one must assume that all three variables operate in a context of greater fluidity than in oligopolical environments. Defining price in a market occupied by a handful of competitors is comparatively stable and, thus, intrinsically more predictable; however, the dynamics of technological change must be accounted for in the process of projection because a sudden innovation could render an existing price strategy obsolete at any moment.

Market structures determine whether a firm acts as a "price-setter" or "price-taker." In competitive markets, a firm is unable to set its own price in the marketplace. In this instance, the firm behaves as a price-taker, accepting prices defined by the intersection of supply and demand among all producers and consumers. In markets characterized by oligopoly or monopoly forces, the firm tends to behave as a price-setter, thus exercising marginal or significant influence over prices consumers pay for such products. It may be inferred that, historically, firms in the telephony and broadcasting industries have typically acted as price-setters rather than price-takers, albeit under regulated circumstances. The computer industry, in both hardware and software development, has evolved under competitive conditions and has precipitated a set of price-taking firms (although proprietary developments in operating systems have narrowed the entrance for emerging competitors). In any case, the Act was intended to supplant regulation with the price-taking dynamics of competitive market structure.

3.3.3 Description, Prescription, and Projection of Pricing Strategy

Descriptive pricing involves profiling the market in which products are introduced. Description is intended to outline the characteristics of the market, thus ensuring that any adopted pricing strategy conforms to its contours. Strategy that results in high pricing relative to competitors is not permissible in a market characterized by dynamic competition. Accurate descriptive methodology allows practitioners to determine the context in which pricing strategy should be designed: competitive, monopolistic competitive, or oligopolical.

Prescriptive pricing is the process of designing a strategy that maximizes competitiveness and market share. To prescribe pricing strategy is to select among alternatives to identify a price that meets market testing—satisfying customers, expanding market share, and building value—adjusted for time horizons. Adaptability to prescriptive pricing quickly differentiates the effectiveness of competitors' strategic decision making.

Projected pricing is the forecasting of price movements necessary to sustain and build market share over the long run. This step is the most difficult and delicate procedure associated with pricing strategy. To elaborate one's own pricing options, it is necessary to simultaneously (1) predict the individual pricing strategies adopted by competitors and (2) anticipate the dynamic interaction of all competitors in the market as they influence the course of prices over the long run. Econometric models exist that facilitate this objective, but they are tenuous in design and application. Indisputably, the forecasting of pricing, with the multiplicity of entrants moving into and out of the industry, constitutes the greatest potential contribution to strategic decision making if models could be derived to anticipate such contingencies. Communications firms will be exposed to models adapted to their industry, with no assurance as to their useful application or results.

In developing a firm's strategic pricing model, all three procedures—description, prescription, and projection—should be included and adjusted for changes in market structure. The core determinants of telecommunications competitive structure, now agitated by deregulation, will shift over time, but such change also should be construed as market opportunity. Every change constitutes opportunity as well as threat.

3.4 Strategic Decision Making and Market Planning: A Methodological Alternative

The material thus far reviewed in this chapter represents traditional thinking about market environments, strategic decision making, and strategic pricing.

In a standard interpretation of market dynamics, it has always been presumed that markets at any moment represent fixed numbers of consumers and that competitors fight with one another to control market share. The outcome of this "war" defined the successful and the unfit. Regardless of market structure—competitive or otherwise—such interaction produced a "zero-sum" game, in which every winner was matched by a corresponding loser [17].

There is a differing, emerging view of market structure that may impinge directly on the outcome of deregulation's effect on the communications industry. This view, now being taken with utmost consideration, is the notion of "coopetition." Conflict, argue exponents of coopetition, will give way in the future to the logic of cooperation blended with the rudiments of competition. Two leading experts, Adam Brandenburger and Barry Nalebuff, present their observations [18]:

> Friend or foe. . . . A player is a competitor if customers value your product more when they have the other player's product than when they have your product alone. A player is your competitor if customers value your product less when they have the other player's product than when they have your product alone. A player is your complementor if it's more attractive for a supplier to provide resources to you when it's also supplying the other player than when it's supplying you alone. A player is your competitor if it's less attractive for a supplier to provide resources to you when it's also supplying the other player than when it's supplying you alone. Customers and suppliers play symmetric roles. Competitors and complementors play mirror-image roles. *There are both win-win and win-lose elements in relationships with customers, suppliers, complementors, competitors . . . think coopetition.*

Brandenburger and Nalebuff surmise that coopetition will surface in various industries where it becomes increasingly clear that the synergies of complementary assistance benefit every player in preparing for competition. We had earlier established (see Chapter 1) that deregulation will impel greater competition within both the domestic and international arenas. Speed to market with new products will punctuate transitions in market development. We had also confirmed the significance of the Act in highlighting the great fact of the post-regulated era: No firm today can independently launch a "bundled" package—or "one-stop shopping" for all communication services. Cooperative synergies will naturally envelop the industry; this is the only way to enhance strategic positioning at maximum speed.

In juxtaposing the Brandenburger/Nalebuff thesis to telecommunications development, it has now become clear that the strategic use of five techniques will be necessary to sustain competitiveness:

- *Mergers:* incorporating two or more firms to create a new entity;

- *Acquisitions:* horizontal or vertical, aimed at gaining immediate access to new markets;

- *Partnerships:* equity interests of two or more firms in launching new or veteran product lines;

- *Alliances:* nonequity interests of two or more firms launching new or veteran product lines;

- *Affiliations:* promotional or other cooperative efforts, establishing new distribution channels.

The various advantages and limitations of each of these tactics are explored in Chapter 8, but it should be underlined here that these developments are consistent with a "coopetitive" telecommunications market environment. The authors of this method contend that such ventures will proliferate and reconfigure all industries, including telecommunications [19]. Content providers thus intersect telephony, broadcasting, and computer producers in all aspects of the industry. To facilitate efficient implementation of their theory, Brandenburger and Nalebuff advocate the use of "game theory," a tool with quantitative and qualitative applications [20]. Game theory is a technique long applied in the military, but its use in contending with the vicissitudes of cooperative ventures remains in its infancy. Certainly, strategic decision makers will want to examine the prospective value of game theory as a tool to design plans and set prices. More importantly, "coopetitive thinking" expands the range of alternatives open to communications firms in setting long-run policy. If one adheres dogmatically to traditional strategic decision making, the range of potential ventures is seriously impoverished.

3.5   Strategic Decision Making and Market Planning: A Conceptual Challenge

Author George Gilder has challenged the two conventional approaches to strategic decision making and the development of market planning for telecommunications [21]. His provocative views are based on an implicit faith in profound future advances in three critical areas: bandwidth, digital compression, and efficient new uses of electromagnetic spectrum. Gilder envisions a metamorphosis in telecommunications convergence: high-speed data transmission, supported by improvements in digital compression, lead to unlimited multimedia applications. In short, Gilder contends that individuals and institutions will be able to move information in all forms—voice, text, and image—in unrestricted fashion at nominal price [22].

With respect to radio transmission, Gilder contends that digital compression supported by more flexible uses of spectrum will lead to abundant applications consistent with his predicted bandwidth breakthroughs. If correct, Gilder's scenario implies that the "anytime, anywhere" ubiquitousness of the utopian information revolution might become a reality. If individuals possessed full information access—retrieval, transmission, and storage—a new dynamic would be introduced in the economy. Prices would fall dramatically, every person would eventually control these tools on their person, and information portability would speed transactions of all kinds.

Apart from the contentiousness implicit in Gilder's technical assessment, the notion of market planning during a period of radical telecommunications transformation should be entertained. Even modest, incremental change in communication services imparts competitive advantage to individuals and organizations who learn to integrate and exploit such benefits. However, the same axiom holds for firms that are able to plan their markets to convey these advantages to their clientele. In other words, telecommunications market planners must educate their customers in identifying the connection between communications innovation and their own competitive survival. In the present environment, this is an important key to a communication firm's success.

Gilder's scenario, if realized, would represent the beginning of a new era in telecommunications market planning and could precipitate, when combined with sophisticated database management, "mass customization" of services for every individual. Communications firms, if technologically enabled to supply unlimited information support for every consumer, would paradoxically eliminate intrinsic communications advantages among those customers. The extension of Gilder's logic and conceptual challenge to traditional market planning signifies a new market paradigm.

A counterpoint to Gilder's position is held by a number of critics who contend that economic, technical, and social costs associated with deregulation offer few advantages to the average consumer. The costs of massive upgrades required to produce state-of-the-art communications architectures, for example, will inevitably be borne by households and institutions. Similarly, even if Gilder's utopian vision were realized in the near term, social interaction—the communication between Americans of diverse backgrounds and interests—would be seriously retarded by the intrusion of such technologies. Much of this criticism hinges on the degree to which such technologies encourage or inhibit interactivity among citizens, groups, and institutions. Whatever conclusion one draws about these developments, the economic influence of telecommunications seems unabated and potentially immense.

3.6   Telecommunications Market Planning: New Markets and Products

The American economy has undergone significant structural changes during the 1990s, and a note must be added regarding these changes relative to prospective communications demand [23]. Some of these changes were alluded to in Chapter 1, but the most significant can be distilled as *telecommuting, home-based businesses, entrepreneurial ventures,* and *outsourcing.* In isolating these trends, we can augment the traditional view of market structure and demand, ordinarily divided between residential and business consumption of communication services.

A note should be added regarding these four emerging structural economic changes. Nine recessions have punctuated the post–World War II period. The last American recession, extending from 1990 to 1991, was unique in the American experience. Economists debate the origin and cause of this recession. Some contend it was the inevitable product of excessive debt accumulation during the 1980s, others believe it was a function of problems in the financial and banking system, and still others link it to problems in the global economic environment [24]. Whatever the cause, what made this recession uncharacteristic in the American experience was the restructuring that followed the restoration of economic stability. After 1991, many corporations operated at lower levels of employment; that is, many firms simply did not hire back a significant fraction of the employees they had laid off [25]. As a result, many individuals opted for alternative means of making a living.

We can verify empirically that the number of home-based businesses has risen dramatically in recent years. These firms range from services to retailing and are supported in some cases by inventories that are managed by other supporting enterprises. The commercial advent of the Internet has encouraged the creation of such firms, and new accounting and financial methods have diminished the costs of running such enterprises. New entrepreneurships, firms independently run outside of the home, and taking a physical community presence, are similarly on the rise. The Small Business Administration now estimates that nearly 800,000 new entrepreneurial ventures are formed each year, up from 650,000 a decade earlier.

Outsourcing, the transfer of employment to independent contractors, has also been increasing sharply since the 1990–1991 recession. Outsourcing diminishes the cost of corporate management while providing a new source of employment to individuals seeking greater autonomy over their profession. Whether such developments come at the expense or benefit of traditionally hired employees remains a matter of contention, but the trend is unmistakable. When combined with traditional consulting, outsourcing represents a significant and growing market development now affecting all sectors of the economy.

Telecommuting, a phenomenon dating to the 1980s, now affects the work of millions. Debates have ensued about the precise number of Americans who telecommute—that is, employees who spend two or more days working at their home or satellite work site—but some have estimated the number at greater than 30 million. Like outsourcing, advantages of reduced corporate costs inspire new working relationships. Social benefits, such as diminished transportation and environmental costs, accompany the shift to telecommuting.

The significance of these developments lies in their influence over the creation and management of new communication product lines. We can estimate that, based on a population of 265 million, 130 million Americans are working at any given moment. Of this number, perhaps 50 million are actively engaged in telecommuting, home-based businesses, outsourcing, or emerging entrepreneurial ventures. Commercialization of the Internet is the key to sustaining productivity in all four of these arenas. In meeting market demand, communication firms must anticipate changes in technology, changes in consumer preferences, and changes in intersecting demand for services between these four segments. The restructuring of the American economy will be monitored very closely by both wireline and wireless firms as they plan and develop new product lines.

References

[1] Colander, David, *Economics*, Chicago, IL: Irwin Press, 1995, pp. 591–599.

[2] Arnst, C., and M. Mandel, "Telecom's New Age: The Coming Telescramble," *Business Week*, April 8, 1996, pp. 44–51.

[3] Collier, S., "Economic Deregulation and Customer Choice: Lessons for the Electric Industry," Washington, D.C.: The Brookings Institution and Center for Market Processes, 1996.

[4] Archey, William T., "Cyberstates: A State-by-State Overview of the High-Technology Industry," *American Electronics Association*, Jan. 1997.

[5] "MCI-Gallup Survey Reveals Technology and Responsiveness," *PRNewswire*, May 2, 1997.

[6] Lynch, Karen, "Crowe's Equation," *Tele.com*, Dec. 1996, pp. 65–67.

[7] Case, K., and R. Fair, *Principles of Economics*, Englewood Cliffs, NJ: Prentice-Hall Inc., 1994, pp. 244–270.

[8]   Colander, *op. cit.*, pp. 593–594.

[9]   *Ibid.*, pp. 603–605.

[10] Branham, Michael, "Measuring What Matters," *Marketing Tools*, April 1997, pp. 12–17.

[11] Frost and Sullivan Inc., *New Strategic Research on Public Data Services*, May 6, 1997.

[12] "New IDC Report Highlights Outsourcing Activities," *PRNewswire*, May 7, 1997; "Communications Preferences Survey Highlights Opportunity to Deliver Packaged Communications Services to Deregulated Marketplace," *BusinessWire*, March 31, 1997.

[13] Evagora, Andrea, "The Dash for Cash," *Tele.com*, March 1997, pp. 52–60.

[14] Tice, William, and William Shire, "One-Stop Telecom: The New Business Model," *Telecommunications*™, April 1997, pp. 63–66.

[15] Holland, Royce, "Future Vision," *Telephony*, May 20, 1996, pp. 36–38.

[16] Barnett, John, and William Wilsted, *Strategic Management*, Boston, MA: PWS-Kent, 1988, pp. 136–168.

[17] Brandenburger, Adam, and Barry Nalebuff, *Coopetition*, New York: Doubleday Press, 1996. Complementary material on coopetitve strategy may be found in Tapscott, Donald, David Ticoll, and Alex Lowy, *Digital Capital*, Boston: Harvard Business School, 2000.

[18] *Ibid.*, pp. 36–40.

[19] Flanagan, Patrick, "The Top Ten Hottest Technologies in Telecom," *Telecommunications*™, May 1996, pp. 29–36.

[20] Colander, *op. cit.*, pp. 604–607.

[21] See Gilder, George, *Telecosm*, New York: Simon and Schuster, 1996.

[22] Gilder, George, "Bandwidth Tidal Wave," *Forbes*, Dec. 5, 1994, pp. 48–56.

[23] Colander, *op. cit.*, pp. 659–682.

[24] For a critical treatment of the restructuring of the labor market, see Dent, Harry, Jr., *The Great Jobs Ahead*, New York: Hyperion Press, 1994.

[25] Dent, Harry, Jr., *Job Shock*, New York: St. Martin's Press, 1995, pp. 4–38.

4

Economic Tools for Telecommunications Professionals

The purpose of this chapter is to survey the primary economic tools destined to influence telecommunications market planning during the era of deregulation. We may generally divide the array of these "economic tools" into two categories: microeconomic and macroeconomic data analysis. Macroeconomic analysis relates the determinants of aggregate market demand to the business cycle. Microeconomic analysis involves the evaluation of supply and demand and market equilibrium, the computation of costs and profits, and estimation of economies of scale and projection of pricing. The principal concern of telecommunications firms during the early stage of deregulation will be pricing strategy aimed at establishing and expanding market share.

We infer that the initial years of deregulation are likely to be turbulent based upon the pattern of historical experience seen in large industries previously deregulated. We also presume that a key dynamic in telecommunications deregulation will be the impetus to competition instigated by the computer industry. As one element of the telecommunications triad, the computer sector remains untouched by the specter of regulation; the economics of the industry may change as computers supplant traditional services offered by its broadcasting and telephony cousins. Innovations in this industry will fuel change in telephony and broadcasting in addition to new approaches to customer relations.

The synergies of technological innovation and relaxed regulatory barriers preventing cross-ownership further reinforce the implicit uncertainties associated with identifying market winners and losers in the years ahead. Moreover, should the computer industry come to assume growing influence in such matters as privacy, security, obscenity, and other noneconomic areas, the paradoxical prospect of gradual deregulation in telephony and broadcasting may be accompanied by government intrusiveness in the third sector. Such noneconomic contingencies may nevertheless have far-reaching economic consequences for the entire industry over the long run. Telecommunications professionals must therefore assert a collection of economic tools that benchmarks emerging competition with its consequent impact on future pricing.

For purposes of this discussion, the application of economic tools is focused on telecommunications market planning. The use of microeconomic techniques to dissect market demand and provider response is emphasized in this presentation, although the costs of providing universal service and interconnection remain implicit and embedded in such a review [1]. Web site economic research tools on telecommunications infrastructure are provided in Appendix C. Of primary concern in this segment is the linkage between economic opportunities resulting from passage of the Act and the market research methodologies presented in Chapters 3 and 5 to 7. Those engaged in strategic planning, market forecasting, and sales management require a collection of tools that will enable them to seize upon opportunities evolving from competitive market interactions.

4.1   Overview of Macroeconomic Analysis

4.1.1   Macroeconomic Tools for Telecommunications Market Planning

Although the primary focus of communications planners during deregulation will be the microeconomic analysis of veteran and emerging product lines, a note should be included here regarding the macroeconomic characteristics of the industry. The forecasting of broad long-term demand for all goods and services is tied directly to the course of employment, interest rates, and cyclical changes in the economy. With variance as to sector and composition of product line, the correlation of these factors varies from industry to industry [2].

In the case of the telecommunications industry, the influence of macroeconomic forces is manifested in the degree to which demand is dependent upon such factors as aggregate employment, income, and debt accumulation. We find in many communications products a relatively low correlation between these variables and resultant demand [3]. If we dichotomize veteran from emerging product lines, we note that the use of the telephone is largely impervious to cyclical variations in the economy. We verify this by evaluating market penetration rates. Stable rates—penetration rates that remain unaffected by transitions between prosperity and recession—suggest that these products and services remain ingrained fixtures.

As time has passed, cable television subscription has come to resemble these same characteristics [4]. Rates of computer adoption for businesses and households imply the same phenomenon in recent years. The most profound post–World War II example of high, sustained rates of adoption seemingly unaffected by turns in the business cycle is television: in the nine recessions punctuating this era, television sales sustained a geometric rate of adoption independent of recession or inflation. A dramatic recent example of the same dynamic is reflected in the videocassette industry [5]. In each case, we note that nominal pricing, coupled with broad business and social acceptance, led to ubiquitous and habitual use.

Telecommunications is thus differentiated from many other industries whose demand is connected directly to movements in the business cycle. The housing, steel, durable goods, and automobile industries all anticipate demand, in part, as a result of broader transitions in the economy. For veteran product lines, communications planners are concerned less with such empirical tests. It should be added, however, that to the extent that demand increases in these "cyclical" arenas, consumption may be expedited for communications services. The lasting empirical lesson from these data is clear and concrete: Given time, individuals and organizations come to rely on such products and services, unless alternative superior products are introduced.

The introduction of new products and services, however, does bear a comparatively close relationship to shifts in the business cycle. When economic output grows rapidly, unemployment falls and real wages rise, and the number of new adopters increases proportionately. It is in this environment that new communication services are most effectively introduced to market. Conversely, market contractions or recessions diminish the number of "experimenters" inclined to purchase newly introduced services, particularly if they are exotic in nature. Marketers have long known this market characteristic and adjusted their presentation of new products accordingly.

The two critical macroeconomic variables to examine in considering prospective demand for communications products and services are employment and income. The decision to purchase, and continue to use, communications services is chiefly a function of these variables. To this extent, the business cycle is an important tool in estimating long-term demand; if this method of analysis is applied, however, the length of current and future business cycles must be predicted. Generally, the lengths of prosperity have averaged four to five years; most recessions have averaged under one year [6]. However, the precise timing of recession and prosperity remains elusive, with recessions occurring in 1949, 1954, 1958, 1960, 1970, 1973–1974, 1980, 1981–1982, and 1990–1991. The cycles have thus been irregular and governed by a confluence of events often unanticipated or surprising. Nevertheless, once the transition from recession to prosperity is clearly evidenced (see Figure 4.1), the impetus generated by lower unemployment and rising income is a significant aid to aggregate demand. Telecommunications planners, even if committed to long-term capital construction, cannot be completely oblivious to the dynamics of business cycle forecasting; the point in time at which a new product is introduced can be fortuitous or disastrous for a firm, especially during periods of accelerated competition.

4.1.2   Other Issues in Business Cycle Analysis

One issue historically divides strategic planners in estimating aggregate demand for their products. This matter concerns the philosophy of adapting the business cycle as a macroeconomic tool to unfolding demand for new goods and services [7]. One philosophical view holds that national or regional demand for all services, telecommunications staples included, is a function of national income. In other words, if a forecaster anticipates rising national income, the presumption is that national aggregate demand will thereafter rise proportion-

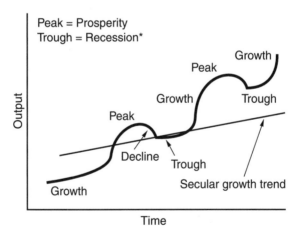

Figure 4.1 Transitions in the business cycle. Note that in the post–World War II era the average expansion has lasted approximately five years. The longest periods of expansion unpunctuated by recession are 1961–1970 and 1982–1990, which were marked by geometric growth in telecommunications infrastructure and services, as defined by convergence. The secular growth trend represents the average annual growth rate in the American economy (which has averaged between 3.0% and 3.5% since 1945).
*Recession is defined as two or more consecutive quarters of negative growth in the Gross Domestic Product.

ately. An alternative outlook contends that there exists a "family of business cycles," in which one must select the appropriate cyclical model to predict probable demand industry by industry. This perspective divides cycles by unique characteristics or attributes and categorizes the following [8]:

- Seasonal analysis, in which demand follows the contours (measured monthly or quarterly) of residential and business demand throughout the year;

- Secular analysis, in which data is examined to elicit persistent underlying trends independent of national, regional, or other forces;

- Three-cycle schema, in which three long-term forecasting waves are adjusted for 40 to 50 years (Kondratieff), 40 months (Kitchin), or Juglar (dividing the Kondratieff wave in six increments). These cycles are superimposed over one another to illustrate emerging short-term developments.

The last of these techniques has been accepted in many academic circles as appropriate business cycle theory; its credibility and application to telecommunications may be a matter of dispute or conjecture, but a macroeconomic grasp of long-term aggregate demand could facilitate capital construction decisions regarding infrastructure. As suggested in Chapter 7, the key to enhancing the probability of success in forecasting lies in applying all methods of analysis, and then searching for areas of common agreement based on intersecting inferences.

Another issue of concern raised by analysis of the business cycle concerns the influence of psychology on consumption. As illustrated in Figure 4.1, incomes rise as the economy ascends to its peak. This ascendancy influences psychology—the attitudes, feelings, and confidence—of consumers and emboldens them to emulation or experimentation, depending on their position on the S-curve (discussed in Chapter 5). Significantly, data reveals that psychology plays a role in determining the timing of initial purchases (i.e., their first experience with a new communications product). Thus, the state of the business cycle at any moment indirectly influences the timing of such experimentation and can accelerate or decelerate consumption accordingly. This factor is of less concern to telecommunications providers—tied to long-term building commitments—but of keen value to content developers, who must consider such information when delivering new products. Psychology is profoundly difficult to quantify, though attempts have been numerous, and its value in illuminating consumer behavior in telecommunications remains unclear at present [9].

4.2 Overview of Microeconomic Analysis

4.2.1 Microeconomic Tools for Telecommunications Market Planning

Microeconomic tools are essential for communications market planners. The application of these techniques varies considerably in terms of infrastructure provision versus content development. While those firms engaged in the management of infrastructure—principally those involved in the telephony and cable television industries—must allocate large sums of capital over time to sustain a customer base, many content developers can readily shift product lines to accommodate changes in market preferences. To this extent, content developers extract competitive advantages vis-à-vis their own competitors by exploiting network economies of scale and scope [10]. We noted in Chapter 2 that Sections 251 and 259 of the Act were written in such a way as to ensure that proliferating networks would become integrated through complete and fair interconnection at pricing designed to assure universality. The translation of this feature of the Act into practical economic terms has meant that many firms (e.g., Internet providers) currently assume a "quasi-free rider" position via deregulation.

The following sections survey the fundamental techniques used to estimate supply and demand, costs and profits, equilibrium and elasticities, and the dynamics of pricing. While the application of these tools applies uniformly to both service providers and content developers, the costs uniquely borne by network managers differentiate their planning agenda from those of other telecommunications firms.

4.2.2 Principles of Telecommunications Supply and Demand

A market consists of buyers and sellers. Resultant transactions are measured in price. Prices are defined by the interaction of supply and demand in markets. Some markets are local in nature, while others are regional, national, or international in scope. The geography of markets is an important characteristic in the evolution of telecommunications networks and the services they support.

If the intent of the Act is to encourage competitive markets, then we would identify three important overlaying characteristics to market supply and demand.

1. The number of buyers and sellers would be such that no entity could independently influence levels of price.
2. Fluctuations in pricing levels would have to remain unimpeded by government sanction, regulation, or intervention of any kind.

3. Movement of buyers and sellers would remain mobile over time (in other words, buyers would be free to select alternative providers at any given moment). Suppliers, too, would be free to move in and out of veteran and emerging product lines to secure their clientele.

As noted in Chapter 3, there exists a spectrum of competition—gradations in the degree to which markets exhibit these characteristics. In perfectly competitive markets, optimal numbers of buyers and sellers moving in great mobility often produce substantial fluctuations in pricing. In any event, it is not government that influences the course of pricing over the long run.

We have established that in the telecommunications triad of the telephony, broadcasting, and computer sectors there exists significant variation in competitive structure. While the computer industry remains fully unregulated, its counterparts are significantly regulated even after enactment of deregulation. Within telephony and broadcasting, some elements of the Act have modified or relaxed regulations (as elaborated in Chapter 2), but government influence over local telephony has remained entrenched. The commitment by federal and state governments to universal service has perpetuated an unfettered interference in market transactions. Whatever the underlying public interest spirit or motive, the fact remains that market supply and demand are influenced by government edict. We must therefore diagram intersecting supply and demand according to the *type* of market activity in which that enterprise is engaged. In areas denoted by competitive market transactions, as illustrated by the principles above, we diagram prevailing supply and demand as illustrated in Figure 4.2. We may refer to the setting of prices in such environments as "competitive market price determination" [11].

The interaction of supply and demand defines price, with supply and demand curves ever changing. Thus, price will ordinarily remain a constant if government asserts its regulatory powers to that end. Price stability, or predictability, will be sacrificed in the short run if deregulation is manifested in the way its sponsors intended, but greater congruence between market supply and demand will be attained in the long run. The effect of this congruence would, in theory, lead to *equilibrium price*, the state in which buyers want to buy the same quantity that sellers are prepared to sell. Therefore, if a seller sets a price higher than equilibrium, surpluses occur; if a seller sets a price too low, excess demand ensues and shortages inevitably follow. Surpluses, in short, beget subsequent declines in price. Shortages induce higher prices to the point of equilibrium.

We may generally conclude that the standard application of supply and demand analysis is appropriate to pricing for content providers. Those communications firms that supply content for infrastructure—telephony, television,

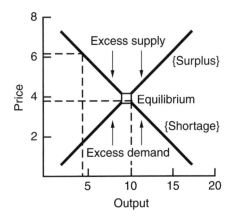

Figure 4.2 Dynamics of market supply and demand. Prices set above the equilibrium point generate surpluses, while prices set below precipitate shortages. Price reductions in the form of "sales" lower excess supply to the equilibrium point over time; price increases mitigate excessive demand until equilibrium is attained. Shifts in lines of supply and demand tend to be more dynamic in instances of excessive supply or demand, with competitors entering and leaving the marketplace at accelerated speed.

and computers—thrive, survive, or die by the prevailing dynamics of supply and demand. Those who manage, control, design, or create networking infrastructure operate, however, in a fundamentally different environment. Costs of generating content for infrastructure tend to be labor-intensive; costs of developing that infrastructure are capital-intensive and typically require a substantial time horizon to gain profitability. In short, the cost of labor is decisive in content development; the cost of capital is key in infrastructure development. Thus, economies of scope (cost savings through multifaceted service provision) coupled with economies of scale (the decline in per-unit costs as production rises) are essential for network managers. Reducing network costs per subscriber while exploiting profit potential from an expanding array of value-added network services is critical to the emerging deregulated environment (see Figure 4.3).

Large network providers obviously hold at least a short-term advantage during the early stages of deregulation. These firms are best able to exploit the fundamental principles of economies of scope and scale; they are best positioned financially (see Chapter 8), as a result, to engage in vertical integration by absorbing competitors [12]. Moreover, there exists an intrinsic comparative advantage for large network providers; the addition of each subscriber increases the value of the network for veteran users [13]. We thus note the economic formula for success in network provision: the exploitation of economies of

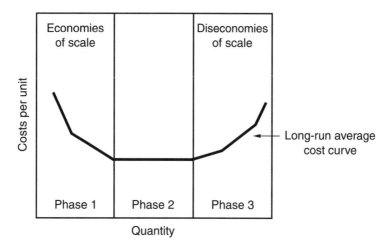

Figure 4.3 Hypothetical telecommunications services model: costs per subscriber. In the long run, the average cost curve for servicing telecommunications subscribers falls due to economies of scale; later, costs of servicing subscribers rise in response to diseconomies of scale (that is, as the firm expands, monitoring costs increase while costs of managing and sustaining employee productivity also rise). In response, the firm seeks to exploit economies of scope by designing a new product line and generating added revenue. This phenomenon takes hold in phase 3 of the long-run average cost curve.

scope and value combined with value-added multiple services. When these factors are juxtaposed to vigorous marketing, thus leading to expansive market share, a strategy is set in motion that relegates short-term pricing strategy to a secondary concern. What matters to network providers in the short term is market share and competitive threat; the law of supply and demand takes over in the long run to define market winners and losers. Despite the comparative advantage of size that some firms enjoy, the risks of committing long-term capital while network innovations advance rapidly are considerable.

Closely allied to capital risk is the problem of user-churn, the frequency with which consumers move from provider to provider for their communication needs. With the prospect of proliferating competition, with new technologies supplanting veteran product lines, user-churn generates both threat and opportunity in the Act's aftermath. The importance of user-churn as related to market share and strategic planning is discussed in Chapter 6, but one must simultaneously contemplate risks of capital and user-churn when projecting long-term plans. The need to satisfy consumers at their basic level of need has now assumed a significance never before entertained by common carriers. In the absence of predictable cash flow, it is not feasible to raise or borrow the capital required to build and maintain state-of-the-art infrastructure. It is for this reason that customer-led customization—one tool for satisfying the consumer's basic communication requirements—is now regarded as a strategic priority in some telecommunications firms (see Chapter 10).

4.2.3   Price Elasticity of Demand in Telecommunications

The measurement of changes in quantity demanded in relation to changes in price is called price elasticity. Simply, firms need to estimate how responsive consumers will be with respect to fluctuations in pricing. Elasticity is calculated by dividing the percentage change in quantity taken by the percentage change in price. In defining elasticities at various pricing points, a firm is able to compute gross revenues in relation to the number of units sold. Thus, analysis of elasticity serves two goals: (1) to gain a keener sense of the sensitivity of consumers to current and prospective changes in price and (2) to facilitate forecasting revenues in the face of competitive pressures in pricing.

Figure 4.4 diagrams the relationship between price changes, demand elasticity, and total receipts. The calculation of elasticity will yield one of three results in each application: a value greater than 1, equal to 1, or less than 1. A value less than 1 indicates inelasticity, meaning demand is not responsive to changes in price. A value greater than 1 defines elasticity, suggesting that demand fluctuates with changes in price. Unitary elasticity occurs when the computation equals 1. If price elasticity of demand yields a value greater than 1, then a rise in price will generate less revenue. If the demand for a product is said to be inelastic, then an increase in price will produce more revenue. In situations where demand is unit elastic, the percentage change in quantity equals the percentage change in price.

Obviously, there is spectrum of demand for communications products that is variously elastic or inelastic. Telephone service for most people is generally inelastic, since it is regarded as a staple of modern living. Cable television service has come to assume the same role for many Americans in recent years. For workers involved in occupations where mobility is crucial, cellular telephones are regarded as a necessity. On the other hand, interactive television service remains comparatively exotic, and thus pricing is highly elastic. The inelasticities associated with many staples represent potential opportunities for firms to increase prices. Yet, under conditions of emerging competition, the capacity of firms to raise prices while controlling market share is severely constrained. In part, unfolding inelasticities of demand for certain services inspired framers of the Act, who sought to mitigate higher prices that would accompany monopolistic or oligopolistic market structures.

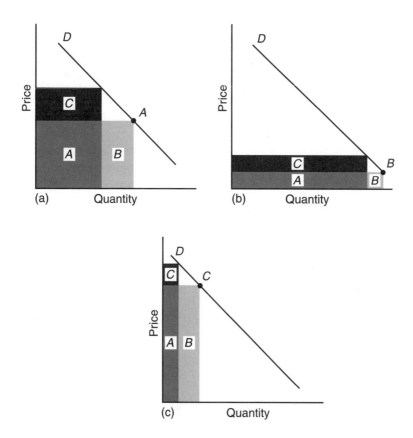

Figure 4.4 Price elasticity and total revenue. The illustrations indicate that as price is raised aggregate revenue expands by rectangle *C* and decreases by rectangle *B* under conditions of (a) unit elastic, (b) inelastic, and (c) elastic market environments.

4.2.4 Profit Maximization in Telecommunications

A communications firm, like any enterprise, maximizes profit by increasing sales and minimizing cost. The difference between revenue and costs constitutes its profit. In competitive market structures, a firm would compute its marginal revenue and marginal costs and identify that level of output at which the two are equal. Marginal revenue is defined as the change in total revenue precipitated by a one-unit change in output level. Marginal cost, or incremental cost, represents the increase or decrease in total costs a firm bears resulting from the output of a unit more or less. Simply, prices are defined from the intersection of supply and demand, and the calculation of marginal costs and revenues identifies appropriate output levels. The extent to which a firm sells its outputs maximizes its revenue; the difference between revenue and costs thus determines profit.

The measurement of profit is vital to evaluating a firm's strategic vision. The computation of profit, the demonstration of profitability, is essential in conveying a firm's credibility to stockholders and potential investors. We will note in Chapters 8 and 9 that profitability means "survivability" in the long-term murky waters of deregulation.

The computation of profit margin provides communications firms with a valuable tool in justifying their strategies to potential investors. Profit margin is the ratio of income to sales and is measured in two formats. *Gross profit margin* is reflected by the percentage return that a company earns over the cost of goods or services sold; it is computed by dividing gross profit (sales minus costs of goods sold) by total sales. The gross profit a firm earns essentially covers operating expenses, including administrative expenses, taxes, and interest. *Net profit margin*, or return on sales, reflects the percentage of net income generated by each sales dollar. The standard computation of net profit margin results from dividing the income statement figure for net income after tax by total sales [14]. Both computations are illustrated in these expressions:

$$\text{Gross Profit Margin} = \text{Gross Profit/Sales}$$
$$\text{Net Profit Margin} = \text{Net Income After Tax/Sales}$$

Both calculations provide investors and competitors with important clues about the value-added desirability of communications products. In an environment of rising competition, the firm that generates a product that provides value-added attributes is likely to generate a higher profit margin. In this sense, the determination of profit margins is significantly more important than gross profits.

4.2.5 Determinants of Supply and Demand

Empirical evidence consistently pinpoints the following criteria as "determinants" or independent variables driving consumer and business demand for communications services. These include (1) income, (2) tastes and preferences, (3) prices of complementary and substitute goods, (4) future expectations regarding market prices, and (5) the number of buyers. These criteria, coupled with current market prices, determine aggregate demand for a product [15].

In the case of supply, the following criteria, apart from price, drive the supply providers are willing to introduce to market. These include (1) cost, (2) technology, (3) number of sellers, (4) prices of other products (or substitutes) that sellers could introduce, and (5) future expectations about market price.

Methodologically, a firm would isolate each of these variables in relation to any given product and apply correlation analysis to determine the extent to which each is influential. A linear equation would then be designed to express the demand function appropriate to that product. From the point of providers, similar equations can be designed to estimate the supply generated by competitors and the resultant impact on pricing. Demand and supply functions are used in all industries and have gained renewed consideration in telecommunications following passage of the Act.

4.3 Telecommunications Demand and the Substitution Effect

The degree to which consumers select from among alternatives to satisfy their communications needs may be described as the "substitution effect" [16]. The measurement and forecasting of dynamic substitution will grow in importance in the telecommunications industry as deregulation unfolds. Multiple factors define the dynamics of substitution, but we may generally conclude that the differentiation of elastic from inelastic preferences for segmented markets reveal much about the future of communications needs.

Few will disagree that food and milk are, in principle, inelastic staples for most consumers. It is reasonable to conclude, also, that video cameras and vacations represent preferences best characterized as elastic in nature, with resultant wider price fluctuations. Consumers essentially do not reduce their consumption of necessities, regardless of changes in price. Price elasticity is substantially related to variation in income for luxury items. In the communications palate, the issue of substitution becomes a complex and nebulous one amid rapid technological change. The need of individuals and organizations to communicate rapidly has grown exponentially in recent years, but consumers will always search out lower cost alternatives—substitutes. How does one measure and project the course of substitution over time? Too many unknowns make such forecasting highly complex, but in differentiating each communications product or service, we may deduce the determinants of market demand as outlined in Table 4.1 [17].

In distinguishing "luxurious" from "necessary" communications services, there will exist substantial variation among income, occupational, and other stratified groups (although these differences are likely to diminish over time, for reasons outlined in Chapters 1 and 5). Where market demand is elastic, consumers will have multiple alternatives available; this might eventually mean, for instance, using Internet telephony as a substitute for local or long-distance

Table 4.1
Determinants of Market Demand

Elastic Market Demand	Inelastic Market Demand
Luxury	Necessity
Many substitutions available	Little or no substitutions available
Price represents large fraction of income	Price represents small fraction of income
Long-run time horizon	Short-run time horizon

telephone service. In markets characterized by a range of available substitutes, the price consumers bear tends to represent a higher fraction of income, particularly as they seek higher value-added services. The longer the time horizon, the greater the propensity toward elasticity; as time passes, new alternatives are introduced to market, and consumers are able to select the most desirable alternative. Consumers alter their tastes and preferences, as well, in response to emerging substitutes. The use of e-mail, for example, has diminished the need to communicate by letter [18].

It has been demonstrated, too, that longer versus shorter time horizons play a significant role in fulfilling the social benefits of elasticity and substitution. For example, in the short run only larger organizations could afford computers, fax machines, and related communications equipment. The long-run time horizon had the effect of reducing the cost of these items, which made them more readily available to first, small businesses, and later households. The dispersed benefits of these technologies cannot be measured exclusively in economic terms and have meant substantial improvement in the quality of life for many people [19].

The economics of "substitution" can be expressed through *cross-price elasticity*, or the degree of responsiveness of consumers to changes in price for a particular product relative to changes in price for available substitutes. For purposes of this measurement, available substitutes may be either identical or complementary in nature. As noted earlier in this chapter, it is obvious that technological innovation threatens veteran product lines; therefore, forecasting cross-price elasticities will assume a heightened priority for telecommunications firms in the future. There is expansive empirical evidence that consumers today explore the desirability of substitutes in a way that their predecessors did not. Therefore, it behooves the modern communications provider to maintain close scrutiny on those firms instigating such changes, especially in the area of computer innovations.

4.4 Telecommunications Indifference Curves

Deregulation, even if not fully realized in the manner anticipated by sponsors of the Act, is likely to have two primary effects on telecommunications supply: (1) there is likely to be a rapidly expanding, possibly explosive, growth in the number of veteran and emerging product lines, as technological innovation is substantially encouraged; and (2) there is likely to be added network capacity, anticipating growing consumption. The two forces are interactive and will instigate profound shifts in the scope of services available. Where a single cable television provider was formerly entrenched, the presence of DBS, MMDS, and cooperative community ventures will expand choice and diminish price. Where a single provider assumed control over local telephony, multiple providers may emerge, perhaps even from industries historically unrelated to telecommunications (e.g., electric utilities). Where television content had been presided over by broadcasting networks, cable television, computer, and other concerns may intervene to provide alternative programming. In these instances and other prospective cases, consumers may experience astonishing choices previously unimagined.

As consumers brace for choice and newly formed substitutions, indifference curves will inevitably become an important tool for communications firms. Both content and infrastructure providers will seek out the benefits of measuring the extent to which consumers satisfy their communications needs by choosing from among proliferating choices. The *indifference curve* measures bundled choices from among alternatives to satisfy needs and desires. In other words, for each dollar a consumer commits to satisfying basic communications needs (see Figure 4.5), how will he or she allocate that sum for telephony, television, and other services?

Consumers may opt for a simple bundle of telephony and television service, or shift to the Internet for delivery of all services, or select a single provider that can supply a menu of options unique to his tastes. The dynamic interaction of the determinants of supply, as outlined, guarantees no simple answer as to the extent to which consumers will augment or simply substitute their preferences for communications.

The problem is a profoundly complex one when innovations are introduced so readily to market. As cellular phones, Internet service, interactive television, and other services augment veteran product lines, how will the consumer reallocate his communications budget? Or, is it conceivable that the need and desire for information management and transmission are so ravenous that today's emerging services will become tomorrow's staples? A general response to these questions cannot be formulated, but the measurement of reallocated communications budgets can be diagrammed through the use of

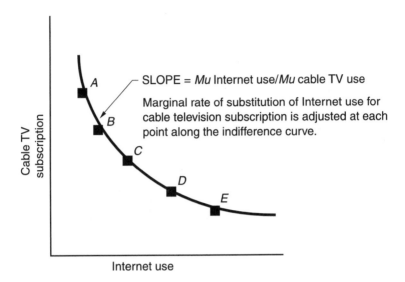

Figure 4.5 Hypothetical indifference curve, which measures the hypothetical combinations of alternative services that satisfy one's telecommunications requirements. An implicit question posed by the curve concerns whether consumer demand for communications services will continue to rise or remain relatively constant, compelling different choices along the indifference curve.

indifference curves; a firm would design indifference curves for each targeted market segment. Indifference curves gain a renewed importance in this industry as deregulation is implemented.

4.5   A Comment on Economic Tools for Telecommunications

Limitations of space herein prevent a fuller treatment of the economic tools that are appropriate to evaluating and predicting the course of the telecommunications industry during deregulation. Appendix C lists a full array of resources, including Web sites, that will permit the reader to track the corporate applications of these tools. Seminal research in the field is now under way, precipitated by the enactment of deregulation, and organizations such as the International Communications Forecasting Conference sponsor innovative microeconomic forecasting methods [20].

It can be stressed, despite the uncertainties and constraints of price measurement and prediction, that an appropriate complement of economic tools for communications managers and planners would include [21]:

1. Business cycle analysis, with special emphasis on the relationship between the business cycle and consequent national income and its relationship to communications consumption;

2. Supply and demand analysis, with emphasis on the design of demand functions appropriate to market segments;

3. Computation of market price equilibrium points;

4. Application of substitution analysis;

5. Calculation of elasticities, particularly with respect to emerging product lines.

Particularly in the case of firms that previously operated in monopoly markets, a fresh perspective on equipping managers and strategic planners with these skills would greatly improve organizational communication and camaraderie. The need to sensitize all members of the firm to the economic realities of deregulation has asserted itself. There is now a manifest requirement to convert the esoteric jargon of economics into a common organizational language.

References

[1] See Heldman, Peter, Robert Heldman, and Thomas Bystrzycki, *Competitive Telecommunications: How to Thrive Under the Telecom Act,* New York: McGraw-Hill Co., 1997.

[2] Valentine, Lloyd, and Dennis Ellis, *Business Cycles and Forecasting,* eighth edition, Cincinnati, OH: Southwestern Press, 1991, pp. 130–168.

[3] See Gasman, Lawrence, *Telecompetition: The Free Market Road to the Information Highway,* Washington, D.C.: Cato Institute, 1994.

[4] Belisle, Patti, "Cable Television," *Communication Technology Update,* Boston, MA: Focal Press, 1996, pp. 35–45.

[5] Brown, Dan, "A Statistical Update of Selected American Communications Media," *Communication Technology Update,* Boston, MA: Focal Press, 1996, p. 350.

[6] Valentine and Ellis, *op. cit.,* pp. 59–105.

[7] Case, Karl, and Ray C. Fair, *Principles of Economics,* Englewood Cliffs, NJ: Prentice-Hall Inc., 1992, pp. 586–629.

[8] Valentine and Ellis, *op. cit.,* pp. 106–128.

[9] Whiteley, Richard, and Diane Hessan, *Customer Centered Growth,* Reading, MA: Addison-Wesley Press, 1996, pp. 146–190.

[10] Egan, Bruce, *Information Superhighways Revisited: The Economics of Multimedia,* Norwood, MA: Artech House, 1996, pp. 166–168.

[11] Colander, David, *Economics*, Chicago, IL: Irwin Press, 1995, pp. 548–568.

[12] Dolan, Edwin, and David Lindsey, *Microeconomics*, Chicago, IL: Dryden Press, 1992, pp. 420–423.

[13] Egan, *op. cit.*, pp. 166–167.

[14] Argenti, Paul, *The Portable MBA Desk Reference,* New York: John Wiley & Sons, 1994, pp. 321–322.

[15] Frank, Robert, *Microeconomics and Behavior*, New York: McGraw-Hill Inc., 1991, pp. 134–158.

[16] Colander, *op. cit.*, pp. 502–506.

[17] Frank, *op. cit.*

[18] See Tansimore, Rod, "In Its Image," *Telephony*, April 21, 1997, pp. 64–70; and Ernst, Daniel, "Consumer Services," *Tele.com*, Nov. 15, 1996, pp. 57–62.

[19] See Strassman, Paul, *Information Payoff: The Transformation of Work in the Electronic Age*, New York: The Free Press, 1985, for a discussion of social (and quality-of-life) benefits arising from integrated use of information and communication technologies.

[20] Consult Web site references listed under the University of Michigan (see Appendix C) to identify institute resources committed to price and market share forecasting.

[21] See Hawkins, D. I., Roger Best, and Kenneth Coney, *Consumer Behavior: Implications for Marketing Strategy*, Homewood, IL: Irwin Press, 1996, Section Four, for a discussion of the integration of microeconomic techniques into marketing and strategic planning.

5

The Economics of Consumer Demand for Telecommunications

The liberalization of regulatory constraints has led to a reexamination of the determinants of consumer demand for telecommunications goods and services. Many factors contribute to consumer decision making and consumption. Thus, the industry faces the challenge of identifying those independent variables most responsible for adoption and habitual use of communication services. Firms must now forecast and plan for demand at a time of technological turbulence, proliferating competition, consumer uncertainty, and global interaction. Without question, the post-1996 competitive environment has refocused the industry's energy on the criticality of consumer behavior.

This chapter provides a foundation for the tools, evaluated in the following two chapters, required to forecast demand and construct strategic plans. It should be stressed, as one proceeds through this material, that a period of gestation will be necessary for consumers to identify the services they require. While firms can accelerate consumer adoption by encouraging an ambiance of experimentation, there is no substitute for the passage of time to underscore the value of such products. Before enactment of the Act, consumers had little or no choice in deciding which communications products were appropriate; after enactment of the Act, confusion has mounted proportionate to the number of new providers and the choices afforded. With this caveat in mind, this chapter elaborates the categories of consumer behavior associated with an enlarged menu of options, the determinants of aggregate demand, and future issues associated with an informed consumer market.

5.1 Determinants of Consumer Behavior

5.1.1 Situational Influence

A determinant (or independent variable) of consumer behavior is any factor that influences decision making at the time of purchase. Such factors include motivation, personality, lifestyle, income, educational attainment, and other social and group influences. Among these important determinants is *situational* influence [1].

Situational determinance involves physical surroundings, social surroundings, temporal perspectives, task definition, and antecedent circumstances. In situational influence we note the impact of atmosphere in setting household demand for communications services. The atmosphere of physical surroundings—whether based on a desire for education, knowledge, entertainment, or other needs—is decisive in defining consumption. Physical settings in business and home life, as noted in subsequent chapters, are critical factors in setting the agenda of consumer demand in the consumption of telecommunications services.

5.1.2 Problem Recognition

Problem recognition constitutes a critical variable in consumer behavior after initial exposure to a new communications product or service. Substantial empirical evidence in other industries has revealed that consumer decision making becomes more refined and extensive as purchase involvement increases. In other words, as consumers gain experience with communications services—and thus become more sophisticated in the use of product lines—their ability to differentiate products heightens [2]. Problem recognition—distinguishing the value of products in relation to their optimal use—defines the category of consumer decision making. Some researchers categorize consumer behavior at the time of product adoption as consisting of *habitual, limited,* or *extended* decision making.

Habitual decision making is defined as that category of consumer behavior in which customers repeatedly purchase the same product; such purchases reflect little desire to search for alternatives to veteran, predictable product lines. Memory and experience linked to branded product lines are the pivotal sway at the time of purchase. Resistance to price and dissatisfaction with service or quality can undermine consumption ordered by habit. Predictably, limited decision making surfaces as the product of unease with the status quo: consumers now seek additional knowledge before buying, often depending on word of mouth before making a decision. Extended decision making envelops consumers when they become uncomfortable with existing product lines, when price sensitivity magnifies differences among multiple alternatives, and when post-purchase evaluation becomes a concrete concern. It should be noted that all consumers at various times exhibit all behavioral identities, depending on the product or service in question. The ability of a communications firm to draw these distinctions is central during deregulation, where branded loyalty was formerly taken for granted by marketing and sales staff [3].

5.1.3 The Information Search

Consumers apply information gleaned from two sources to make their adoption decisions: internal and external. Internal information—or stored information—defined by personal experience, memory, and passive recall from comments made by others, is of particular relevance for corporations with established reputations in the telecommunications industry. External information, or knowledge resulting from individual interaction with other consumers, as well as advertising, can also be gleaned from contact with consumer groups, government databases, direct product inspection, and other personal sources. Customer pursuit of external sources of information, after problem recognition, will be an important long-term concern of the communications industry. One may reasonably infer that the early transition to deregulation will be one in which firms will attempt to reinforce the value of branded reputations.

What emerges from a serious analysis of a consumer's information search, and what is particularly salient for communications firms, is that the pending proliferation of products—from basic telephony to Internet usage, from wireless services to interactive television—will overwhelm and confuse many consumers in the short run. As time passes and as an atmosphere of acceptance is manifested, telecommunications companies can expect customers to invest additional time in external information searches. Just as significantly, for those firms that can design appropriate bundled packages of services, the early stage of deregulation will provide a ripe time to optimize market share based on consumer anxiety [4]. This important factor is given additional elaboration as to its strategic implications in Chapter 7. Inevitably, however, bundled packages based on standardized strategies may be effective in the initial years following deregulation; however, consumer tastes, tending to greater fastidiousness as competition intensifies, will compel firms to design strategies increasingly sensitive to external information channels [5].

5.1.4 Lifestyle and Attitudes

Among the dominant factors influencing consumer behavior are lifestyle and the attitudes associated with the human lifecycle. Identifiable independent variables influence consumption with the passage of age, and the lifestyle one adopts. Empirical evidence confirms that as one ages lifestyle generally transitions to a more conventional, conservative pattern. Conformity to social norms and dictates of family life defines social, political, and cultural attitudes [6]. Such transformations are accompanied by rising consumption, as noted in Chapter 1. Personal consumption peaks at about age 49 and generally subsides thereafter [7]. Lifestyle and attitudes define the content of consumption. One cannot separate age, lifestyle, and attitudes from the pattern of telecommunications consumption likely to be seen in the next decade.

To gauge the impact of lifestyle and attitude on future configurations of telecommunications demand, market researchers apply psychographic methodology [8]. Psychographics is a tool that estimates consumption according to categories of shared attitudinal characteristics. By grouping consumers into such classifications—classifications based on psychological need, age, peer groups, and other schemes—firms can designate the core of demand. In an era of intensifying competition, those firms able to accurately apply these methodologies will retain a distinct market advantage. A visual description of psychographic methodology is provided in Figure 5.1.

5.1.5   Motivation, Emotion, and Personality

Much of consumer behavior, and thus purchase and post-purchase decision making, is a function of economic intangibles. Principally, these intangible, often indefinite, but always influential variables include individual motivation, emotion, and personality makeup [9]. Voluminous literature has been compiled to document empirical tests relating the variance with which motivation determines consumption. A. H. Maslow's seminal research suggests that all human beings shift through five stages of motivation before achieving full satisfaction in life: these include, in order of transition, physiological, safety, belongingness, esteem, and self-actualization [10]. These transitions, in turn, are connected to varying tastes and rates of consumption. Complementary research by others has identified the importance of curiosity, self-expression, affiliation, and modeling. Examined in totality, human beings—despite their general desire for independence and self-assertion—are highly influenced by the actions, motives, and behavior of others [11]. This phenomenon is pertinent to an analysis of the adoption rates of new services now anticipated in the telecommunications industry.

The use of surveys, focus groups, and individual and group testing allows a firm to ascertain the extent to which motivation drives particular product choices. Emotion, a subset of motivation, can drive the purchase of goods and services but is less likely to influence purchase and post-purchase decision-making processes. The ambiance likely to be characteristic of telecommunications consumption in the age of deregulation is not a function of discretion

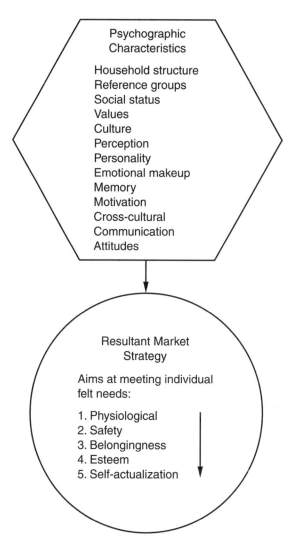

Figure 5.1 Psychographic determinants of consumer behavior and decision making.

drawn from emotion; it is the result of felt need and reliance on continued innovation. Thus, personality will be given a higher priority in market research in the future.

Personality is the collection of traits forming the individual. For reasons discussed in greater detail in Chapters 6 and 7, the economy and industry are transitioning to an era of customized menus. Advanced technologies now afford a firm the luxury of designing a communications palate specialized and attuned to the taste of individuals. Passage of the Act effectively accelerates the competitive thrust to meet such demands. Those firms unwilling or unable to meet this challenge are likely to suffer declining market share and possible extinction; the link between personality and consumer demand grows increasingly apparent as a result of deregulation. Furthermore, new market research efforts in "data warehousing" and "data mining" facilitate competitive opportunities scarcely known previously in the industry [12].

5.1.6 Evaluating Alternatives

The process of judging a product's traits—price, quality, and other objective characteristics—constitutes objective criteria. Other characteristics, such as reliability, dependability, and prestige, are more subjective in nature but nevertheless influence consumer behavior. Under such circumstances, it is common for consumers to depend on branding—or corporate image and reputation—to differentiate alternatives. How do consumers decide which product or service to adopt when inexperienced with available competitive alternatives? The effectiveness with which firms address this issue will be an important factor in attaining long-term success. In a world of proliferating choice in telecommunications, particularly as firms align in forming partnerships to deliver services, the identification of evaluative criteria becomes an immediate economic and methodological question.

We can assert a general methodology to the process by which consumers evaluate alternatives: the measurement of qualitative and quantitative information that consumers use to differentiate choices. How consumers perceive alternatives for each product on the basis of branding, image, reputation, and prestige constitutes the first and foremost area of inquiry for market research. Second, the comparative importance of each of these four variables—as well as others—defines the use of evaluative criteria necessary to construct a marketing plan. In this regard, the material supplied in the following two chapters discusses forecasting methods appropriate to assessing the process by which consumers distinguish product alternatives.

5.2 Household Consumption

Among data compiled by telecommunications firms are household numbers detailing the patterns of aggregate household consumption. These data are formulated in cyclical, seasonal, and other forms. These are firm, reliable data about which credible forecasts can be made as to regular and periodic demand for communication services. Before passage of the Act, such forecasts were often generated with a high degree of confidence, even certainty about prospective aggregate demand; with competition about to proliferate, with new technologies pending implementation, and with underlying economics now favoring industry convergence, analysis of household consumption becomes a delicate craft.

Consumption of telecommunications services generally falls within three definable areas of adoption: consumer applications, household lifecycles, and organizational (business/industrial/governmental/nonprofit) demand. With respect to household lifecycles, there exist certain driving variables that impinge on demand. These variables include cultural and social class variables, patterns of socialization, family structure, role specialization, segmented decision making, and the changing stages (relative age) of the household lifecycle [13]. In previous eras, the decision to adopt communications products would have been relegated to a single member of the household. Deregulation promises to complicate the process of forecasting demand amid proliferating choice since many emerging services have a unique value to each family member. In an era where telephony and basic cable television service constituted essential communication services, aggregate consumption was comparatively streamlined and straightforward.

It is reasonable to anticipate that emerging wireline and wireless services, particularly in terms of interactive applications, will be customized to the need of each member of the household. Proliferating competition assures that every individual will approach the key economic decision maker in the household for those services of greatest personal value. The household decision maker, therefore, will have to contend with multiple choices and the heightened demands of others to formulate a menu of services. For marketers, it will be increasingly necessary to focus on the habits of each individual to estimate household consumption patterns over time [14].

5.3 Organizational Consumption

Decisions regarding communications product adoption were formerly the domain of a single individual or committee in many organizations. Today, the convergence of the telephony, computer, and broadcasting industries, and the industries that support them have had the effect of encouraging individuals throughout an organization to investigate product alternatives. We thus note a parallel situation in contemporary organizational purchase decisions: Proliferating alternatives have overwhelmed even telecommunications experts in many firms, while individuals without such expertise are nonetheless cultivating expertise in a single technology that they deem necessary to their work. Inevitably, deregulation will indirectly encourage friction between management layers who cannot cooperate or communicate effectively on matters of communications integration.

In designing marketing strategies to meet the consternation of organizational buying during the initial stages of deregulation, communications might contemplate the following tactics.

1. Isolate purchase decisions in this fashion: (a) new task adoption, (b) straight rebuying, and (c) modified rebuying. The most complex of these decisions involves the deployment of new communications services—adopted to solve emerging problems—while replacement decisions ordinarily are less provocative in their organizational implications.

2. Identify the organizational style of a firm: styles can generally be classified as *command, cooperative,* or *competitive* [15]. In firms where command decisions are made, product adoption is simplified from the viewpoint of the vendor; on the other hand, post-purchase servicing is typically more facile and agreeably sustained in organizations that adhere to cooperative styles of decision making. Telecommunications firms clearly must consider the strategic and tactical implications of the organizational style governing product and service agreements. This issue becomes a difficult one with which to contend in organizations that have reputations for shifting from one style to another over time.

5.4 Monitoring Consumer Behavior: Standard Versus Emerging Techniques

In monitoring, measuring, evaluating, and forecasting consumer behavior relative to the telecommunications industry, firms may integrate a set of traditional and state-of-the-art tools. Traditional methods of monitoring changes in consumer preferences, tastes, and patterns of consumption include a set of qualitative and quantitative applications. Standard procedures, including surveys, focus groups, historical analogy, and others, are described in the following two chapters. The use of such techniques remains a fixture in basic market research and in the extrapolation of data to estimate prospective demand. These techniques have recently been augmented by sophisticated new methodologies based on data warehousing, data mining, and interactive applications [16].

Data warehousing is the process of accumulating information about all aspects of consumer behavior. Database management and the use of object-oriented technology to integrate diverse aspects of multiple market databases now permit organizations to identify nearly every detail impinging on consumer demand. Data mining is the methodological procedure of filtering voluminous data in a warehouse; by isolating and applying the patterns of behavior unique to each consumer, unprecedented inferences can now be drawn about current and prospective demand. In short, the stage has been set for the marketing of customized telecommunications services to individuals, households, and organizations. Monitoring consumer behavior is thus more effective than ever, despite the complexities of expansive competition and enhanced services.

5.5 Diffusion Rates for Communications Services

The pattern by which veteran and new product lines are introduced to and adopted by a market is defined as a diffusion rate. Diffusion rates describe the rapidity with which consumers purchase communication services. Estimation of diffusion rates is a critical link in predicting aggregate future demand for a firm's products and is performed in assessing individual, household, and organizational lifecycles.

Diffusion rates generally fall into one of three categories for telecommunications and will come to take on increasing significance in the near future. We note *rapid diffusion, standard diffusion,* and *slow diffusion* as typical descriptions of unfolding rates of adoption. Figure 5.2 identifies market diffusion rates. Obviously, there are important strategic reasons associated with a firm's identification of its prospective product market opportunities. When viewed in the context of unfolding consumer behavior, diffusion illuminates several important strategic considerations.

1. The entry to market by competitors is effectively accelerated after product introduction in cases of rapid diffusion. In other words, rapid consumer demand quickly inspires competitors to emulate, simulate, or substitute for a successful product line.

2. The appropriate strategy to exploit consumer demand is to inaugurate a new product and rapidly improve its quality as it "rides" the diffusion wave amid rising competition. In this fashion, the firm establishes a reputation for both creativity and innovation in emerging product lines.

3. Each of the three illustrations describe an S-curve phenomenon. That is, consumer adoption, and therefore diffusion, is slow at the outset as a new product (accompanied by high price) gains acceptance;

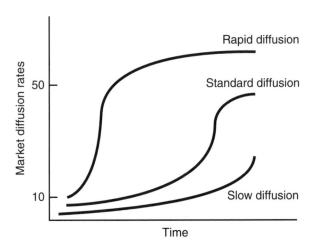

Figure 5.2 Diffusion rates associated with new product introduction. Rapid diffusion rates are defined by new products that quickly gain 10% market acceptance; thereafter, the time necessary to secure a 90% penetration rate approximated the time required to secure the initial 10% increment. Determinants of rapid diffusion include high observability in the marketplace, extensive marketing effort, strong felt need, low complexity, easy trials, positive word-of-mouth, large relative advantage, low risk, and "ubiquitous" pricing.

thereafter, growth becomes exponential as consumers influence non-users to adopt the same product or service lines.

4. Generally, rapid diffusion is contingent on new product introduction. Standard and slow diffusion rates are characteristic of innovative product lines.

5. In rapid and standard diffusion rates, the time required to generate 10% market penetration is comparable to the period necessary to shift from a 10% to 90% rate. In other words, market acceptance is slow at the beginning and accelerates rapidly when nonusers see the benefits gained by those who experiment.

S-curves, as tools used to assess prospective consumer behavior, have various limitations and contingencies. We may infer that their most effective application occurs in those instances in which broad practical and social acceptance seems likely, in which economies of scale can reduce manufacturing costs, and when such services perform a manifest need or otherwise substitute for existing product lines in efficient or cost-effective ways. Additional information on S-curve forecasting, as applied to the wireless telecommunications sector, is supplied in Chapter 7.

5.6 Emerging Issues in Consumer Behavior

5.6.1 Categories of Consumer Behavior and the S-Curve

Much research has been performed in recent years to categorize, classify, differentiate, and segment consumers [17]. The effort here is aimed at determining what marketing strategy is necessary to appeal to every "type" of consumer. The presumption is that certain groups of consumers share characteristics that then enable a firm to standardize its approach to the market. Marketing is defined as the pricing, promotion, and distribution of goods and services. For communications firms, the initial market research question following the advent of deregulation concerns the application of the S-curve to forecast demand in the context of explosive competition.

A starting point for the description and analysis of consumer behavior involves the placement of consumers along the S-curve. Early adopters are those interested in experimenting with new products; these individuals have the discretionary income to engage in such experimentation and enjoy establishing precedent or creating new images. Early majority consumers are those who perceive benefits tangibly demonstrated by those who initiate such purchases but who are unwilling to be the first to adopt such innovations. Late majorities, riding the S-curve to 90% market penetration rates, join the early majority as prices drop and applications become obvious. Laggards only come into the market after social acceptance of new products has been established, prices approach nominal levels, and the stigma of being "left out" becomes palpable [18].

It should be noted that many firms further segment these categories—early adopters, early majorities, late majorities, and laggards—in many additional segments. However, all such breakdowns can be crystallized into these four typologies. The crucial point here is that exponential growth in sales can only be achieved when a firm meets the criteria outlined above and when product pricing falls to accommodate each successive category coming into the market. One may also presume that improvements in quality and service rise relative to pricing as consumers ride the S-curve, although there is substantial variance from firm to firm in this regard.

5.6.2 Consumer Behavior and Bundled Communications Services

The initial years following a deregulated communications market are likely to be characterized by a consumer desire to "bundle services." Simply, in moving from a monopolistic market—in which clear lines were drawn between the telephony, broadcasting, and computer industries—to a market-based system, consumers may be so inundated by choice that consternation could immediately follow. The lack of experience in deciding upon a local telephony provider, for example, could conceivably instill in consumers a desire for simplicity and predictability. By extension, it is logical that a firm that can provide all fundamental services—telephony, television, Internet, and wireless access—can readily secure market share. Such expediency now accounts, in part, for the wave of mergers and acquisitions that followed enactment of the Act.

As underscored several times in this volume, it is reasonable to infer consumers will gain increasing sophistication in sorting through communications alternatives. As a result, a bundled strategy holds effective tactical implications for the short run but cannot substitute for competitive pricing, quality, and servicing over the long term. For segments of the early adopters, early majorities, late majorities, and laggards, "bundled" market tactics represent an expedient measure, but the competition that unfolds thereafter could instigate new innovations in product design and delivery. Consumers, five years hence, are likely to become savvy purveyors of telecommunications as they link market knowledge with personal need.

5.6.3 Consumer Behavior and Particle Marketing

Particle marketing, discussed in terms of its future implications in Chapter 10, represents the pricing, promotion, and distribution of services to customers on a one-to-one basis. In other words, to fully engage particle marketing a firm must customize services for every customer in its defined market. This is an unprecedented development and has particular significance for the telecommunications industry. Because the communications industry depends on the provision of intangible (nonphysical) goods, instantaneous transmission of information is a realistic objective. When consumers exploit the dynamics of particle marketing, they are empowered to provide feedback to providers in a way unimaginable in most industries.

A contentious inference would hold that an informed consumer contingent, once sensitized to proliferating options, would be willing to abandon the stability of bundled packages for personalized services adapted to changing tastes. One can only speculate about the efficacy of sophisticated market networking to build and sustain market share; it seems likely, however, that high-end consumers (those whose incomes permit expansive consumption) will explore options that permit personalized menus. Thereafter, a "learning curve" for other customer segments could well dictate future communications services. Particle markets, using methodological options now in their infancy, are predicated on a pattern of deregulation reducing market concentration. The development of an industry oligopoly would minimize the probability of this scenario. In the meantime, a competitive environment encourages a firm to exploit its organizational resources to advantage; strategic positioning is thus aimed at establishing an image of accommodation to residential and business need. The strategic and tactical implications of particle marketing are discussed in the following two chapters.

It should be emphasized that a firm's ability to monitor changes in consumer behavior is crucial to exploiting the emergence of particle markets. We may reasonably conclude that such factors as strategic marketing, organizational restructuring, process reengineering, and the realignment of customer support services will be vital in superseding the challenges of competitors who perceive the same opportunities.

5.7 Consumer Demand and Changing American Tastes

Social and cultural values decide, to a significant extent, the varying patterns of economic consumption. Therefore, consumer demand projected for telecommunications services must be examined in light of changing tastes. Changes in tastes—or consumer preferences—are influenced by shared social and cultural norms among a variety of segmented groups. To fully anticipate changes in demand, and to forecast with precision their impact on sales and post-purchase servicing, one must (1) define social and cultural values that distinguish each segmented group and (2) link these values to patterns of demand for each service [19].

Mechanically, one approaches the problem of forecasting changes in consumer preferences by stratifying groups according to age, sex, gender, ethnic background, socioeconomic status, and other characteristics [19]. By applying correlation analysis to the identification of "strength of relationship" between each of these independent variables and resultant demand, we gain a keener sense of prospective consumption. A crucial element in the post–World War II period affecting the communications industry is the transformation of American behavior precipitated by dynamic changes in technology, politics, economics, and social interaction. As a result of dramatic change, consumer behavior has been altered fundamentally because conflicts emerge between and among stratified groups; that is, differences in social and cultural values magnify differences in consumer attitudes. It is therefore vital to determine the marketing strategy appropriate to *changing*, segmented groups. A paradoxical economic opportunity implicit in this dynamism concerns the potential for greater understanding and acceptance among the same groups. Such understanding may hold an important economic ramification as private alliances might emerge between stratified groups who learn to employ improvements in communications and information infrastructure to generate new business.

An important footnote to this discussion concerns the economic influence of women in American society during the past two decades. An increasing fraction of the workforce consists of "working women," a contingent whose need for advanced communications services parallels that exhibited by men. With outsourcing, telecommuting, and home-based businesses now transforming segments of the American economy, one cannot overemphasize the significance of such restructuring. It will therefore be important for a firm to forecast demand for each stratified group according to preferences exhibited by men and women. In other words, there are likely to be differences between men and women as to their consumption of services within a wide range of bundled alternatives.

5.8 Demand for Telecommunications Services in the Post-Act Era

The great economic question posed by the passage of regulatory reform concerns underlying consumer and business demand for communications services. Expressed plainly, how will prospective burgeoning supply affect unfolding demand? Has primary consumer and business demand already been satiated at some nominal level? Does innovative supply create its own demand through attractive marketing efforts? Will increases in demand, if any, be marginal with respect to patterns of consumption in the past, or does there exist pent-up demand for a broad array of new services? Researchers are now contending with this murky area of inquiry as vast capital flows are poised for infrastructure development [20].

Empirical evidence postulated in Chapter 3 suggests that the natural instinct of most consumers will be to "substitute" rather than "augment" communications usage when presented with evolving value-added alternatives. However, opinion is divided as to potential aggregate demand for emerging services. Such analysis is largely limited to conjecture since we have not yet experienced the blossoming of competition previously promised in local telephony [21]. Seminal research now being conducted points to the "bundling" (or packaging) of services as the key determinant to the expansion of aggregate consumption [22]. Favorable pricing accompanied by a simplified, self-directed menu of services appears to be the most expedient method for carriers to encourage consumer demand. This straightforward strategy represents the communications firm's point of view in the post-Act era.

Projecting aggregate demand from two perspectives, that of the consumer and the provider, may be a bit myopic. One might wish to engage in a third dimension—environmental—as outlined in Chapter 1. The intersection of economic, political, and demographic forces seemingly impel a greater need for communications services: As competition intensifies, the need to transmit, receive, record, and synthesize information intensifies proportionately. Superimposed over each of these variables is the stimulus of global competitive reconfiguration.

There exists a clear linkage between wealth and advanced communications infrastructure. Studies indicate that the influences of these two variables are interactive, with both factors exercising independent and dependent roles.[1] In other words, the efficient manipulation of information generates wealth, while material progress facilitates further improvement and innovation in required infrastructure. One induces wealth by being productive; the role of communications, therefore, lies in its capacity to make all factors of production—capital, labor, and technology—more efficient. Such efficiencies will characterize the most productive individuals, organizations, and nations in the future. This is not speculation, but historical extrapolation.

5.9 Consumer Demand and Changing Global Tastes

Just as consumer tastes change with the passage of time and the dynamic interactions described previously, so too do consumer preferences vary throughout the world. We can identify two overriding considerations in behavioral change affecting the demand for global communications services. The first concerns the cultural influence of American entertainment, education, technology, and public policy. The United States is the clear and uncontested largest exporter of intellectual property in its broadest form: the contribution of ideas and knowledge to the design of new enterprises [23]. As a result, when intellectual property is fused with advanced communications, American influence on global consumer preferences sustains a profound impact. Many people throughout the world simply emulate, in varying degrees, American tastes and attitudes. The exercise of such influence is connected directly to the adoption of emerging technologies aimed at linking diverse peoples to American dissemination of information. The demise of Cold War political structures has, of course, accelerated this trend.

---

1. Evidence points to communications infrastructure as being both an independent and dependent variable with respect to wealth creation. Review the Web site selection under the University of Michigan (see Appendix C) for database listings on international empirical research conducted in this field.

Concurrent to this phenomenon is the growing influence of "globalization"—one market, not many, in which arbitrary geographic distinctions are simply ignored by consumers. Despite governmental attempts to control, dictate, or regulate production and employment within their own boundaries, consumers—now equipped with sophisticated forms of transmitting and retrieving information—assert their independence in purchasing decisions. Chapter 9 discusses developments in deregulating and privatizing markets in each area of the globe; these political decisions further encourage the movement toward greater consumer discretion.

References

[1] Lidz, Theodore, *The Person*, New York: Basic Books, 1968, pp. 70–92.

[2] Hawkins, D., R. Best, and K. Coney, *Consumer Behavior: Implications for Marketing Strategy*, sixth edition, Homewood, IL: Irwin Press, 1995, pp. 90–184.

[3] Tice, W., and W. Shire, "One-Stop Telecom: The New Business Model," *Telecommunications*™, April 1997, pp. 63–65.

[4] "Inquisitive Middle Class? Lost Cause?" Survey, *PRNewswire*, April 14, 1997.

[5] Tansimore, Rod, "In Its Image," *Telephony*, April 21, 1997, pp. 64–70.

[6] Lidz, *op. cit.*, pp. 509–522. See also Maier, N. R. F., *Psychology in Industry*, New York: Houghton-Mifflin, 1970, and Hall, C. S., and Gardner Lindzey, *Theories of Personality*, New York: Wiley Press, 1974, for additional seminal information and historical analysis in this field.

[7] Dent, Harry, Jr., *The Great Boom Ahead*, New York: Hyperion Press, 1993, pp. 21–34.

[8] Hawkins, Best, and Coney, *op. cit.*, pp. 284–341.

[9] *Ibid.* See also Bourne, L., and B. Ekstrand, *Psychology: Principles and Meaning*, New York: Holt-Rinehart-Winston, 1991, for additional background and analysis.

[10] Drucker, Peter, *Management: Tasks, Responsibilities, Practices,* New York: Harper & Row, 1990, pp. 195–216.

[11] Kotler, Philip, *Marketing Management*, eighth edition, Englewood Cliffs, NJ: Prentice-Hall Co., 1996, pp. 416–484.

[12] Baker, Dan, and Brian Washburn, "Bill of Sales," *America's Network*, April 1, 1997, pp. 17–22.

[13] Dent, *op. cit.*, pp. 28–46.

[14] Russell, Cheryl, *The Master Trend*, New York: Plenum Press, 1993, pp. 55–62.

[15] Arlen, Gary, "Choosing the Right Path," *Convergence*, May 1996, pp. 39–48.

[16] Edelstein, Herb, "Mining for Gold," *Information Week*, April 21, 1997, pp. 53–70.

[17] Bernhardt, K., and T. Kinnear, *Cases in Marketing Management*, New York: McGraw-Hill Co., 1997, pp. 349–416.

[18] Dent, Harry, Jr., *The Great Jobs Ahead,* New York: Hyperion Press, 1995, pp. 65–92.

[19] Bernhardt and Kinnear, *op. cit.*, pp. 514–645.

[20] Wilde, Candee, "Telecom Analysts Take Stock of the Market," *Tele.com,* May 15, 1997, pp. 113–132.

[21] Flanagan, Patrick, "The Ten Hottest Technologies in Telecom," *Telecommunicatons*™, May 1997, pp. 25–38.

[22] Salak, John, "Delivering the Goods," *Tele.com,* May 1, 1997, pp. 56–65.

[23] Edvinsson, Leif, and Michael S. Malone, *Intellectual Capital,* New York: Harper Business Press, 1997, pp. 65–74 and 101–138.

6

Telecommunications Market and Competitive Research

Deregulation now prompts, urges, and cajoles telecommunications firms to engage in the most sophisticated techniques of market and competitive research. We may generally surmise that in the new deregulatory environment, new product development, improved customer service, and market access will differentiate the successful from the unsuccessful. Entrenched bureaucracy, regulatory constraints, and transitions in strategic planning limit the capacity of established telephony and other enterprises to quickly adapt to the new deregulatory environment. The competitive marketplace will compel the adoption of and adaptation to new market and competitive research techniques.

This chapter surveys the incorporation of both standard and emerging techniques to sythesize and evaluate data on marketplace and competitor developments. The object of this research is to facilitate both short-run and long-run strategic planning. The convergence of the telephony, broadcasting, and computer industries, supported by deregulation, has precipitated the most dynamic industry in the world. As noted in Chapter 1, the dynamics triggering competition in this field create profound structural change in all other sectors of the economy as enterprises learn to exploit such technologies. To the extent that the telecommunications industry innovates its planning mechanisms, we can expect to see other sectors transforming their strategic contingencies [1].

6.1 Market Versus Competitive Research

One must distinguish between two critical research mechanisms that necessarily precede a discussion of forecasting and strategic planning. *Market research* is the collection of data relating consumer behavior to prospective demand for goods and services. Market research, at its core, juxtaposes elements of behavior—attitudes, socioeconomic background, personality, and so on—with changes in consumer preference. The focus in this regard emphasizes the tracking of market segments (as discussed in Chapter 5) and their unfolding changes in tastes, preferences, and fundamental needs. To engage in market research is to examine unceasingly every dynamic influencing customer demand [2].

Competitive research, also referred to as "surveillance" by some practitioners, is the review and evaluation of strategic planning and performance promulgated by one's competitors. Competitive research, in its totality, includes quantitative and qualitative treatment of competitor performances, monitoring of emerging alternative technologies, and assessment of consumer interaction with other firms relative to price, quality, service, and post-purchase satisfaction. Those who employ competitive research techniques seek to determine the extent to which competitors' strategies alter prospective market share and the effect this will hold for the future of the firm [3].

6.2 The Methodology of Telecommunications Market Research

6.2.1 Traditional Tools of Market Research

The basic, or traditional, tools of market research include *segmenting, targeting,* and *positioning.* Segmenting the market means, in essence, identifying and dividing customer bases for promoting a product. Market segments could expand or contract, depending on changes in consumer preference as evidenced by customer surveys, focus groups, and other tests. Targeting involves pinpointing current and future market opportunities, that is, identifying prospective congruence between a firm's resources and its ability to satisfy a market segment. Positioning is the procedure by which a firm sets its product line apart from those of competitors, which is especially vital in industries where impending competition is about to proliferate [4].

6.2.2 Procedures for Market Segmentation in Telecommunications

Traditional market segmentation calls for the application of the following sequential procedures:

1. Identification and measurement of the market: This procedure demands that researchers group consumers within market segments and apply appropriate methodologies to confirm the extent to which a firm's product line conforms to consumer needs. Can the firm measure a segment's characteristics, desires, and needs?

2. Determination of segment profitability: Is the market segment of sufficient size to justify a firm's commitment of financial and other resources?

3. Market communications and distribution: Can the firm introduce its product to market and effect channels of communication to monitor prevailing consumer preferences?

4. Elasticity of response: Does the market respond to changes or differences in market strategies? In other words, how can the firm ensure that the effectiveness of its market strategy can be scientifically evaluated?

5. Market segment stability: Do forecasts reveal stability—therefore predictability—of market segments, or is dynamic change expected? This factor is critical in determining a projected modification of market and strategic plans.

The basis upon which market segments are defined includes the manipulation of both descriptive and behavioral platforms. Descriptive platforms consist of variables, the integration of which provides a profile of groupings within defined market segments. Such variables include age, sex, gender, nationality, income, occupation, religion, family size, household lifecycle, and education. Behavioral platforms are comprised of variables including brand loyalty, readiness to buy, social class, lifestyle, usage rates, user status, and other identifying characteristics focusing patterns of behavior impinging on demand. The creative assimilation of descriptive and behavioral platforms is what distinguishes firms in terms of their relative success in market research [5].

Established telecommunications firms are likely, during the first five years of deregulation, to apply historical analogy in defining and projecting their approaches to market segmentation [6]. Historical analogy is a methodological technique aimed at isolating relevant patterns of business experience found either inside or outside the same industry. Telephony and broadcasting firms will survey, in light of the Act, the effectiveness of marketing plans devised by firms with comparable competitive experiences in other industries. In this way, errors in market segmentation will be avoided and the probability of marketing success will be enhanced. Historical analogy is discussed in greater detail in the next chapter.

6.2.3   Procedures for Market Targeting in Telecommunications

The selection of target markets is characterized by the identification of groups whose consumption would result in greatest profitability. In other words, ordinarily a firm chooses to introduce its product to those groups within market segments who not only can afford to purchase that item but where profit margins can best be extracted in a competitive environment. Thus, there exists a ranked-order methodology to targeting; the largest and most profitable groups are introduced to new product lines before extended market absorption is attempted [7]. This procedure is both traditional and sensible in the eyes of many, but new marketing techniques may obviate its effectiveness in light of unpredictable competitive threats emerging from technological substitution.

The identification and evaluation of target markets lead to strategies appropriate to exploit such groupings. Typically, marketers will employ either *concentrated* or *differentiated* strategies to exploit emerging opportunities. A concentrated strategy focuses on a single, most profitable group of consumers within a market segment. A differentiated strategy, sometimes described as "multisegment strategy," fully develops all niches within a segment and promotes its product to each differentiation [8]. Particle marketing constitutes a third approach to the identification of targets: each consumer becomes a market unto him- or herself. Some observers refer to particle marketing as "one-to-one" promotion or, alternatively, "atomization." The trend toward narrowly defined targets is unique to this decade, but some speculate that the next progression in targeting may be "subatomization": the identification of the predominant independent variables responsible for a customer's behavior. These subatomic units can vary over time and thus require constant monitoring and measurement to be useful tools in the targeting process [9].

We can infer an important underlying impetus to target research for communications firms during the next decade. For established companies previously constrained by the rigors of regulation, it is probable that they will ease into traditional concentrated versus differentiated methodology before embarking on the delicate and sophisticated mechanisms of target atomization. Organizational inertia and operational size prevent large, regulated firms from engaging customers on a one-to-one basis. Those communications firms previously untouched by regulation, and who remain more nimble by virtue of smaller, decentralized marketing units, are better suited to apply such atomization.

6.2.4   Procedures for Market Positioning in Telecommunications

Market positioning incorporates a number of organizational and product characteristics to gain a foothold in either established or emerging markets. The emphasis is not on image but on competitive advantage: The issue at hand concerns gaining comparative leverage over all other competitors. In the communications market, the procedures by which positioning is accomplished are especially critical because they must be utilized in a context of great volatility and uncertainty.

Those engaged in positioning telecommunications firms pinpoint:

- *Product attributes:* The association of advantageous characteristics vis-à-vis competition (cellular phones, for instance, are superior to paging services because of value-added characteristics).

- *Solution specification:* This service solves or satisfies a basic consumer or business need (wireless data transmission enables mobile workers to integrate e-mail and Internet services into their daily routine, for instance).

- *Product user:* Firms have positioned themselves by linking personalities (athletes or film stars) to their firm or product line.

- *Quality:* Studies reveal that an increasing fraction of the population is increasingly sensitive to product quality and reliability; communications firms producing hardware, for instance, are attempting to exploit this sensitivity.

- *Multifunctionalism:* Communications firms are now defining their market positions by designing products with versatile functionality. A single tool that can solve multiple problems retains great customer appeal during the early era of deregulation when uncertainty prevails in the minds of many consumers.

Segmenting, targeting, and positioning are the operative words of contemporary telecommunications market research. Passage of the Act urges the incorporation of all three techniques to gain comparative and competitive advantage. Rapid adaptation to these strategies leads to the development of new markets, facilitates expansion of market share, and mitigates the costs associated with retention of customers. Firms formerly regulated must now exploit state-of-the-art marketing strategies, as cultivated in other industries. Firms operating without a history of regulatory constraint now pursue their strategies in the face of prospective capital infusion by their larger counterparts; funds designated for market and competitive research can be expected to grow exponentially in the years ahead.

6.2.5 The Advance of Competitive Research

Competitive research focuses on the analysis and evaluation of competitors' strategic plans. The stress in this regard is on the segmenting, targeting, and positioning of other firms seeking a larger market share. In this fashion, those who employ competitive research on an ongoing basis seek definition of "market path": the future pattern or road of the market is regarded as a function of strategic plans laid at present.

Apart from gaining a clearer sense of future market structure, the object of competitive research is to "benchmark" a firm's own record relative to others operating within comparable markets. Such comparisons of performance provide a practical method of measuring progress while increasing marketing accountability. Such data, if carefully and properly evaluated, are likely to be a significant force in creating strategic plans during the early stage of deregulation. The use of benchmarking techniques is provocative because considerable debate exists regarding appropriate methodologies for its application and interpretation. The primary methodology, at present, involves the use of accounting and statistical methods applied across a spectrum of firms operating under comparable conditions (new tools are now in development throughout the industry and have drawn inspiration from the manufacturing sector).

For the telecommunications industry, competitive research (sometimes referred to as competitive surveillance) has emerged as an essential tool of strategic planning [10]. Two dynamics make competitive research in communications profoundly difficult in today's environment: the proliferation of competition and technological innovation. Deregulation has fueled change in both dynamics. In relaxing regulatory constraints, the government has simultaneously encouraged not only new competition but also technological innovation and substitution precipitated by converging industries. Competitive research has become a multilayered process of investigation, embracing unfolding strategic and tactical plans formulated by competitors.

6.2.6 Transformation of Telecommunications Market and Competitive Research

During the 1950s, mass marketing techniques were aimed at generating demand for single-brand markets; one kind of automobile, one kind of cosmetic, and one kind of appliance were promoted for large markets, geared to a period of rising consumption. This pattern continued into the 1960s, with some differentiation in market segments (or market segmentation); a single product was subdivided into a set of four or five niches, distinguished by variance in consumer income, attitude, or taste. The advent of more sophisticated methodologies brought further subdivisions, sometimes referred to as "niche" marketing during the late 1970s and 1980s. A serious beginning was made during this era toward designing market research techniques aimed at producing customized products for limited customer segments.

The 1990s, as suggested in Chapter 5, signified a period ripe for the development of "market customization," that is, the creation of product lines tailored to the specific needs of individuals, households, and businesses. The array of database procedures, combined with technological innovations, now makes this process feasible. We may distinguish two primary forms of customization. "One-to-one" marketing is the distribution of services and products to customers based on the use of business geographics (GIS) systems and is accomplished by carefully monitoring changes in consumer preference based on "substitutes" introduced to market by competitors.

The second, and extraordinary, form of market customization is "particle marketing"—product lines designed by customers themselves. This embryonic form of marketing, potentially of profound significance to the communications industry, is best accomplished when distribution channels are outlined such that customers provide immediate feedback as to their needs. Substantial marketing costs are often built into the system on the "front end," but this approach beckons many firms because it can be exploited to sustain market share. It is generally agreed that the costs of recruiting a new customer exceed by a factor of five times the investment necessary to maintain an existing customer base. In theory, there is a greater probability that customers will stay with a firm that designs its product menu such that changes in preferences are immediately accommodated. Customers, in short, dictate what the firm produces on a one-to-one basis. Telephony and software firms, via data transmission, storage, and retrieval, are innovating services that facilitate precisely this opportunity; it is therefore logical to expect the most aggressive members of the industry to move in this direction.

6.3 Current Issues in Telecommunications Market and Competitive Research

A number of cutting edge issues quickly surfaced for marketers within months following adoption of deregulation. Among the most vital concerns is managing customer lifecycle marketing. Simply, firms wanted to know: (1) how do we expeditiously seize market share; (2) how do we maintain that market share in the face of rising competition; (3) how do we implement a lifecycle strategy (satisfy customers over the long run); (4) how do we build effective customer retention programs; (5) how we do exploit greater profitability for each market segment; (6) how do we forecast emerging customer needs; (7) how do we identify the most profitable customer lifecycle segments; (8) should we diversify early or later across geographic boundaries; and (9) how do we efficiently employ state-of-the-art database strategies to track customer recruitment, retention, and satisfaction?

Lifecycle marketing had become a pivotal concern for communications firms because capital costs compel the formation of customer bases that build regular cash flows. Costs of capital construction can only be justified by predictable cash flows generated month to month. In regulated industries, such predictability was a foregone conclusion; in the face of competition, capital construction necessarily became a riskier venture. The key to industry success became gaining a keener grasp of the forces driving consumer behavior. While lifecycle marketing became crucial, new techniques of leveraging data to extract further profitability were held as an equal priority [11].

To contend with these developing issues, communications firms in some instances sought unconventional strategies to differentiate themselves from competitors. Strategically, redefinition of marketing channels, cooperative alliances, mass-customization techniques, new affiliations with related industries, and other unconventional methods were explored with increasing seriousness [12]. Above all, customer loyalty became a consuming obsession for firms in both wireline and wireless services [13].

6.4 Linking Market and Competitive Research to Strategic Planning

Strategic planning is the culminating step in a three-stage process aimed at defining the future of the firm. The planning of strategy to identify a firm's mission, goals, and objectives, discussed with respect to its economic implications in the following chapter, represents the difference between success and failure in today's environment.

A firm cannot develop its strategic plan unless it precedes this process with a two-step procedure. The first step involves environmental scanning, isolating the technological, social, economic, political, and cultural variables impinging on customer demand. The second step, forecasting, is the manipulation of qualitative and quantitative methods to ascertain the significance of data accumulated through scanning. These measures are elaborated with respect to their linkage in Chapter 7, but a note must be inserted here as to strategic planning and "market fit." A strategic plan cannot succeed unless it is congruent with changing market structure. Those who design a firm's plan predicate its integrity on the basis of credible data filtered through the scanning and forecasting processes.

This linkage between scanning, forecasting, and planning illuminates a vital aspect of market and competitive research: the importance of verifying the integrity of data and monitoring its continued relevance for strategic planning. Therefore, the work of market researchers during the age of deregulation holds great significance for the future success of the firm. The ascendancy of credible market data now directly impinges upon the very existence of the company itself.

6.5 Current Studies on Telecommunications Market Demand

Within a year following adoption of the Act, a series of studies produced by academia, private research firms, government agencies, and business groups began to pinpoint prospective demand for telecommunications products and services [14]. In and of itself such studies were hardly unprecedented, but this was the first time such research had been conducted within a context of deregulatory interaction. The value of this research was as much a function of its effort to gauge the effect of industry convergence as it was an attempt to forecast demand.

The studies now treated the computer industry in all its manifestations as an equal player to telephony and broadcasting in setting the agenda for telecommunications. The studies were especially intriguing because their results intersected in a way that supported some, but not all, suppositions of industry experts. At the same time, the public seemed impervious to developments that are likely to influence the way in which they live and work in decades to come. Several studies in this regard are worth recapitulating.

In an industry review in May 1996, three months after the commencement of deregulation, a major study reported that the following "hot" technologies would influence the course of the communications industry in the early stage of deregulation [15]:

- *Java programming language:* stirring greater accessibility to and commercial use of the World Wide Web;

- *Voice over frame relay:* speeding data and voice communications via a new generation of fast packet switches;

- *Virtual LANs:* independent of physical infrastructure, virtual local area networks now facilitate improved and flexible communications systems;

- *Cable modems:* vastly superior to telephone modems, given their capacity and ability to speed delivery of Internet and other multimedia services;

- *Gigabit LANs:* high-speed local area networks extraordinarily more effective than standard network architecture and effectively "leapfrogging" over existing infrastructure, thus enhancing productivity while minimizing communications costs;

- *Internet appliances:* the use of network computers to facilitate Internet access—holds the promise of reducing communications costs, thus promoting possible ubiquitous use of such technology;

- *Personal satellite phones:* new hardware designs lead to easy access to satellite transmission while reducing the cost of voice communications (85% of the world's land mass is not covered by traditional cellular service);

- *Intranet:* the creation of internal communications networks, fused with Internet access and new security characteristics of virtual data networks;

- *Automated network management:* new advances in artificial intelligence can now be translated into management of the entire network system, fueling new efficiencies while reducing network costs;

- *Data mining:* a technique for extracting information from a "data warehouse"—stored organizational information—that provides great potential for organizational effectiveness by generating new information to answer questions pertaining to management, marketing, finance, accounting, and information systems.

There have been serious attempts to measure the past, current and prospective impact of new enterprises as spurred by telecommunications deregulation and technological innovation. One manifestation of projected new business opportunities is an annual list of 10 emerging communications opportunities published by *Telecommunications*™ magazine. The following table enumerates the annual lists of new telecommunications business opportunities in the years immediately preceding and following the 1996 statute.

While this list may represent a somewhat fashionable composition of emerging technologies, a list whose components will vary with time, what is significant is the dramatic evidence it displays of industry convergence. Deregulation has already encouraged the integration of computer advances—from telephony to network management—to traditional methods of telecommunications. A spur to telecommunications innovation is thus confirmed, in part, by the passage of legislation. Evidenced is a wave of mergers, acquisitions, alliances, affiliations, and partnerships (discussed further in Chapters 8 and 9) that promises intensified integration.

Tied directly to the list of technologies for which there is projected demand are recent studies that have specified the dynamics of consumer behavior in the industry. Two early studies, issued amid deregulation, have confirmed the nature of two distinct behavioral trends: (1) consumers, desiring simplicity in communications menus, prefer "bundled" packages; and (2) a "new media

Year's Ten Hottest Technologies in Telecommunications*

Rank	1994	1995	1996	1997	1998	1999	2000
1	Digital Wireless	Plug/Play Internet Use	Java Prgmmg	Web Broadcasting	Multi-Layer Switching	Service Mediation	VoDSL
2	Remote Networks	DBS	VO Frame Relay	Remote Acc Servers	Erbium Metro Doped Amps	DWDM	Time Division Duplexing
3	ATM	PCS	Virtual LANS	Extranets	Packet Voice	OCE	Tunable Lasers
4	Gigabit Networks	Carrier Frame Relay	Cable Modems	Internet Telephony	LMDS	HDSL2	ODSI
5	LAN Switch Hubbing	ATM/LAN	Gigabit LANS	ENDS	WLL	Unified Messaging	Synchronization
6	Super-Interconnect (ODSI) Intelligent LAN	Videosavers	Internet Appliances	Website Management	Virtual Pvt Networks	Policy-Based Management	IP VPNs
7	Voice/Data Integration	Wireless Data	Personal Sat Phones	IP Switches	Web Call Centers	MPLS	10-Gigabit Ethernet
8	Reengineered Security Systems	Data Crypts	Intranets	WDM	XML	SmartAntenna	Very-Short-Range Optics
9	Neural Netks Management	Digital Cash	ANMs	Digital Sub Lines	TMN Technology	Optical Switching	Soliton
10	Multimedia Info Delivery	Real-Time App Sharing	Data Mining	POTS Modems	Java Management	NetWireless Location	Passive Optical Networks

*Note: Projections drawn from a variety of experts throughout the telecommunications industry, including consultants working in the field of specialization indicated. Listing employs qualitative methodology and culls expert opinion each May during the respective years identified above.

The assimilation of data reveals both convergent and divergent opinion about those communications technologies most likely to impel future change in the industry. Significantly, the year-to-year listings elaborate the growing emphasis on Internet usage, wireless services, and technical apparatuses required to facilitate storage and rapid distribution of information. What is most striking, perhaps, is the extraordinary proliferation of new technologies; the dramatic rapidity of innovation renders discontinuity throughout these lists. No single technology retains its place or prominence on a year-to-year basis. One can discern obvious technological propensities in wireless and wireline data traffic, Internet commercial usage, multimedia development, computer programming, and similar trends, but experts have great difficulty in forming a consensus about dominant state-of-the-art tools. The absence of agreement reflects precisely the uncertainty embedded in current strategic management processes. Clarity of planning begins in successfully forecasting these emerging methods of communication.

The data arrayed do not suggest that technological innovation, in the short run, has been significantly influenced by adoption of the 1996 Act. What has changed in the period immediately following passage of regulatory reform is the nature, number, and design of new business formations intended to exploit those technologies. The Telecommunications Act of 1996 is regularly denounced by some as having fallen short of the goals initially projected for it. This debate will surely endure through the coming decade, but it now appears irrefutable that the statute has intensified the complexity of modern strategic management for the industry.

Source: Telecommunications™.

household" engenders interest primarily from high-income consumers in the early phase of deregulation [16].

A bundled package is defined as a set of communications services for which consumers select a single outlet offering such provision at a single (often discounted) price. Industry consternation and the proliferation of choices are expected to lure many consumers to the simplicity of single providers. This development, if realized, may or may not represent a long-term industry trend. These empirical results suggest that those firms that are successful in targeting markets blend effective pricing with service. This analysis also indicates that it is incumbent on single providers to quickly analyze the extent to which their targeted strategies succeed in satisfying customers. After price-sensitivity, these studies confirm that intangible factors (such as faith in a firm's reliability) influence consumer choice.

A number of households are aware of the new possibilities of emerging communications technologies to enhance the quality and productivity of their lives [17]. These innovators—or experimenters—are curious about such technologies and foresee obvious application in education, health, entertainment, and general information value. Although comparatively small in number (representing perhaps 3% of household consumption), this contingent's experimentation is necessary to launch the wave of early-majority, late-majority, and laggard segments characteristic of the consumer S-curve elaborated in Chapter 5.

Tracking behavioral change in service adoption can now be accommodated by manipulating billing data [18]. Thus, immediate answers to these questions can now be accomplished, which was unimaginable just a few years before. The tracking of consumer behavior—the application of state-of-the-art techniques to formulate marketing strategy—may well distinguish the effectiveness of strategic plans in the near future. For smaller firms unable to internalize the tracking of such data, new outsourcing strategies are available; firms specialized in the evaluation of such data can provide interpretation in an expedient fashion [19].

6.6 Multiorganizational Tracking of Consumer Demand

Because effective market and competitive research is tied today to uncovering the mysteries embedded in customer data, a seminal development in monitoring consumer demand is the cooperation of firms whose skills are complementary in nature. Such synergies are referred to as multiorganizational tracking; that is, the evaluation of customer data by firms who "team" in order to efficiently implement billing and customer care strategies. A firm interested in the value of customer data for its objectives might be willing to perform customer analysis for another company willing to part with its database. Strategic planning for both enterprises is engineered in a way that maximizes cost effectiveness.

There is a delicate balance to be observed here between proprietary hold on data versus the need to gain expertise at minimal cost, but the early glimmer of such an industry is in evidence [20]. We note again the slow movement toward complementary and supportive alliances characterizing strategic planning during the initial stage of deregulation. No single firm in the industry can independently supply all of its needs while pursuing every available avenue of opportunity; "coopetition," discussed in the next chapter, may provide the key to future market and competitive research.

6.7 Adaptation to Competition and Change: Established Versus Emerging Firms

The integration of market and competitive research points decisively to the importance of organizational response to economic and technological change. In such an environment, what firms hold an edge in gaining market position and market share? A variety of factors overlay the answer to this question, but in dichotomizing established from emerging firms, we are provided with a clue to a profound question in deregulation.

The established firm, which relies on experience, acumen, and sagacity, possesses the attributes of stability, caution, and calculation in projecting and responding to developing trends. The emerging firm (or start-up), which is unencumbered by bureaucracy and/or entrenched organizational interests, is nimble, adaptable, and opportunistic. In such a struggle, who succeeds? In the year following passage of the Act, we have had an explicit demonstration of market intent: established firms have sought merger/acquisition routes to the expansion of market share (see Chapter 8). New firms have simultaneously pursued the capital markets for funding new ventures, particularly with respect to wireless enterprises.

We have had a glimpse of the short-run practical economic effect of deregulation in telecommunications. It may be, however, that both large and small firms have neglected an obvious yet incisive counterpart to their strategies: adopting a "mixed" strategy, in which the small firm replicates structural elements associated with their larger cousins. Likewise, the established players can emulate the most positive attributes of effective start-ups. In essence, such a change in mind-set would mean a serious effort to minimize the limitations intrinsic to both small and large enterprises, while exploiting the clear advantages of both systems. A discussion of the future implications of this issue is provided in Chapter 10.

References

[1] Ekdahl, Lyle, "Playing for Customer Keeps," *WB&T*, Feb. 1997, pp. 36–39. See also Howard, William, and Bruce Guile (Eds.), *Profiting from Innovation*, New York: Free Press, 1992.

[2] Mowen, John C., *Consumer Behavior*, Englewood Cliffs, NJ: Prentice-Hall Inc., 1995, pp. 188–221.

[3] Kefalas, A. G., *Global Business Strategy*, Cincinnati, OH: Southwestern Press, 1990, pp. 257–280.

[4] Higgins, James M., *The Management Challenge*, New York: MacMillan Publishing Co., 1994, pp. 231–252.

[5] Hiam, Alexander, and Charles D. Schewe, *The Portable MBA in Marketing*, New York: John Wiley Press, 1992, pp. 207–216.

[6] Pettis, Chuck, *TechnoBrands*, New York: American Management Association, 1995, pp. 201–212.

[7] Hiam and Schewe, *op. cit.*, pp. 219–226.

[8] *Ibid.*, pp. 204–208.

[9] Whiteley, R., and Diane Hessan, *Customer-Centered Growth*, Reading, MA: Addison-Wesley Press, 1996, pp. 18–51.

[10] Naisbitt, John, *Global Paradox*, New York: William Morrow Press, 1994, pp. 53–102.

[11] Daniel, Larry, "Combining Technologies to Support the 1:1 Agenda," *Business Geographics*, Nov.–Dec. 1996, p. 18.

[12] Edelstein, Herb, "Mining for Gold," *Information Week*, April 21, 1997, pp. 53–70.

[13] McCartney, Laton, "Customer Service Calling," *Information Week*, March 31, 1997, pp. 37–44.

[14] "Communication Preferences Survey Highlights Opportunity to Delivery Packaged Services to Deregulated Marketplace," *BusinessWire*, March 31, 1997.

[15] Flanagan, Patrick, "The Ten Hottest Technologies in Telecom," *Telecommunications*™, May 1996, pp. 29–38.

[16] "Inquisitive Middle Class? Lost Cause?" *PRNewswire*, April 14, 1997. See also Tice, William, and Willow Shire, "One-Stop Telecom: The New Business Model," *Telecommunications*™, April 1997, pp. 63–64.

[17] Dholakia, R., N. Mundorf, and N. Dholakia (Eds.), *New Entertainment Technologies in the Home*, Mahwah, NJ: Lawrence Erlbaum Associates, 1996, pp. 157–172.

[18] Martin, Eric, and John F. McClure, "Using Billing Data for Telecom Marketing," *Telecommunications*™, Sept. 1996, p. 66.

[19] Guy, Sandra, "Putting Words into Action: Bundled Services Are in the Cards for 1997," *Telephony*, Jan. 20, 1997, pp. 36–38.

[20] Brandenburger, Adam, and Barry J. Nalebuff, *Coopetition*, New York: Doubleday Press, 1996, pp. 110–158 and 198–233.

7

Forecasting Telecommunications Market Trends

A necessary precondition for the creation of strategic plans is the application of forecasting methodology. Serious attempts to predict market developments, however complex, provide a context in which firms can forge competitive response. In the absence of forecasting, it simply is not possible to define organizational mission, goals, objectives, and future contingencies. In applying forecasting methods, a firm does not obsess with absolute predictive accuracy. To prognosticate market outcomes is to fashion a firm's current mind-set about planning alternatives; this, after all, is the overriding and essential significance of organizational forecasting.

For the telecommunications industry, effective forecasting has never been more important. Deregulation has sped new competition to market, encouraged a rapid proliferation of mergers and partnerships, engendered new incentives for product substitutes, and heightened the anxiety of consumers inexperienced with the proliferation of alternatives from multiple suppliers. It is precisely dynamic market environments that invite firms to engage in the most sophisticated forms of forecasting methodology.

The field of telecommunications forecasting is undergoing extraordinary change, particularly with respect to establishing systems aimed at anticipating changes in consumer preference and the vagaries of consumer loyalty. This field of endeavor is too vast to review in the limited coverage of this space, but this chapter discloses primary applications of forecasting and suggests possible future applications in light of deregulation. This analysis also provides a review of the principal tools used to forecast product adoption rates, which is a critical issue for both established and emerging firms. You will note the inclusion of a methodological treatment for the wireless communications industry and a set of forecasts for the elaboration of diffusion rates. It is hoped that these illustrations will reveal the potential of forecasting for the purpose of clarifying strategic planning options amid the profound structural changes now enveloping the industry.

7.1   Telecommunications Forecasting

7.1.1   Standard Tools and Applications

The collection of telecommunications forecasting methods hereafter outlined is used to anticipate market developments in two essential endeavors: (1) generating general market data on prospective consumer preferences, sales, emerging competition, and other priorities as defined by the firm; and (2) specifying the patterns of technology adoption precipitated by variations in product diffusion, mortality, and substitution. You may recall the economic significance of adoption rates in material discussed in Chapter 5.

With respect to the assimilation and evaluation of data on general market developments, two dozen qualitative and quantitative techniques are employed in the business environment. The application of these tools holds universal value in all industries [1]. The array of qualitative (or nonmathematical) tools includes historical analogy analysis, visionary forecasting, Delphi methodology, panel consensus, and pre- and post-testing of focus groups [2]. The virtue of qualitative methodology lies in its capacity to treat information not susceptible to traditional statistical treatment. Useful information generated through environmental scanning, discussed later in this chapter, is often anecdotal or intuitive in nature, and such methods provide a mechanism to transmit the intangibles of market research. While some may dismiss the subjective nature of these tools, practitioners have used them to great advantage, particularly during periods in which the rapidity of change supersedes the value of data rendered quickly obsolete. The instinct of qualitative methodologists is to identify patterns of consumer behavior that are inherently repetitive—and therefore cyclical—in nature. Forecasts are thus contingent on patterns of behavior presumed to be repetitive in terms of lifestyle, attitudes, household age, or other subjective standards [3].

Quantitative methods include the use of: moving average, regression analysis, exponential smoothing, Box-Jenkins tools, X-11 time analysis, trend projections, traditional input-output analysis, systems theory, anticipation surveys, econometrics, lifecycle analysis, leading indicator composites, and diffusion indexes [4]. Other quantitative methods, traditional and emerging, are sometimes applied by firms in the forecasting process, but these can generally be regarded as falling into one of these categories. Quantitative methods, with their often arcane and intimidating scientific foundation, are often preferred by strategic planners and marketers. As is the case with qualitative methodology, however, these techniques have their limitations; always, their value lies in the credibility of data inserted into the appropriate method [5].

For communications firms, all of these traditional techniques hold potential value in gaining competitive advantage. As a result, two principles guide the application of forecasting methodology: (1) practitioners should identify every tool appropriate to the data or information at hand; and (2) the intersection of results, when the methods have been applied, should be treated as the key to effective predictions. Expressed differently, prognosticators should employ every method congruent to the objective and supporting evidence and presume that the greater the consistency in results, the greater the probability of producing an accurate forecast. It is assumed that the intersection results from the application of diverse techniques that leads to accurate predictions.

By incorporating traditional qualitative and quantitative methodology, we can summarize the main advantages and attributes of each forecasting technique and connect these to telecommunications market research. Table 7.1 lists and defines each tool, discusses its value, and comments on the relative accuracy of each. You will note the comparative advantages and limitations of each tool. An additional, admittedly contentious, commentary is also provided in this table as to the most effective tools to be applied in forecasting during the early stage of deregulation. A large body of reference material is available to guide the process of applying each technique (consult the references in this chapter as well as Appendix C for additional information).

7.1.2 Adoption, Diffusion, and Mortality

In an industry driven by technological change, with competitive threats implicit in innovation, it has become incumbent upon prognosticators to isolate the determinants of market penetration. Arguably, it is in this area that forecasters have had their greatest success in outlining the future direction of the communications market [6]. New product introduction and resultant adoption rates have become the overriding priority of telecommunications forecasting in today's market environment [6].

In predicting the fate of new telecommunications products and services coming on-line, forecasters differentiate three economic forces:

1. *Product adoption:* the rate at which consumers—residential and business—purchase new products. Adoption, for new products destined

Table 7.1
Telecommunications Forecasting Methods

Qualitative Methods			
Forecasting Method	**Accuracy Rate**	**Applications**	**Information Required**
Delphi method	Fair to very good	Forecasts of new-product sales	A sequence of questionnaires is used, with several tests conducted
Market research	Excellent	Forecasts of product sales	Two sets of reports over time based on market data
Panel consensus	Poor to fair	Forecasts of new-product sales	Information gleaned from a panel of experts is presented in group meetings to arrive at a consensus forecast
Visionary forecast	Poor	Forecasts of new-product sales	Scenarios are elicited from group of experts
Historical analogy	Poor	Forecasts of new-product sales	Use of historical parallels to project future

Quantitative Methods			
Forecasting Method	**Accuracy Rate**	**Applications**	**Information Required**
Moving average	Poor to good	Inventory Management	A minimum of two years of sales history
Exponential smoothing	Fair to very good	Production Control	Identical to moving average
Box-Jenkins	Very good to excellent	Production Control	Identical to moving average
X-11	Very good to excellent	Monitoring sales	Three years preceding sales history

Quantitative Methods			
Forecasting Method	Accuracy Rate	Applications	Information Required
Trend projections	Very good	New-product forecasts	Preceding five years' sales history
Regression model	Good to very good	Forecasts of sales by product line	Several years' quarterly history of sales
Econometric model	Good to very good	Forecasts of sales by product line	Identical to regression
Intention-to-buy and anticipation surveys	Poor to good	Forecasts of sales by product line	Several years' data are necessary to evaluate company sales by applying indexes
Linear programming	Fair to good	Prediction of short-term veteran product sales; demand function forecasts	Extensive data on consumer demand and competitor response
Economic input-output model	Poor to good	Company sales for households and businesses	The same as for a moving average and X-11
Diffusion index	Poor to good	Forecasts of sales by product line	The same as an intention-to-buy survey data
Leading indicator	Poor to good	Forecasts of sales by product line	The same as an intention-to-buy survey data
Lifecycle analysis	Poor to good	Prediction of new-product sales	Annual sales of the product considered or comparable

Note: Consult Appendix C for a listing of statistical and forecasting tools; these Internet sites provide instruction as to appropriate application of these techniques.

for success, is visually captured through the construction of an S-curve, tracing the phenomenon of consumers who gradually enter the market as a product gains market acceptance. Often, new product adoption represents an effort by consumers to "substitute" improved products for older, less efficient ones.

2. *Product diffusion:* a descriptive tool employed to itemize "types" of consumers likely to purchase a new product. This itemization is discussed in detail in Chapter 5; you will recall that "innovators," "early adopters," "early majority," "late majority," and "laggards" constitute the primary components of the diffusion process. The significance of this forecasting method lies in the fact that early adopters do not enter the market until innovators have established purchase precedent; acceptance by early adopters is necessary to trigger consumption by early majorities, and so forth. It should be added that many firms further divide these diffusion typologies depending on the unique characteristics of their products and the consumers they attract [7].

3. *Product mortality:* product breakdown, not obsolescence, drives "product mortality." In the case of certain telecommunications products, the motive to purchase is prompted by product failure. The age of existing product lines—standard phones, for instance—is linked to a desire for replacement rather than product innovation. Only when a product fails do consumers seek a replacement.

In today's deregulatory environment, with innovation triggering product obsolescence at an accelerated rate, predicting adoption and diffusion rates becomes ever more critical. An examination of both the opportunities and limitations of such research is evidenced by the following case study, which illustrates forecasting methodology in the wireless telecommunications industry.

7.2 Wireless Telecommunications: A Case Study in Forecasting

The wireless communications industry continues to enjoy explosive growth as the new century unfolds. The industry, whose initial burst of growth has been prompted by the moderation of pricing in the cellular arena, is now poised for spectacular expansion. A growing paging sector has experienced market growth in recent years exceeding that attained by cellular providers. Accompanying this growth are the specialized mobile radio (SMR) and personal communications services (PCS) industries. Together, the four wireless competitors are poised for rapid growth that may touch virtually every American before 2010 [8].

The cellular industry, which emerged in 1983, has benefited from the absence of direct competition in the "voice" market. In the two decades following the advent of cellular communications, only enhanced paging services have directly impinged on the growth of the cellular market and its resultant penetration. In recent years, the paging industry has produced market growth that has exceeded that attained by its cellular counterpoint, characterized by a vigorous pricing strategy.

The cutting edge of uncertainty in the industry today concerns the prospective development of SMR and PCS.

- Are these dynamic new players capable of eventually supplanting cellular technology as leading providers of voice and other value-added wireless services?

- Are these emerging technologies, or rather, merely complementary providers, able only to capture niche markets?

- What of the role of the paging industry?

- Is this merely a "bare-bones" technology, destined to serve consumers who simply cannot afford the higher pricing associated with cellular providers?

- Is, perhaps, paging poised for continued spectacular growth based on incremental improvements in its technology?

These are among the ambiguities implicit in data now surfacing from private researchers, corporations, and government agencies.[1]

7.2.1 Data and Forecasts

One of the more intriguing elements of the wireless industry in recent years has been the variance at which forecasters have drawn their statistical inferences. The divergence of forecasters and their predictions has been historic since the early 1980s. Both AT&T and the FCC, for example, predicted in the early 1980s that fewer than 1 million Americans would be willing to pay for wireless voice communications. In fact, the explosive growth that has unfolded in the industry has astonished even the most enthusiastic prognosticators; in cellular technology alone, 35 million Americans now use analog or digital phones. In 1995, majority sentiment among marketers held that not more than 40 million cellular phones would be in use by 2000 [9]. Today, that statistic

---

1. The most unpredictable factor in forecasting market growth for these technologies remains engineering and technical improvement. For purposes of this analysis, it is assumed that paging is a complementary service that will not approach the value-added functionality associated with its three competitors. Should a major technological breakthrough occur such that paging is enhanced to offer unlimited voice communications, for instance, the industry's S-curve and ascending concave arc will force a profound industry shakeout well before 2010.

has been revised upward by some forecasters to as many as 90 million domestic users, and this growth is independent of rising competition by the SMR and PCS industries.

Mounting empirical evidence suggests that virtually all Americans, without regard to age or other significant independent variables, are prospective candidates for wireless services. As noted in the body of data and other information herein, the industry has all of the intrinsic advantages associated with dynamic growth industries based on emerging technologies, namely:

- Underlying value;

- Inherent need;

- Ease of use;

- Positive mass psychology.

We may reasonably infer from the data that the key determinant for optimal market penetration is pricing; thus, the competitive positioning of the four key players will ultimately be a function of that technology that sufficiently lowers price to accommodate the broadest consumer base. Within reasonable parameters of quality and service, price will be the key independent variable responsible for market penetration. The public wants and needs the value-added services—instant communications, security, access, information, mobility—that only this industry can promote.

7.2.2 Comparative Advantages and Disadvantages

No serious discussion about wireless providers and their pending competitive positioning could be entertained without noting differences in their technical capabilities and attributes. The seminal wireless technology—cellular—has sufficient bandwidth to accommodate large numbers of consumers and operates within a duopolistic competitive framework. The cellular industry, despite continued technical improvements and digital enhancements, cannot accommodate, however, the enormous demand estimated for the industry at the turn of the century. It is for this reason that the FCC has continued to allocate—at first through essentially "free," now auctioned licensing—electromagnetic spectrum.

SMR, which originated in 1974 as essentially dispatch designated spectrum, has grown modestly in recent years, but is now poised for geometric growth. Major industry players—Nextel, OneComm, and various regional providers—have accumulated SMR licenses throughout the country to form networks potentially competitive to cellular providers. The absence of sufficient spectrum and continuing technical problems have hampered the growth of the industry, but one certainly cannot rule out rising prominence for these competitors should technical problems be relieved.

The enormous spectrum allocated to PCS, dwarfing the combined bandwidth of SMR, cellular, and paging, underscores the greatest potential competitive threat to the voice duopoly. The multiplicity of PCS license holders in each metropolitan market, however, along with serious issues of capital availability, probably means many years will be required to seriously erode cellular's present "leader" status. The variance in competitive advantages and disadvantages may be seen more concretely in Table 7.2.

In sum, the cellular, SMR, and PCS industries will divide and shift market share in response to comparative pricing strategies. Pricing will depend very heavily on the established competition unique to each urban, suburban, and rural area. With the recent completion of PCS auctions (a handful of PCS and SMR licenses remain to be auctioned), we are gaining a clearer sense of the positioning of the players. It is comparatively more expensive to build out and maintain cellular and PCS systems as opposed to their SMR and paging counterparts. However, barring major technological improvements in paging and SMR, these latter providers will remain niche players, despite implicit pricing advantages (they control an insufficient spectrum to capture substantial market share in the short term). Despite financial and technical constraints, each wireless technology will nevertheless influence price and resultant market share. Taking these advantages and disadvantages in totality, cellular retains the greatest competitive advantage at present. The "head start" it has gained during the past two decades of development has produced a seminal technical and financial grasp of the market.

Table 7.2
Wireless Services: Competitive Advantages and Disadvantages

	Cost of Business Per Subscriber	Bandwidth	Pricing	Competitive Structure
Cellular	High	Moderate	High	Duopoly
Paging	Low	Minimal	Low	Multiple
PCS	High	Extensive	Moderate	Three to six*
SMR	Low	Minimal	Low	Multiple*

*Depends on location.

7.2.3   Recent Market Forecasts

Among the more robust forecasts of the wireless industry market are those generated by Paul Kagan Associates [9]. In 1995, this firm predicted that cellular subscribers would mushroom to 75 million in 2001 from its current level of approximately 40 million. The firm also estimated that SMR subscriptions would expand from fewer than 1 million to more than 3 million during the same period. PCS platforms, just now being erected, would generate 8 million customers in 2000, exploding to 37.5 million six years later. By 2006, it was estimated that enhanced specialized mobile radio (ESMR) would garner 12 million customers while cellular providers would maintain a base of more than 93 million. The wireless "voice" market would thus generate 143 million subscribers by 2006. This statistic excludes paging providers, whose long-term positioning remains a matter of ambiguity and conjecture. One may reasonably assume that the paging industry would have a primary market base of some 40 million or more subscribers. For paging, the pivotal issue concerns whether technical refinements will permit it to compete effectively with its highly value-added cousins.

A visual depiction of these data underscores the enormous growth projected by Kagan and others (see Figures 7.1 and 7.2). We note that cellular's growth continued unabated through 2000. Thereafter, its growth rate is projected to diminish as PCS and SMR take hold. The concurrent growth in paging remains a matter of debate and contention. We cannot estimate precisely the role of a lower priced competitor during an era of rising competition between three dynamic players. Is paging poised to capture a "low-end" market, or will it simply be supplanted by a competitor that performs a multiplicity of functions at a declining price? The market penetration rate graphed in Figure 7.1 for paging presumes continued growth that probably will not diminish until well after 2000. In any event, some 60% of the American population could be using one or more communications devices sometime between 2004 and 2006.

7.2.4   Applying S-Curve Analysis

The use of a product S-curve, which is a method of forecasting prospective market share based on the introduction of technologies aimed at vast consumer markets, has been in use for decades. The S-curve gained prominence for high-technology marketing in 1970 when it was first used to predict the value of research and development in terms of its contribution to productivity [10]. In essence, the S-curve, as applied to measuring market growth potential, asserts that:

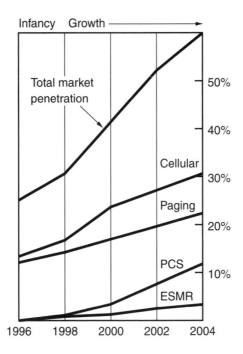

Figure 7.1 Comparative market performers: penetration measured as a fraction of the population. Note that all four technologies are projected to grow until market penetration exceeds 60% sometime between 2004 and 2008. At current projection rates, a minimum 48% market penetration will be attained by 2004 with respect to voice communications. If enhanced paging technologies are added to this mix, and assuming this technology sustains a subscription rate in excess of 30 million units, as many as 60% or more Americans will own a wireless communications device by 2004–2006.

- The time required to gain a 10% market share is approximately the same as that required to move from a 10% to 90% penetration rate.
- The shift from a concave to convex curve represents a diminishing growth rate in a rising, competitive mass consumer market.
- Product innovation and improvement should become the prime concern of market participants as they move from growth to mature markets.
- Competition compels market leaders to behave in this fashion.

In practice, the S-curve is a strategic tool that pinpoints the time required to introduce and capture the bulk of a ubiquitous consumer market. Because the wireless communications industry fits this criterion, the S-curve has obvi-

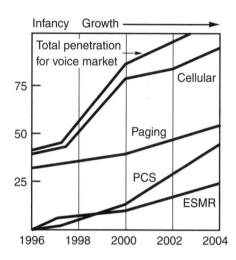

Figure 7.2 Comparative market performers: measured by projected growth rates. Subscription rates are denoted in millions. For purposes of this analysis, enhanced specialized mobile radio, cellular, and personal communications systems are distinguished from paging counterparts by virtue of their ability to offer unlimited voice communications. Total market penetration for all four technologies will probably exceed 175 million by 2006.

ous application. We note, for example, that between 1993 and 1995, the cellular industry generated a market penetration of 12.9%. In short, a decade had passed before a 10% penetration rate had been established. We thus infer that a comparable period—approximately 10 or so years—would produce 90% market acceptance.

By this estimation, as many as nine out of 10 eligible consumers would secure a wireless communications device sometime during the period running from 2006 to 2010. There is some implicit imprecision in this forecast—technical problems could impede market growth, for instance—but the dynamics of market acceptance underscore a credibility to the prediction. Even in the absence of significant contributions by SMR and PCS, the industry has already attained a 25% penetration rate when paging is factored with cellular subscriptions [9].

In Figure 7.3, we can isolate an S-curve that suggests 90% market penetration for wireless services no later than 2012. A comparable penetration rate could be achieved earlier, of course, if price dropped sufficiently to expand the market base. There are ample precedents for similar and dramatic S-curves throughout history: radio, television, and the videocassette recorder all achieved 90% market penetration defined by a time frame consistent with the S-curve. The years required to gain a 10% market share approximate the number of years required to move from a 10% to 90% market share.

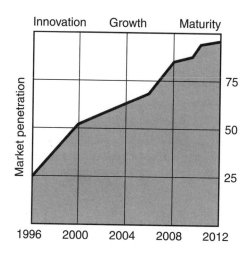

Figure 7.3 Wireless S-curve analysis: cumulative impact. Following the initial development of cellular technology in 1983, a penetration rate in excess of 10% was attained by 1995. By this reckoning, and taking into account a standard S-curve adoption rate, a 90% penetration rate could be achieved by 2012. Such a phenomenon would effectively mean that nearly all American households would acquire one or more personal communications devices of some form within the next 15 years.

When the S-curve for wireless communications is adjusted to exclude paging services (see Figure 7.4), which is a tactic employed to specify the potential market for voice communications, we note a series of "steps" that market participants are likely to embrace. As the industry continues to reconfigure in the years following the enactment of deregulation, we identify the origin of radical innovation in quality, pricing, and service for wireless services. As market share expands, an industry shakeout is inevitable. Before 2008, and perhaps as early as 2004, we are likely to see a substantial number of mergers, acquisitions, and strategic alliances that will only diminish as we approach 90% market penetration. It appears probable that the impending chaos of these consolidations will only abate when we move from 2008 to 2012 [11]. The period of maturity that follows thereafter will be characterized by established markets and predictable, sustained market shares. There will be a place for all three primary voice technologies with reliable hardware retaining versatile functionality.

Should this S-curve scenario unfold in a manner characteristic of other mass consumer markets, we cannot identify with certainty the "ultimate victor." The winner, in the era of mature and established markets, most probably will be either cellular or PCS. Their control of spectrum imparts, with or without major engineering refinements, is a strong advantage in this environ-

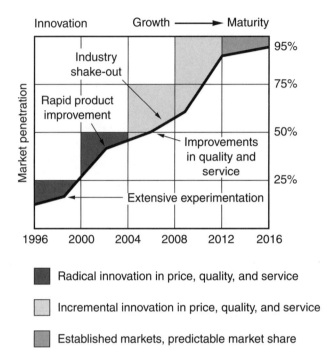

Figure 7.4 Projected wireless industry behavior. This illustration excludes paging applications. Should the wireless communications industry adhere to traditional S-curve developments characterized by comparable technologies, we can expect major product innovation to differentiate the PCS, cellular, SMR, and paging industries sometime between 2006 and 2008. In other words, an industry shakeout, distinguishing winners from losers, is likely to occur well before 2008.

ment. In this case, the consolidations that follow radical innovation during the next eight years will probably determine the dominant market technology. By this time, it may well be immaterial which technology "wins" such a battle. We may well have networks that are indifferent to the hardware that sends or receives wireless messages. Corporate consolidations resulting in control of spectrum may well be regarded as the most important determinant of market dominance during this era.

7.2.5   The S-Curve and Prospective Technology Substitution

One of the more intriguing strategic issues confronting wireless providers over the next decade will concern technology substitution. In other words, as we move along the path of the S-curve, will there be a move from one technology to another by consumers? This is a more profound question than one might initially expect. A simple focus on pricing is not sufficient to respond to the question. History confirms that market acceptance is often filled with subtlety and even mass psychology.

As previously indicated, that point on an S-curve that represents a transition from concave to convex growth signifies a diminished growth rate. These inflection points signal vulnerability to a market leader in a rising consumer market in which competition mounts. We note such an inflection point for cellular technology in 2000–2002. Although cellular continues to grow at a substantial rate, PCS and SMR technologies accelerate their growth and penetration rates in a rising tide (see Figure 7.5). The period from 2001 to 2006 represents an era of severe competition during which the market leader will have to innovate in response to aggressive emerging technologies. It is improbable that cellular can capture more than one-third of total market share during

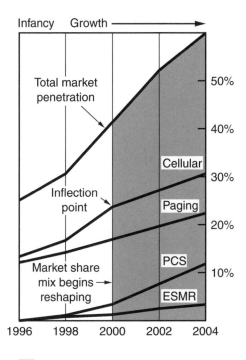

Area indicates intense emerging competition

Figure 7.5 Prospective technology substitution. Note that as each of the four primary wireless technologies develops into the next century, the graph designates cellular technology as the first to reach an inflection point: The shift from concave to convex market penetration implies a market opportunity for the three remaining technologies.

this period, and it is a distinct possibility that PCS (exploiting its bandwidth and economies of scale) could significantly erode cellular market share. Continued technical improvement with commitment to marketing and after-sales service will define market success during this period. It is quite possible that market share will be volatile during this era, with SMR and PCS capturing an expanding market at cellular and paging's expense. It is unclear whether the rising prominence of these alternative technologies will reflect a niche status or serious threat to cellular's long-term survival.

Should business consolidations reduce the number of PCS providers to two or three per market (rather than the four to six participants now licensed), PCS could be the "natural substitute" for cellular telephony. With ESMR relegated to a largely niche dispatch role, a fight between cellular and PCS for rising market share in an exponentially growing market seems inevitable. To succeed as a "substitute," PCS must make its mark (note cellular's inflection point and diminished arc) during the first six years of the next decade, and its most serious obstacle is the presence of so many "sister competitors." Therefore, the extent to which PCS will succeed will depend to a large degree on its capacity to sort out viable, consolidated competition to the enduring duopoly.

7.2.6 Jumping the S-Curve and Strategic Positioning

When technologies in growing consumer markets experience a change of arc in their growth rates, potential market vulnerability should signal major product improvement. In effect, a mature technology can "jump" an existing S-curve and once again experience an ascending concave arc. The logic implicit in this strategy is obvious, and, for the cellular industry, the key turning point is likely to occur before 2002. Yet, in the past, a number of technologies have not been improved to accommodate such a strategy. We note these cases as illustrations of lost opportunity and eventual product decline.

Historically, companies have hesitated in "jumping" to a new S-curve for a number of reasons. Often, product improvement that might have resulted in an elegant move from established to new S-curves has been thwarted by:

- A misinterpretation of market data;
- Reticence in taking on new business risk;
- Overlooking rising competition from alternative technologies;
- Failure to satisfy consumers during periods of expanding numbers of alternative suppliers.

Cellular providers could be undermined by overconfidence during the next decade if they presume that PCS, paging, and SMR providers can achieve no more than marginal market shares. Success often spoils leaders who presume that their early entry to market secures permanent advantage. No such presumption is valid in the wireless industry.

It has often been assumed that the most effective way to capture a growing market share in a rising consumer market is to innovate a technology, expand with the market, and dominate pricing in the early years of minimal competition. It is this tactic that has characterized the growth of the cellular industry since the 1980s. The primary disadvantage associated with this tactic is, of course, the extraordinary expense associated with product innovation and "consumer education." The front-end costs are seen as being fully justifiable during an era of an expansive S-curve. Although PCS and ESMR are different forms of wireless telephony, they are essentially identical in their ability to offer consumers a similar multiplicity of services. Therefore, increasingly, the notoriety of being the market innovator will be less significant, and it is possible through incremental improvement to compete with this established technology. PCS and ESMR have the opportunity to learn from mistakes made by the cellular industry: Technical, financial, and market lessons painfully learned in cellular can lead to sustained, incremental improvements for these upstart technologies.

It has become axiomatic in the computer industry that "first to market acceptance" is substantially more important than being first to market. Seventy-five percent of Americans do not have access to wireless communications at present, and it matters little to them which technology they employ. Consumers are interested in procuring that form of wireless telephony that, predictably, is predominant in pricing, quality, and service.

The ascending concave arc now characteristic of the industry S-curve affords competitors an exceptional and, perhaps, unprecedented market opportunity. It is unlikely that such an opportunity will return after 2006.

7.3 Lessons of Telecommunications Forecasting

The preceding case study demonstrates both the strengths and limitations of telecommunications forecasting as it pertains to the prediction of consumer behavior and demand. Unfolding enthusiasm now in evidence for wireless communications eluded some of the most sophisticated prognosticators in the industry. We conclude from this experience that any communications technology for which there is prospective practical application, and in which ease-of-use is a definable attribute, holds the potential for market penetration. At the moment these two characteristics coincide, pricing becomes the critical determinant to ubiquitousness.

It should also be emphasized that forecasters reviewing prospects for wireless technology relied almost exclusively on quantitative forecasting techniques to draw their predictions. As a result, they lost the benefit of exploiting the value of qualitative methods that might have unlocked the latent needs and desires of consumers. The importance of applying multiple forecasting methods and investigating the intersection of findings is therefore underscored by this study [12].

7.4 Forecasting Competitive Bidding: The Case of Electromagnetic Spectrum

Responding to the intent of the Omnibus Budget Act of 1993, the FCC launched a policy of "auctioneering" electromagnetic spectrum in 1994 [13]. The FCC adopted a policy of competitive bidding for several reasons. Principally, this method of allocating radio channels would theoretically expedite the buildout of a variety of wireless services, thus promoting economic and job growth in new industries. The presumption here was that only a company seriously interested in building out its system would bid for available spectrum. Secondly, the policy of "free radio" licenses (in prior years the FCC had granted licenses through a "first come, first served" application process or, alternatively, through lottery procedure) had promoted speculation and "hoarding" of radio frequencies by many individuals. Auctioneering would effectively prevent, in theory, such manipulation. Finally, and perhaps most significantly, auctions would generate vast new revenues for the federal government (some estimates had indicated that in excess of $80 billion could be raised if all unused spectrum were to be allocated on the basis of competitive bidding) [14].

The first auctions were held in August 1994, and yielded in excess of $1 billion for two-way paging and interactive video and data services (IVDS) licenses [15]. In a complicated series of auctioneering rounds, investment syndicates and corporations applied gaming theory to the estimation of license value and bid accordingly. We note in these experiences—and subsequent auctions held for PCS, SMR, and other spectrum—a critical value of forecasting methodology for the future of telecommunications.

Forecasting is assuming a multidimensional role for telecommunications planners: Its importance can be seen in predicting the behavior of consumers and caliber of competitors, in defining the intrinsic future value of their own organization, and now in estimating the worth of prospective licenseholdings. With respect to public policy, one might question the desirability of holding auctions for public assets, but it is hardly deniable that the strategy has clarified the intentions of communications firms.

A footnote should be added regarding the course of FCC auctions during the initial three-year experimental stage. The federal government collected large sums for the public treasury during the first two years of competitive bidding—with much smaller amounts generated in late 1996 and early 1997. Recent spectrum auctions for satellite transmission rights, for example, have produced fewer bids and less revenue than anticipated by both industrial and governmental experts. This experience implies a keener grasp of the inherent initial value of the licenses, perhaps, as well as a greater recognition that technological innovations may render some licenses highly valuable but others significantly less so. The experience suggests, too, that participants have learned to adjust their strategies to the procedures that government authorities have designed to maximize revenue while minimizing collusion.

7.5 Environmental Scanning, Forecasting, and Strategic Planning

As indicated in Chapter 6, the contribution of forecasting in clarifying strategic planning options can only be made on the basis of sound data assimilated through the process of environmental scanning. Environmental scanning synthesizes both quantitative and qualitative information according to a set of variables that includes technological developments, social and cultural trends, economic change, political factors, and other elements. At this stage the transformation of data leads to the application of forecasting methodology; resulting in a set of predictions that, in turn, is transmitted to planning models.

Data warehousing, data mining, and other state-of-the-art methods now lead to an intriguing possibility for the creation of telecommunications strategic plans. If the plethora of data on consumer behavior were used to greatest advantage, the phenomenon of "mass customization" could theoretically arise in the industry [16]. In other words, it may well be possible for large corporations to provide highly personalized service of a kind only known in today's smallest firms. Personalized service occurs when the customer dictates adjustments in customized menus at his own discretion and the organization accedes to that design. It has been noted earlier that customized menus of telecommunications services are very likely in the near future. Fusing abundant data and sophisticated forecasting measures may yield personalized services that reinforce customized menus. Corporate executives should now take this possibility seriously as they construct their strategic plans.

References

[1] Valentine, Lloyd, and Dennis Ellis, *Business Cycles and Forecasting*, Cincinnati, OH: Southwestern Press, 1991, pp. 462–515.

[2] See the seminal work, Spencer, E., C. Clark, and B. Hoguet, *Business and Economic Forecasting*, Homewood, IL: Irwin Press, 1961, for an examination of the array of forecasting methods. A simplified portrait of these methods is elaborated in "Forecasting Business Trends," *Harvard Business Review*, June 1975.

[3] Whiteley, Richard, and Diane Hessan, *Customer-Centered Growth*, Reading, MA: Addison-Wesley Press, 1996, pp. 265–285.

[4] Schneider, K., and C. R. Byers, *Quantitative Management*, New York: John Wiley Press, 1979, pp. 3–52.

[5] Valetine and Ellis, *op. cit.*, pp. 59–80.

[6] Vanston, L. K., and J. H. Vanston, *Introduction to Technology Market Forecasting*, Austin, TX: Technology Futures, Inc., 1996, pp. 1–8.

[7] John C. Mowen, *Consumer Behavior*, Englewood Cliffs, NJ: Prentice-Hall Inc., 1995, pp. 234–266.

[8] Paul Kagan Associates, "Wireless Communications Industry Forecast," *PRNewswire*, Oct. 5, 1995. This study assimilates data on prospective demand through 2006. Data cited hereafter reflect the summary findings of this industry forecast.

[9] Paul Kagan Associates, *op. cit.*

[10] Asthana, Praveena, "Jumping the Technology S-Curve," *IEEE Spectrum*, June 1995, pp. 49–54.

[11] O'Brien, Meghan, "Cellular Telephony and PCS," *Communication Technology Update*, fifth edition, August Grant (Ed.), Boston: Focal Press, 1996, pp. 290–299.

[12] Maynard, Jr., Herman B., and Susan E. Mehrtens, *The Fourth Wave: Business in the 21st Century*, San Francisco: Berrett-Kohler Publishers, 1993, pp. 127–170.

[13] Guy, Sandra, "The Next Frontier," *Telephony*, Sept. 16, 1996, p. 74.

[14] Salters, Harold, "E9-1-1 Coming to CMRS," *Radio Resource Magazine*, pp. 11–45, Oct. 1996.

[15] Dorsey, Michael, "Unplugged," *WPI Journal*, Spring 1994, pp. 22–27.

[16] See Brandenburger, A. M., and B. J. Nalebuff, *Coopetition*, New York: Doubleday Press, 1996; and Whiteley, Richard, and Diane Hessan, *Customer Centered Growth*, Reading, MA: Addison-Wesley Press, 1996, for a discussion of prospective trends in personalized, customized service.

8

The Economics of New Business Formations

Passage of the Telecommunications Act of 1996 has spurred a vast new flow of capital to the telecommunications and its ancillary industries [1]. New capital formations always reveal the future intent of organizations and entrepreneurs, as the private sector seizes upon a conception of unfolding opportunities. The shift from regulated to deregulated industries, as noted in Chapter 1, inspires a reconfiguration of strategic planning and business development.

For the telecommunications industry, unfolding deregulation prompts contemplation of these salient developments:

- Review and evaluation of the economics of "communications convergence";

- Adaptation to new strategies of horizontal and vertical integration;

- Tactical consideration of mergers and acquisitions;

- Investigation of prospective strategic alliances and partnerships;

- Earmarking of future venture capital;

- Likely emergence of a telecommunications business opportunity paradigm (BOP);

- Coming tertiary economics of telecommunications;

- Unfolding of National Information Infrastructure (NII)/International Information Infrastructure (III) business networking.

8.1 The Economics of Convergence

In anticipation of deregulation, telecommunications strategists have adopted the word convergence in describing pending business development. Although this term is often used in several contexts, often confusing the general public as to its meaning, "convergence" may be defined as the *interaction and evolving relationship of the television, telephony, and computer industries.* In other words, the dynamic, continuously changing involvement of these three innovative enterprises now denotes a major restructuring of both domestic and international commerce [2].

The significance of unfolding convergence, coupled with technological advance, lies in the fact that the telecommunications industry is likely to transform itself via these mechanisms:

- Increased merger and acquisition activity, as noted below, fostering consolidation in both domestic and international markets;
- Unfolding "strategic partnering," in which telecommunications service providers seek favorable agreements with content developers;
- Strategic "alliances" and "affiliations," in which telephony, television, and computer enterprises join in bilateral and multilateral agreements that enhance their value-added capabilities.

The dawn of convergence in the communications industry represents the tangible recognition that no firm can fully participate in all sectors of the industry—transmission services, content generation, value-added enhancements, and hardware development—in the absence of acquiring, merging with, or cooperating with complementary firms. The industry is so vast and expanding at so rapid a rate that no enterprise can successfully accommodate consumer demand while relying exclusively on its own resources. In the era preceding divestiture, the enormous size and capital availability of AT&T and other international monopolies would have permitted such a contingency, but today even the largest concerns cannot hope to pursue diversified strategies without complementary support.

It should be stressed that some observers speak of convergence when discussing potential agreements among content providers; or refer to cooperative relationships between telephone providers and motion picture studios; or emphasize the significance of emerging software development for Internet usage, and so forth. The multiplicity of such developing relationships has muddled a concrete definition of economic convergence. In essence, individuals who apply the term "convergence" in different contexts should define with precision their application of the word. As a practical matter, the seminal definition of telecommunications convergence links current and prospective relationships between the computer, telephone, and television industries. The discussion elaborated in Chapters 8 and 9 will strictly adhere to this working definition.

8.1.1 Telecommunications Mergers and Acquisitions

A merger may be defined as a combination of two or more firms, the purpose of which is to establish a singular identity for a resultant enterprise. Mergers are often precipitated by a shared vision that holds that enlarged size results in expanded market share; that expanded market share gives the newly defined company greater leverage over its competitors; that rising profit margin may result from the transaction; and that expansive capital availability will permit the firm to innovate at an accelerated rate. Mergers typically are facilitated by exchanges of stock or accounting "good-will" techniques, in which the larger partner passively assumes the difference in assets and costs of transaction. Because a merger can be negotiated as a tax-free pooling of assets, stockholders often benefit greatly from such agreements [3].

An acquisition is defined simply as the purchase of one firm by another [4]. Multiple motives characteristically surround acquisition decisions. Often, an acquisition is prompted by the desire of one firm to engage in vertical or horizontal acquisition. As suggested in Chapter 2, strategic planning aimed at these forms of business integration are viewed with concern by the Anti-Trust Division of the Department of Justice.

Vertical integration is defined as the ownership of networked production or distribution of a product or service [5]. In essence, a firm chooses to purchase or merge with those enterprises that support its delivery of products or services. If, for example, a fast-food franchise specialized in the production of pizza elects to buy complementary firms engaged in the production of cheese or tomato sauce, it has designated vertical integration as a strategic course of action. Vertical integration may be expressed as "forward" or "backward": If the pizza franchise buys a bread company that facilitates the production of its crust, its strategy is supportive and therefore backward; if the same company purchases a frozen-food distributor, its strategy is identified as forward, since the firm now enters a market not otherwise available.

Horizontal integration occurs when a company purchases a competitor in order to broaden and extend its product or service distribution [5]. General Motors recently purchased Saab, for example, in order to extend its automotive line to other markets. Horizontal integration, as a practical matter, can result in a firm's greater influence over price, since competition is often effectively reduced. The motive to acquire competitors can also involve enhancing internal expertise by absorbing the most talented employees working within the industry.

For obvious reasons, the Justice Department views with suspicion the prospective vertical and horizontal integration strategies that are established by the largest firms within an industry. The Sherman Anti-Trust Act, the Clayton Act, and similar antitrust mechanisms can be applied to block prospective mergers or acquisitions if they are deemed to threaten competitiveness through the manipulation of market share. As described in Chapter 2, the philosophy and political machinations of the Anti-Trust Department typically mirror the presidential administration of that era.

A paradox of antitrust application is that the administration that files an antitrust suit is often replaced by one that refuses to prosecute or otherwise allows prosecution to go dormant. Changes in the White House can shift antitrust priorities, thus extending by a factor of many years final approval of a proposed integration. Clearly, large, capital-intensive telecommunications firms will have to factor this consideration into their long-term strategic planning.

As one reviews the behavior of telecommunications firms immediately before and after the passage of telecommunications reform, a pattern becomes obvious. Many companies have completed negotiations aimed at creating mergers of horizontal or vertical character [6]. Deregulation has lubricated the regulatory climate to facilitate many of these transactions.

Among the major firms completing such negotiations are AOL, AT&T, Verizon, SBC, British Telecommunications, WorldCom Inc., Hughes Electronics, and Frontier Corporation. Major targets of these takeover attempts include Time Warner, Pacific Telesis, MCI, MFS, US West Media Group, Sprint, TCI, and PanAmSat Corporation. As was the case with the aviation, trucking, and natural gas industries, we note an emerging wave of merger activity likely to precede similar developments during the next decade. The objective of this strategic planning is *consolidation*: a tactic aimed at expanding market share, extending industry influence, and securing marketing and other expertise, while expediting entry to new markets.

Whether these transactions will be challenged by the Justice Department remains a matter of considerable conjecture, with billion-dollar firms gambling the vicissitudes of market timing and product introduction. Against this background is the continuing specter of government involvement at any point. Typically, government action is precipitated by concern over disproportionate market share or the implicit barriers to entry of firms who cannot compete with the economies of scale enjoyed by merged entities. Nevertheless, many mergers are deemed to be consistent with the public interest, and the Federal Communications Commission predicted well before passage of telecommunications reform that mergers, in many instances, would be a desirable by-product of deregulation.

It should be noted that heightened merger and acquisition activity, quite apart from the continuing debate about the desirability of such market dynamics, in no way assures market success. While many mergers and acquisitions, particularly those cutting across international boundaries, may eventually determine market winners, it can be reasonably inferred that some of these transactions will produce losers as well. Those integrations that do not result in producing telecommunications products or services that the consumer desires are destined to fail.

The historic justification for mergers, despite antitrust concerns, lies in the efficacy of economies of scale: price per unit of a product or service drops as the costs of capital, labor, and technology diminishes. The underlying presumption of a benign merger or acquisition is that the firm can reduce the cost of these three factors of production when its size proportionately expands (note the discussion of economies of scale in Chapter 3). There are historic examples of a firm's failure to measurably enhance economies of scale through a planned integration. Moreover, through recent advances in software development and inventory management, the cost of producing a good or service at reduced scale is decreasing; in other words, in many industries it is now possible to produce an item in lower quantity while approximating the cost advantages associated with higher levels of output [7]. Economies of scale as the key determinant to competitive pricing are thus eroding. Small, boutique enterprises may emerge as major competitors in all sectors of the telecommunications industry (further elaboration on this scenario is provided in Chapter 10).

In sum, we crystallize the recent proliferation of telecommunications merger/acquisition activity in Table 8.1.

Horizontal and vertical integration is likely to continue for approximately a decade, with market consolidation the key objective. If the natural gas, aviation, and trucking industries present a reasonable historical guide, the front end of merger and acquisition (M/A) activity will principally involve major providers and distributors of services; the back-end, final wave of M/A strategy will be aimed at the integration of service and content providers. Whether this scenario comes to dominate industry performance is a matter of great dispute and contentiousness, as hereafter discussed.

8.1.2　Strategic Alliances

Rapid technological change promotes an ambiance of unpredictability in today's telecommunications markets. If one reviews telecommunications forecasts of the past decade, particularly those involving the advent of wireless

Table 8.1
Projected Horizontal and Vertical Integration, 1997–2007

| | Objective/Motive | | | |
| | Front-end (1997–2002) | | Back-end (2002–2007) | |
M/A Partners	Horizontal	Vertical	Horizontal	Vertical
Service providers	Immediate expansion of market share	Procurement of marketing expertise to support sales	Diminution of competitor market share	Diversification (global and domestic)
Content developers	Access to networks and technical support	Procurement of complementary expertise in emerging markets	Long-term competitive positioning	Access to global markets

communications, interactive television, and Internet usage, one is startled by, alternatively, the exaggerated under- and overestimates of industry experts. In retrospect, gross underestimates projected for mobile communications usage stand in contrast to extravagant claims for both wireline and wireless interactive television services [8]. In both cases, forecasters had failed to fully anticipate both advances and inadequacies of emerging technologies and resultant competition. Additionally, changes in consumer preferences, as specified in Chapter 6, compounded the difficulty of forecasting unfolding substitutes of emerging technologies for veteran product lines. The glib pronouncements of forecasters, particularly the marketing staffs of telecommunications firms, often give way to the grim intrinsic mystery of household and business consumption.

In this environment, it becomes increasingly evident that *flexibility, adaptability,* and *customer responsiveness* dictate future success. In light of today's market realities, to rely exclusively on mergers and acquisitions to expand market opportunities impoverishes a firm's range of strategic options. Deregulation has prompted a number of telecommunications companies to seek speedy entry to new markets via supportive alliances: agreements, often temporary in nature, that typically join service providers with content developers to seize upon emerging market trends. In late 1996 America Online set temporary agreements, for instance, with Netscape and Microsoft to facilitate access to larger consumer markets[1]; these agreements last only as long as the interests of

---

1. Public affairs announcements specifying these contingencies were made by AOL in October and November of 1996; Steven Case, chairman of AOL, also noted that these agreements were subject to chance and dependent upon customer responsiveness demonstrated by both Microsoft and Netscape.

both parties are served. Either party may sever the agreement the moment circumstances warrant. A rapid response to changes in consumer preference is expedited at minimum financial risk to the firms in alliance. The virtue of a strategic alliance—limited risk—can be a potential pitfall, however; exceptional long-term opportunities can be forsaken at the whim of either party. Nevertheless, in markets characterized by exponential technological change, this option is a significant component in a company's strategic arsenal. Nimble strategic alliances remain a key element in swift organizational response to anticipated market opportunities.

8.1.3 Strategic Partnerships

For those telecommunications firms that must invest enormous sums to enter new markets and whose resources cannot permit standard merger/acquisition activity, strategic partnerships appear to be an effective alternative. Unlike the limited and temporal character of strategic alliances, partnerships afford long-term "strategic fit": the opportunity to develop a symbiotic relationship in which complementary and supportive expertise and resources enlarge competitive advantage. Such relationships can be formed between any service or content provider in the computer, television, or telephony industries.

Recent strategic partnerships have included Verizon and Pacific Telesis in the formation of Tele-TV; Motorola and Nextel in the creation of a specialized mobile radio network; NetTV and Intel in the development of integrated television/personal computer systems; WRQ and GTE in the delivery of wireless data; and OMNI and LitelNet in the formation of a new electronic software distribution network [9]. In each instance, we detect a systematic theme: the effort to build an enduring relationship while simultaneously leveraging risk. Put differently, the strategic partnership is characterized by a long-term vision of emerging consumer demand and sustained by tacit admission that a single firm cannot create and maintain that market. You will recall the significance of the S-curve (introduced in Chapters 5 and 6) in predicting unfolding consumer behavior. The strategic partnership pivots on a shared organizational view that a long-term commitment is required to proactively seize upon market opportunity.

The strategic alliance also serves the pragmatic interests of two or more firms seeking an alternative to the complexities of standard mergers and acquisitions. Strategic partnerships generally require greater time to consummate than do strategic alliances. However, they can be expedited in a fraction of the time necessitated by planned mergers. More importantly, partnerships do not inspire intervention by the Federal Trade Commission or the Anti-Trust Division of the Department of Justice.

Strategic partnerships often originate in cases where equipment providers seek channels of distribution via service providers or, alternatively, in situations where software providers can greatly aid the productivity of either equipment or service providers [10]. Increasingly, the sophisticated development of software now drives the service and content sectors: enhanced efficiency results in lower price, thus inducing greater numbers of consumers and businesses to experiment with new services or upgrade existing ones [11]. It is therefore reasonable to project that a substantial number of future telecommunications partnerships will result from evolving relationships between telecommunications providers and the most advanced, incisive software developers.

8.2 Venture Capital and the Telecommunications Industry

A critical financial source for the creation of business start-ups is venture capital. Venture capital, sometimes referred to as "risk capital," is a primary source of funding for start-up businesses whose projections of growth typically exceed 25% annually during the first five years of development [12]. Often venture capitalists seek returns of up to 50% annually within the first five years of creation, although they are committed to returning all profits back to the enterprise in the name of establishing and expanding the business.

8.2.1 An Overview of the Venture Capital Industry

Venture capitalists have come to play a vital role in the development of all high-technology industry in the post–World War II period. Ordinary equity financing—the underwriting and selling of stock—is a comparatively conservative undertaking in the creation of new firms whose prospects are based on high risk. Entrepreneurs lack the time, expertise, and finesse so often associated with deployment of stock. Deployment of capital aimed at bringing young firms to market often remains a viable option in a business world otherwise allowing only equity and debt financing.

The venture capital industry has expanded significantly since the passage of favorable federal legislation in 1978 [13]. Reductions in capital gains taxes promoted significant capital diversions to the industry throughout the 1980s. Recognizing the contribution of small business to the expansion of employment opportunities, Congress moved to lower capital gains tax rates in 1997.[2]

---

2. During the 1996 presidential campaign, both major candidates and their parties proposed capital gains tax relief. Capital gains tax rates for long-term investments were lowered to 20% from 28%, effective August 1997.

As a practical matter, favorable tax treatment for this form of investment promises substantial financial support for the future of telecommunications development. Provided standard investor criteria are met for the deployment of venture capital—*high rates of return relative to risk, development of state-of-the-art technologies, anticipation of expansive consumer base, and substantial potential profit margins*—there exists a growing pool of funds for telecommunications enterprises. "Venture capital alleys" in Silicon Valley, Houston, Boston, and other key metropolitan areas are exercising growing influence in defining the future of telephony, computer, and television development.

Inevitably and invariably, venture capitalists provide only seed money in forming start-ups. IPOs are essential mechanisms in generating the capital required to sustain the business 24 months after its birth [14]. For purposes of this discussion, and in the context of the fast-growing, fast-moving telecommunications industry, note the broad purposes of venture capital management [15]:

- *Seed capital:* devoted to financing research and development, providing essential supplies, materials, and initial payroll;
- *Working capital:* aimed at completing the initial stage of development in addition to determining strategic planning, marketing strategy, sales objectives, and essential overhead;
- *Acquisition capital:* applied in those cases in which a start-up must procure the expertise of another firm in order to come to market. In some cases, this means purchasing on a vertical or horizontal basis another small or developing firm that supports the mission of the start-up.

Working capital remains the vital link—and glaring dilemma—confronting the start-up. It is in this stage that a start-up must face its greatest challenge. In this phase a firm often goes to market with its IPO strategy. Significantly, the prevailing winds of the IPO market, which may be characterized as favorable, unfavorable, or stagnant, often dictate success or failure. Spectacularly favorable IPO markets—that is, periods in which investors aggressively sought such speculations—included the late 1970s, mid-1980s, and early to mid-1990s. Decidedly unfavorable eras included much of the 1970s and late 1980s. Thus, investor perception and the general economic climate will be critical variables outside the immediate control of telecommunications firms at any given moment.

8.2.2 Issues in Telecommunications Venture Capital

In designating venture capital deployment for the telecommunications industry, investment firms and individuals are pursuing these concrete goals:

- Desirable rates of return, as defined above;
- Strategic participation, that is, direct influence over company mission and long-term strategic planning;
- Equity share percentages, in which specified ownership of the company's assets are identified from the beginning;
- Board positions, in which capitalists may obtain permanent chairs within the firm's board of directors.

Collectively, these tactics provide sufficient rationale for the venture capitalist to risk his or her capital. Obviously, the entrepreneur will resist the intrusiveness of these actions, particularly with respect to the second and fourth goals. The success of the entrepreneur on this level will be a function of their persistence and prestige as well as the allure of their seminal concept. Among the many critical issues facing both venture capitalists and entrepreneurs in the 1990s are:

- Accuracy and persuasiveness of product projections;
- Volatility and unpredictability of technological innovation;
- Dilemmas of product substitution, linked to unfolding competition.

Forecasting prospective market trends is exceptionally difficult in an industry as transformational as telecommunications. The dynamics of change, both domestic and international, have often rendered traditional forecasting methods obsolete. We have noted in Chapter 7 the array of scientific forecasting tools available and the manner in which they have been applied to produce strikingly accurate long-term predictions. Increasingly, however, the number of inaccurate forecasts are just as striking. The IVDS industry, for instance, had been projected to grow at a spectacular rate by both government and industry analysts; this venture has hardly emerged since its seminal development in 1991 [16]. Demand for cellular communications services, on the other hand, has been consistently underestimated by both government and industry forecasters [17].[3]

Technological innovation in telecommunications has been so spectacular in recent years that some experts estimate that the effective "knowledge half-life" of an electrical engineer is four years. Put differently, half of what an engineer learns after graduating becomes obsolete within four years of his or

---

3. Both the FCC and AT&T, forecasting in 1983, estimated that cellular communications technology would never penetrate more than 1% of the consumer market and would be considered a "luxury market."

her professional employment. Within 24 months following commercialization of the World Wide Web, for example, primitive Internet telephony was introduced. Scarcely an expert had anticipated this contingency. Volatility, unpredictability, and instability must somehow be taken into account before large sums of venture capital are channeled to entrepreneurs.

Similarly, venture capitalists are reluctant to commit major sums to product development before they secure credible estimates as to competitive breakthroughs in the market. In other words, entrepreneurs face the intense problem of persuading capitalists to finance their venture at a time when competition is growing increasingly difficult to estimate—the product of exponential change gripping the global marketplace.

8.2.3 Venture Capital and Telecommunications Fraud

Apart from the strategic challenge of calculating and projecting prospective demand for telecommunications goods and services, the industry has recently experienced a wave of fraudulent telecommunications sales tactics. Unscrupulous firms have marketed a series of fraudulent investment packages—notably, preparation of FCC communications licenses, creation of wireless cable television franchises, and development of specialized mobile radio stations and wireless interactive television services—that have recently distorted the distribution of available venture capital [18].

The Federal Trade Commission estimates that telecommunications investment fraud may have exceeded $80 billion since 1991 [18]. These "venture capitalists" seek to exploit investor enthusiasm and technical ignorance so as to enrich themselves. In fact, investment scams have not only resulted in theft, but have distorted the pooling of future investment capital by instilling profound skepticism in potential investors. It has become increasingly difficult to secure seed money from small investors; entrepreneurs thus searching for venture capital must turn increasingly to managed venture capital firms who will extract greater concessions at the time of investment. Investor fraud and continuing scams have induced the Federal Communications Commission, Securities and Exchange Commission, Federal Trade Commission, and Justice Department to recently seek stronger criminal sanctions in IPO underwriting and financing.

8.3 Telecommunications Business Opportunity Paradigm

The proliferation of mergers and acquisitions, the dynamics of industry convergence, the move to strategic partnerships and alliances, and the criticality of venture capital financing, collectively designate a telecommunications business opportunity paradigm (BOP). The complexity of evolving technological innovation fused with insatiable demand for productive, efficient communications impels a model of industrial development.

Strategic planners can no longer afford to develop a telecommunications good or service in isolation, presuming that a sound product will find its niche merely on its own merits. Three key issues automatically confront telecommunications firms when contemplating the impact of their product on the market—as well as the market's influence on the future development of that product.

First, when a new product captures a customer base in excess of 10% of the market, competition sets in [19]. Success breeds imitation and product substitution. Second, the creation of a successful new product will stimulate tertiary consequences that will result in the development of support services. Third, larger firms will overtly seek to preempt or buy out the successful creator. All three contingencies are central facts that guide the development and evolution of telecommunications firms.

In seeking to synthesize these three themes and connecting them to the realities of a predatory marketplace, it becomes incumbent for entrepreneurs and established firms to design a model that anticipates the effect of its own products on future customer demand. In short, the telecommunications provider must ask and answer these enduring questions.

1. If we introduce Product X and succeed in capturing significant market share, how will our competition respond? Is our product so uniquely conceived that it cannot be immediately imitated? If it can be replicated, how much time do we have before our competitors bring to market effective substitutes?

2. In designing this product, have we created the demand for new firms who will support the sales, maintenance, and efficacy of that product? Simply, will unfolding business opportunities emerge from the creation of Product X and, if so, should we pursue these secondary opportunities as well?

3. In bringing to market Product X, do we invite the predatory instincts of our largest rivals? If we succeed, will the competition usurp our market through cost-effective pricing, manipulation of distribution channels, or any other tactic? Would we be prepared, and would we have the resources to mount a legal challenge to any questionable antitrust practice in pricing, distribution, marketing, or intellectual property infringement?

Abundant evidence exists to confirm that these three interconnected issues often confront the successful telecommunications firm, emerging or veteran. These matters are dealt with effectively when they are forecasted on the front end of product development. We note the model in Figure 8.1, an illustration of the BOP, which asserts the developing market impact associated with the introduction of a new product. The dynamics of new product development are such that we may anticipate, adjusted for an appropriate time horizon, *customer, competitor,* and *industry responsiveness.* One must forecast the short- and long-term environmental consequences indicative of telecommunications innovation and is compelled to do so at a time of turbulence and volatility. If the task were a simple one, a majority of firms would succeed; it is the minority that successfully calculates and anticipates the result of their own actions on the marketplace, and vice versa.

We may infer from the model that the most advanced, sophisticated, progressive telecommunications firms are beginning to seriously address the long-term market consequences associated with their own actions. Where a firm once sought a quick exploitation of a new product or service, serious thought is now applied to long-run ramifications of new product development and the role this will have in defining the future of the firm. This phenomenon

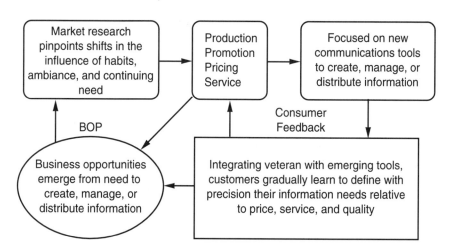

Figure 8.1 Telecommunications business opportunity paradigm.*

---

*Premise: Future telecommunications product/service innovation is driven by customer-led customization; that is, consumers identify with increasing specificity the product applications unique to their own needs. For the first time, consumers are able to exercise direct influence over production, pricing, and service of new products prior to the firm's application of standard market research techniques.

translates into a simple maxim for telecommunications providers: There is no such thing as a telephone, cable, television, or computer company, per se. Instead, today's telecommunications company is a *creator, manager,* or *distributor of information* and should therefore be willing to pursue any avenue that advances that image and purpose. As time passes, dichotomies between these three functions will evaporate, thus setting the stage for revised strategic design and resultant business opportunities in the next century.

8.4   The Coming of Tertiary Economic Analysis

Implicit in the BOP model is the emerging significance of tertiary economic analysis. Based on the application of systems analysis to economic theory and business development, tertiary economics is a mysterious, nebulous endeavor that nonetheless holds an important key to future profits in the telecommunications industry.

In ordinary business activity a firm is committed to bringing its product to market and rarely evaluates the long-term significance of present decision making. American business people, by habit and custom, design and market a product and exploit its profitability (note Chapter 6) until its lifecycle is effectively eroded or ended. However, support industries often emerge inadvertently, and occasionally advertently, from the dissemination of that good or service. History abounds with new technologies whose implementation resulted in necessary economic undergirding.

What surfaces from a consideration of these and other successful technologies is a linkage between *primary, secondary,* and *tertiary effects.* In other words, every new technology has a primary impact on its market and the pattern of household and/or business consumption that follow. A secondary consequence of that adoption is typically the maintenance of that product once purchased. Thereafter, third-, fourth-, and fifth-ordered tertiary consequences are evident

---

Assumptions: (1) The use of advanced telecommunications services becomes increasingly habit forming. (2) The adoption of new services compels associates (friends, family, colleagues) to acquire comparable tools for purposes of communication. (3) The generation, storage, transmission, and retrieval of information represent "inexhaustible" markets; that is, the growth of information precipitates continuous global expansion for each phase of telecommunications management. Unlike other resources, information cannot be depleted.

Inference: Modern telecommunications corporations define themselves as creators, managers, or distributors of information. Mergers, acquisitions, partnerships, and alliances are the business vehicles by which customer-led customization is facilitated. Business opportunities result from cooperative ventures in which complementary expertise is required to enter and build new markets.

in the immersion of that product in the economy (the future business and economic implications associated with tertiary consequences are explored in the following chapter). These are intriguing emerging developments because they represent:

- The absence of control or containment by the initiating enterprise, emerging competitor, or government regulator;
- New business opportunities, separated from initial product introduction only in terms of time and place;
- Impending competition, often international in character, that invariably leads to multinational corporate development.

Without question, telecommunications firms—owing to their growing economic influence—have stimulated a debate among strategic planners, economists, and marketers about the relative effectiveness of traditional forecasting measures in facilitating the prediction of future market opportunities. At present, there is no systematic, satisfactory method to fully anticipate the long-term tertiary impact of the introduction of new telecommunications product. This is not to say that the tertiary effect should be dismissed or ignored in strategic planning; to the contrary, new techniques are evolving to accommodate a greater understanding of appropriate strategy to exploit these opportunities (note the model postulated in Chapter 10). Superior telecommunications firms are "deep-thinking" the ramifications of evolving methodology.

8.5　The Synergies of Telecommunications Development

8.5.1　The NII and International Business Opportunities

The buildout of the NII—wireline and wireless—is being surveyed with great interest by other nations. These perceptions, and their influence in outlining telecommunications development in other regions, form the basis of an emerging synergy that may create an III. The NII/III confluence, when fully extrapolated, allows for a range of prospective international telecommunications partnerships, mergers, acquisitions, alliances, and entrepreneurial opportunities.

Successful international telecommunications ventures—that is, alliances or mergers integrating telecommunications firms originating in two or more nations—cannot be consummated in the absence of two overriding variables: *cultural harmony* and *managerial compatibility*. These are the variables that typify successful multinational corporate relationships in other industries. Cultural harmony, characterized by effective cross-cultural understanding and cooperation, facilitates the kind of rapid decision making that is a part of the competitive landscape. Managerial compatibility, defined as effective organizational communication, is a critical ingredient in setting a mission with supporting strategy, tactics, and goals.

A multitude of driving forces suggests probable international telecommunications alliances and mergers. These may be categorized accordingly:

- Access to new markets with established consumer segments;
- Knowledge of cultural tastes conducive to the growth of market penetration;
- Absorption of technical staff competent to respond to governmental regulation unique to that environment;
- Leverage of capital, labor, and technology required to bring new communications technologies to large markets;
- Competitive access, the compulsion of large American firms to establish market presence and strategic branding/bundling before domestic competitors rise in response to probable deregulation in that region.

Major telecommunications providers in Europe, Japan, and Latin America are now scrutinizing American international strategic planning in light of telecommunications reform. Promulgation of U.S. telecommunications deregulation establishes important precedents in the emerging global arena. Governments throughout the world now contemplating communications deregulation look to the American experience in order to ascertain its effect on economic growth and competitive behavior. Clearly, should the American experiment with deregulation enhance economic growth and employment opportunities, governmental adherence to traditional regulatory formats (explored in the following chapter) will recede [20].

Moreover, mounting evidence suggests European regulatory agencies are concerned that opposition to prospective American telecommunications alliances with that region's providers will result in lost competitive opportunities. Whatever the suspicions may be with respect to unfolding mergers or partnerships, it can hardly be denied that American technology and competitive skill will speed communications services to market. In a world galloping toward heightened competition, a conscious decision by international regulators to prevent such alliances has the practical effect of inhibiting growth in other sectors of their respective economies.

8.5.2 Emerging International Telecommunications Services

Chapter 9 elaborates the short-, mid-, and long-term prospective telecommunications services that may inspire potential American alliances with European and Asian enterprises. These alliances, mergers, partnerships, and acquisitions may be described as developing in a *parallel* environment: that is, mergers and acquisitions are likely to be the engine of initial international affiliations of large corporations. We are likely to see intense M/A activity between American and European counterparts continue through this decade [20]. Simultaneous to these relationships are both temporary and permanent alliances drawn between European, American, and Asian small businesses seeking to cultivate consumer markets in database sharing, retailing, banking services, and electronic commerce [21].

Regulatory reform and competitive thrust are stimulating cutting-edge technologies that now enable small businesses to develop business opportunities formerly the domain of multinational corporations. You will note in Appendix C a section on Internet-based information access to international small business formations.

References

[1] Shaw, James K., "Future Scenarios for the Telecom Industry: A Ten-Year Forecast," *New Telecom Quarterly*, Dec. 1996, pp. 25–26.

[2] Pappalardo, Denise, "The State of Convergence: Study Reiterates Importance of Killer Apps," *Business Week*, Nov. 25, 1996, special insert.

[3] Argenti, Paul, *The Portable MBA Desk Reference,* New York: John Wiley & Sons, Inc., 1994, p. 281.

[4] *Ibid.,* p. 390.

[5] *Ibid.,* p. 215.

[6] Thatcher, B., and R. McNamara, "How Merger Mania Has Redefined the Communications Landscape," *Telecommunications™*, Oct. 1996, pp. 42–44.

[7] Toffler, Alvin, Testimony before United States Congress, Subcommittee on Commerce and Telecommunications, Feb. 12, 1996.

[8] Lim, Kenneth (Ed.), "The Top Ten Myths of Digital Convergence," *Cybermedia 2001*, March 1995, V1.2, pp. 1–3.

[9] Wood, Wally, "Can Telcos Survive," *Telephony*, March 4, 1996, pp. 22–29.

[10] Pettis, Chuck, *TechnoBrands*, New York: American Management Association, 1995, pp. 165–200.

[11] Poletti, Therese, "America On-Line Dominates On-Line Service Business," *Reuters News*, Aug. 23, 1996.

[12] Bygrave, William D., *The Portable MBA in Entrepreneurship*, New York: John Wiley & Sons, Inc., 1994, pp. 184–185.

[13] *Ibid.*, pp. 278–317.

[14] *Ibid.*, pp. 182–187.

[15] Agenti, *op. cit.*, pp. 189–190.

[16] Mills, Mike, "An Interactive Dream Unfulfilled," *Washington Post National Weekly Edition*, July 10–16, 1995, p. 20.

[17] See Shaw, James K., "Wireless Communications and Technology Substitution: What S-Curves Reveal About Pending Cellular Competition," *New Telecom Quarterly*, Summer 1996, for a discussion of underestimation of consumer demand in the wireless arena.

[18] "SMR and Wireless Cable Scams," *Federal Trade Commission Documents*, May 1995.

[19] Dent, Harry S., *The Great Boom Ahead*, New York: Hyperion Press, 1993, pp. 220–245.

[20] Nee, Eric, "The International Information Highway: Two Views," *Upside*, Dec. 1994, pp. 46–51.

[21] Elliott, T. L., and Dennis Torkko, "World-Class Outsourcing Strategies," *Telecommunications*™, Aug. 1996, pp. 47–49.

9

The Economics of International Business Formations

The passage of the Telecommunications Act of 1996 has encouraged other nations to investigate the desirability of comparable deregulatory strategies. With the exception of Great Britain, where experimentation with deregulation and privatization was instigated during the mid-1980s, there has been little movement in this direction on a global scale. The speed with which international telecommunications mergers, acquisitions, alliances, and partnerships unfold will likely be a function of the perceived effectiveness of the American experiment. Indeed, those who manage international capital earmarked for the promulgation of telecommunications services eagerly await a market assessment of the impact of telecommunications deregulation on the American economy.

Should the American economy enjoy robust economic growth and expansive employment, it is likely that precedent will have been established for decisive deregulatory experimentation elsewhere. However, should a narrow elite of oligopolies control market share and minimize competitive benefits while enhancing their own profit margins, it may well be that other nations will eschew such legislation and pursue a strategy of government-sponsored monopoly. Inevitably, a serious discussion of unfolding international telecommunications opportunities pivots on the speed of entry to market: For many regulators, the failure to institute serious regulatory reform could have the effect of providing American companies with a permanent competitive wedge. Such an advantage could paradoxically provide U.S. firms with a sustained economic advantage for decades.

It should also be noted that an underlying presumption of this chapter is that the growth of international telecommunications business opportunities for the next 10 years will hinge on cooperative alliances established by multinational corporate providers. The author assumes that global standards in networking, transmission, and other infrastructure will necessarily precede the stimulus and creativity of international venture capital and its resultant impact on cross-country entrepreneurial activity. This is a somewhat contentious argument but, as postulated in Chapter 8, it remains salient that we "follow the money" in isolating the emerging economics of international telecommunications activity. In this regard, the outline of global capital suggests that multinational corporate planning will be pivotal in the development of telecommunications business opportunities a decade forward.

9.1 The Economics of International Convergence

We defined in Chapter 8 the meaning of telecommunications convergence: "The interaction and evolving relationship of the television, telephony, and computer industries." International convergence is thus defined as the interaction of the multinational television, telephony, and computer industries. Viewed from the perspective of economists, the evolving relationship of these international organizations—and the governments that regulate them—will result from the degree to which management cooperation, collaboration, and communication can be facilitated. This is indeed a challenging task, as common standards of technical, financial, political, regulatory, and cultural interaction must be established [1].

International economic convergence in telecommunications has been expedited by regulatory reform in the United States. Although many nations have been hesitant to introduce competition to state-owned or regulated enterprises, to risk delayed regulatory relaxation would only have the effect of enhancing the competitive positions of American and British concerns. Speed to market with new telecommunications services becomes the predominant objective in the convergence of veteran and emerging technologies across national boundaries [2]. In projecting unfolding international economic convergence, we note the following dynamics.

- The development of mergers and acquisitions is likely to precede international strategic alliances and partnerships, given the need by major competitors to secure immediate market share.
- Fundamental regulatory reform, or deregulation, will be instigated only after success with these reforms has been demonstrated in the United States.

- Necessary preconditions for the earmarking of international venture capital include political stability, predictable regulatory/legal administration, limited capital gains tax rates, and other factors hereafter identified.

- The economic value of strategic partnering, which pivots on the successful maturation of merger/acquisition activity on the front end of deregulation, will not be realized until reciprocity is enacted by all nations through negotiated trade agreements.

- Strategic alliances, critical in establishing the synergies of cooperative agreements, are likely to accelerate international business formations during the next century, assuming the preceding four dynamics are firmly woven in the new fabric of global communications.

The data and other empirical information supplied in Chapter 1 evidence the growth of the American post-industrial economy and the vital role telecommunications provides in spurring that activity. In a larger sense, the growth of telecommunications—and the resultant influence it has on productivity and competitive advantage—will stimulate the same in the global arena. To stimulate American economic growth through telecommunications development is to provoke comparable growth in the international economy. To spark U.S. telecommunications activity is to create supporting industries to follow and otherwise enhance the whole of the American economy. In this sense, it may be said that American telecommunications success is inextricably linked to economic growth in other post-industrial economies (most notably, Japan and Europe); telecommunications advances in these regions lead to the stimulus of economic development in industrial and agrarian economies elsewhere in the world. Deregulation, provided successful, will evitably invite a string of similar efforts throughout the world. Deregulation begets convergence, and convergence inspires economic growth in other sectors of the global economy.

9.2 International Telecommunications Regulatory Overview

In the years preceding American divestiture via AT&T and the liberalization of the telecommunications markets in Great Britain, international mergers and acquisitions hinged on cooperative governmental ventures negotiated through the General Agreement on Tariffs and Trade (GATT) [3]. Because GATT effectively acted as a bridge between competing national interests—and the monopolies they regulated—forging significant cross-national agreements represented a slow, cumbersome process. The instigation of deregulation in the British and American markets thus established a precedent for regulatory relaxation and privatization. There is great variation in the European, Asian, Latin American, Eastern European, and African regulatory experience. The variation in governmental authority is a function of enduring political, legal, and business traditions. However, the fear of the "competitive wedge" (described in Chapter 8)—the intrinsic competitive advantages enjoyed by those nations that permit their telecommunications providers to enter into global agreements—provoked ". . . an inexorable, worldwide trend toward privatization and competition" [4]. To yield competitive advantage places domestic telecommunications enterprises at great competitive risk, with rivals seizing and prospectively sustaining their customer bases. This theme underlines a contemporary examination of communications regulation as discussed elsewhere in this chapter.

9.2.1   Telecommunications Regulation in Europe

The largest economy in Europe is Germany [5]. The German customer base is the greatest in Western Europe; the standard of living is the third highest in the world; and the deutsche mark is the linchpin of regional monetary stability. Mergers between European and other providers can occur only through the competitive dismantling of Europe's largest communications company, Deutsche Telekom. The first step toward German telecommunications deregulation was taken in November 1996 with the privatization of Deutsche Telekom: the company began offering stock to domestic and international buyers. With competition now introduced to the German market, a wave of regional alliances and partnerships immediately followed [6]. It can reasonably be hypothesized that intense negotiations will now take place among primary and secondary communications providers throughout the region [7]. In evaluating the shift to privatization in the German market, we place ourselves in a more advantageous position to forecast probable acquisitions in the near future.

While the dominant engine of telecommunications deregulation, by weight of sheer size, is now the former German monopoly, it is nonetheless salient that British Telecommunications remains the most influential private telecommunications enterprise in Europe. As noted, the British, under the guidance of the Thatcher administration, dismantled a number of domestic monopolies during the 1980s. In the shipbuilding, energy, transportation, and communications industries, the British strategized a vision of increased competition to mitigate inflation and enlarge corporate profitability. This strategy, accompanied by tax reduction, set the United Kingdom as the engine of "deregulatory change" in the European Union (EU). Under terms of the "European 1992" agreement, competition in the telecommunications sector formally commenced in 1998. In response to this agreement and the competitive ambiance already introduced in the region, the Germans, French, Italians, and other members of the EU adopted similar deregulatory procedures.

As is always the case, those governments that adhere to traditional regulation argue that monopolies or limited oligopolies will invariably dictate market price at the expense of consumers. Protection through regulation becomes the only appropriate vehicle in a capital-intensive industry with the unique characteristic of mutually dependent, interacting networks. Even if empirical evidence should prove the contrary, these governments claim that the introduction of competition could not possibly be introduced with sufficient speed to prevent short-term consumer harm. Such remains the mind-set of certain European regulators. This traditional dogma is gradually evolving as new technologies, and EU regulatory edicts, obviate the advantages once exclusively controlled by large entities.

A critical issue to emerging competition in Europe concerns the extent to which the EU will permit the importation—and thus competition—of global telecommunications goods and services [8]. While individual governments control their own tariffs, quotas, and similar import regulations, this is done within broad terms negotiated before 1992. Whether these governments, individually or collectively, seek to construct a fortress in which telecommunications imports are effectively limited is not yet clear. In general, the Italians and French are most entrenched in their resistance to competitive change, while the Germans, Belgians, and Dutch are now pursuing reform with vigor [9].

Paradoxically, Eastern European nations—most especially Slovakia and the Czech Republic, Hungary, and Poland—are more open than their Western counterparts in inviting telecommunications competition. In the wake of the collapse of the Soviet Union, Eastern Europeans are eager to embrace economic change congruent to their political revolutions. American firms seeking access to markets at comparatively cheap prices are therefore eyeing opportunities throughout the whole of Europe. A key stumbling block to such agreements are common legal standards, a necessary precondition to international mergers and acquisitions. Among these concerns are mutually acceptable standards of stock issuance and equity interest, hiring and personnel control, marketing restraints, and similar items where government sanctions can impinge on corporate directive. Already, issues extraneous to acquisition proposals—including the contentious item of intellectual property—have slowed progress otherwise attained.

Nevertheless, Americans and Europeans share a common economic imperative: Much of their traditional telecommunications markets are saturated, and there is little growth in the number of domestic consumers. Following the postwar baby boom—a population explosion experienced in both Europe and the United States after World War II—annual population growth in most

European nations is now less than 1%; indeed, in some nations, aggregate populations are expected to fall in the near future. Population in the United States expands at the rate of about 1% annually.

In this global environment, growth in demand of telecommunications products and services will be a function of two forces: (1) growth in new services that replace existing ones and (2) growth in consumer markets (i.e., consumer growth based on rising populations). For Europeans, in particular, it is crucial that they exploit both of these avenues so as to incite future economic activity. With respect to the former, an expedient method to encourage state-of-the-art technologies is the merger/partnership/alliance route. It is a ready technique to bring new products to market, results in favorable technology transfer, and expeditiously elevates supporting industries through rising productivity [10].

In light of Europe's mixed regulatory environment and the uncertain cross-country political machinations slowly unfolding, we can infer that mergers and acquisitions are likely to accelerate during and perhaps beyond this decade. To "beat" competition to market, in those instances where mergers can be consummated without provoking regulatory and judicial bodies, American and British firms will pursue immediate access to prospective customers; this strategy circumvents any "regulatory fortress" that might later be installed by the EU. Should the EU relax trade constraints and invite international investment, these firms would retain considerable leverage over future global rivals [11]. For reasons discussed later in this chapter, the Japanese are hindered in their attempt to penetrate European markets, particularly in light of delays associated with the deregulation of their own communications industry.

9.2.2   Telecommunications Regulation in Asia

The engine of economic growth in Asia during most of the post–World War II period has been Japan, which maintains the second largest economy in the world, significantly larger than that of Germany, and maintains a sophisticated regulatory environment guiding its telecommunications industry. Inspired by the regulatory system adopted in most large European nations following the end of the war, the Japanese established a biregulatory environment to facilitate telecommunications services. One element of this regulatory strategy was the creation of the government-sponsored Nippon Telephone and Telegraph Corporation [12], guiding development of primary domestic services. The other element involved the creation of Kokusai Denshin Denwa Company, whose legal monopoly status provided for the private management of international communications services. Fusing these two regulatory strategies has presented the Japanese government with several dilemmas as this decade unfolded.

Recognizing that the bimodal regulatory strategy was inhibiting not only telecommunications innovation in Japan but also the nation's international competitive position, the government has committed itself to the gradual relaxation of restrictions since the mid-1980s [13]. Although the relaxation of restrictions has invited some international investment in domestic telecommunications development, capital inflows have not been as substantial as the Japanese had initially projected. Moreover, NTT maintains a very close relationship with politicians who are indirectly responsible for any breakup of this $60 billion monopoly [14]. International business firms are thus effectively discouraged from entering this enormous market. Proverbial "red tape" stifles innovation in the domestic telecommunications market, and consumers effectively pay as much as two and a half times comparable phone rates Americans bear. Consumer consciousness about these inequities has been aroused, as has the Japanese concern that the industry may stagnate in the absence of a competitive international presence.

In late 1996 the Japanese took further steps to liberalize trade restrictions on foreign entrants, and some now believe that its domestic telecommunications industry will enter an era similar to that witnessed in the United States during the late 1990s. However, considerable skepticism remains that the two incumbent monopolies may mitigate the effects of competition by seeking favorable political decisions from senior politicians. Whether this proves to be the case is a matter of conjecture, but the Japanese have limited their own ability to export telecommunications products to both Europe and the United States by resisting reciprocal trade arrangements. Ironically, in seeking to protect their domestic telecommunications industry, the Japanese may have laid the groundwork for diminished future export revenues. One encouraging development in late 1996 was the World Trade Organization (WTO) initiative to simultaneously eliminate all tariffs on information technology imports. This proposal was enacted by the bulk of the world's economies in December 1996. A free-market economy has emerged for information technology products and services. This agreement effectively establishes precedent for the elimination of duties on all high-technology products, including those that are part of the telecommunications industry. This critical development could stimulate substantial economic growth, particularly in the United States, since Japan would be induced to import many emerging technologies.

Elsewhere in Asia, many observers have concluded that the two most intriguing long-term telecommunications markets are China and India. With enormous customer markets now primed for telecommunications demand (China and India are the two most populous countries), all major international telecommunications providers are seriously developing, or have already conceived, strategic plans for these territories. The regulatory structure of China, in principle autocratic, but in practice one managed by local authorities, has gradually liberalized its restrictions [15]. In committing itself to economic expansion, the central government has allowed joint partnerships, alliances, and other financial vehicles to spur telecommunications activity. Many venture capitalists now pursue the Chinese market with vigor, and joint telecommunications enterprises now flourish. The immediate impact of this development was evidenced in coastal regions where communications undergird economic growth. Recently, such ventures have penetrated the central region of the country as well.

In India, where telecommunications penetration is less than 1%, there exist enormous opportunities for international telecommunications providers [16]. The Indian government has always maintained tight control of the telecommunications industry through strict regulatory enforcement. The public monopoly remains a comparatively rigid one, but attempts to encourage private capital investment—both internal and external—have established precedent for international joint ventures. European and American firms in recent years have attempted joint alliances with the Department of Telecommunications (DOT) in order to facilitate the buildout of both wireline and wireless systems. Additionally, Indian entrepreneurs are now seeking private capital to design wireless cable (MMDS) systems in metropolitan regions. The move to regulatory relaxation appears inexorable.

Other Asian nations can be described as fitting into one of two parallel development routes: those countries that have committed themselves to building state-of-the-art telecommunications infrastructures, based in part on outside capital investment; and those nations maintaining traditional monopoly structures, while resisting the intervention of outside concerns. In Hong Kong, Singapore, South Korea, Malaysia, and Taiwan, telecommunications systems are comparatively advanced and approaching advanced access to the citizenry. In Indonesia, Burma, and North Korea a spectrum of extremely limited outside investment to no foreign investment is evidenced. In these nations private, international investment is regarded with great suspicion. It should be noted, however, that the Indonesians are permitting limited ownership, usually contained to 20%, of domestic communications enterprises [17]. Should the Indonesian government conclude that economic growth is efficiently encouraged by the introduction of such private investment, other "restrictive" government may follow suit. A critical determinant of the speed with which international investment seeks out these new markets will be the licensing approval process:

Many foreign firms and venture capitalists are discouraged from investing if they perceive an extended, cumbersome regulatory process. Legal and regulatory reform therefore must precede significant capital investment by foreigners in these emerging markets.

9.2.3 Telecommunications Regulation in Latin America

As is the case in Asia, there is great invariance in the enforcement of government regulation throughout Latin America. In a purely economic sense, the most attractive markets are Brazil and Mexico, owing to their large consumer markets. However, Chile—in political, legal, and regulatory terms—is the nation most immediately attractive to international entrepreneurs. Chile embarked on deregulation of most major industries during the 1980s, with enhanced liberalization and privatization of resources during this decade [18]. Domestic and international capital has flowed into the country, with advanced wireline and wireless systems now under construction. With these developments, Chile is approaching a highly competitive telecommunications market structure. The significance of this experiment, still undetermined in terms of its long-range implications, is that other nations in the region will carefully evaluate this important precedent. The key to forecasting emerging telecommunications opportunities in South America may well lie in the Chilean experience.

Reviewing Chile's experiment with deregulation is of great interest to Brazil. The Brazilian consumer market, with a population of nearly 160 million, is immensely attractive to telecommunications firms. Despite continuing problems of inflation, the Brazilian economy promises long-term economic growth based on natural resource development and a rising standard of living. The Brazilian government, beginning in 1995, introduced incremental privatization in its domestic infrastructure. The Brazilians are particularly interested in instigating the development of value-added services, vital to the maturation of other sectors of its economy [19]. The government has resisted attempts to sell majority interest in its state-owned communications enterprise; in December 1996 the Communications Ministry of Brazil indicated to international investors that it was unlikely to fully deregulate domestic telecommunications for at least three to five years. Nevertheless, the ministry also indicated that through a "phased-in" process, national operators would be permitted to provide service anywhere in Brazil. The reluctance to experiment with immediate deregulation as practiced in Chile has retarded the growth of advanced infrastructure. Still, foreign firms are poised to enter the largest consumer market in South America when conditions are optimized for private investment.

In Mexico, with a consumer base of approximately 1 million, telecommunications buildout is correlated very closely with the relaxation of government sanctions, particularly as they relate to privatization of TelMex, the government-sponsored monopoly. In an elaborate system of shared ownership, including a pivotal block of equity held by employees, TelMex reflects the nationalism implicit in Mexico's regulatory policy; however, privatization has generally led to an improvement in communications productivity accompanied by a higher level of consumer market penetration. Value-added services have been enhanced and competition in other areas of telecommunications, notably in satellite delivery and cable television delivery, have occurred due to the liberalization of foreign investment [20]. One gauge of the effectiveness of privatization in Mexican telecommunications is the exploding value of TelMex stock since the early 1990s; concurrent with rising profitability exceeding that of telecommunications firms in major nations, the Mexicans have established an important guidepost for other Latin American countries.

Throughout Central and South America—in nations ranging from Costa Rica and El Salvador in the north to Uruguay, Paraguay, Bolivia, Peru, Argentina, and Venezuela in the south—the experiences of Mexico, Brazil, and Chile are regarded as crucial. Government resistance to experiment broadly with complete deregulation and the private investment that ensues will only abate in relation to positive returns in employment, government revenue, and productivity. Chile now provides us with unfolding empirical evidence as to Latin American experimentation with deregulation.

9.2.4   Telecommunications Regulation in Other Regions

In Australia, telecommunications deregulation has systematically proceeded concurrent with the experiences of the United States and the United Kingdom. The Australians have opted for a gradual relaxation of government regulation and a fundamental reform of its state-sponsored monopoly, but private investment in new technologies is actively sought [21]. The Australians are especially keen in speeding to market advanced wireless technologies, including MMDS, PCS, and other cellular communications.

In Russia, and the new republics that align this region, there exist vast new opportunities for foreign firms. In many sectors of the economy, reduction, elimination, or modification of government regulations have already spurred private investment. Telecommunications, unlike natural resource development, has not significantly advanced. Complex legal problems have slowed the growth of private investment in communications systems: There is great resistance to majority-held equity interest by foreign firms and a growing difficulty in setting corporate tax rates and tariff schedules. Compounding these dilemmas are the obvious political uncertainties associated with the kind of long-term investment commitment required to build networks. In this environment, foreign firms have an incentive to build out sophisticated wireless systems to sustain business infrastructure. The result is that consumer markets have been starved of capital required to develop fundamental wireline systems.

In Africa, as is the case with most underdeveloped regions, government control through regulated monopolies remains the norm. This is gradually changing in certain nations, but there is little incentive, tragically, for foreign firms to invest in nations whose political systems are often unpredictable and where changes in leadership occur with great frequency. With limited entrepreneurial tradition and standards of living unable to accommodate telecommunications consumption, foreign investment is likely to be delayed in this continent. The instinct of former firms will remain in establishing a presence in large, stable markets, followed by more exotic and risk-taking opportunities found elsewhere. No one can predict with precision how long African nations may have to await such investment, but data supplied by telecommunications analysts imply major investment might be deferred until after 2010.

In the Middle East, telecommunications policy may generally be described as restrictive. In Israel, Saudi Arabia, Iran, Iraq, and most other Mideast nations, there exists substantial government control over the buildout of telecommunications systems. Monopoly control is enforced in the tradition of the European experience. The motive for resistance to possible deregulation varies from economic dogma to political control over the means of information transmission. It should be noted, however, that state-of-the-art telecommunications research via software development is advanced in Israel [22]. In the past several years, notable technical achievements have been introduced by Israeli firms in specialized mobile radio and Internet telephony. While the government maintains vigorous regulatory oversight over domestic telecommunications, Israeli high technology flourishes through explicit governmental incentives.

9.3 International Business Formations

9.3.1 International Telecommunications Mergers and Acquisitions

There has been a marked increase in international mergers and acquisitions since 1994 [23]. The multiplicity of government regulations and antitrust sanctions that dictate successful mergers in the United States are further promoted by comparable cross-national agreements. Whereas domestic agreements can be negotiated between buyers and sellers with some intuitive sense of probable government reaction, global communications agreements must be shepherded through a maze of regulations and legal interpretation adjusted for regional political machinations. Nevertheless, it is anticipated that, for reasons later identified, international telecommunications mergers and acquisitions will proliferate within the next several years, particularly in the developed nations.

In Europe merger strategy has historically been muted. Europeans refrain from the vigorous pursuit of acquisitions, traditionally a hallmark of American multinational corporate policy. For American firms, mergers and acquisitions represent (1) immediate entry to market, (2) absorption of competitive market share, and (3) access to technical expertise housed by competitors. Competitive advantage, despite the drawback of initial organizational disruption, is pursued at all costs. Europeans frequently look upon mergers with a jaundiced eye: mergers represent a "violent and uncivil" strategic approach to gaining market share [24]. The surge in American mergers, both domestic and international, has prompted European telecommunications firms to suppress their reticence to initiate such negotiations. In anticipation of the 1996 Telecommunications Act, American firms, beginning in 1993, pursued mergers with zeal. These mergers and acquisitions cut across services and products, entertainment and hardware, and telephony and television. Content fused with transmission was thus embraced. Europeans have taken the cue.

In 1996, in anticipation of EU regulatory liberalization pending in 1998, mergers were launched between British Telecommunications and Cable & Wireless PLC, and between Bertelsmann and CLT, as well as a number of others [25]. Most significantly, British Telecommunications' proposed merger with MCI has inspired comparable negotiations between European firms and their large counterparts on several continents. The cutting edge of merger and acquisition activity has thus commenced, the product of strategic fear: to delay immediate access to market is to invite competitors to seize available market share. American deregulation has thus prompted Europeans to preempt further gains by U.S. multinational corporations. Mergers consistent with economic convergence, as herein defined, signify the initial phase of global telecommunications synergy.

International mergers, whether the product of firms originating in Europe, the United States, or Asia, must overcome four principal regulatory concerns: universal access, price manipulation, licensing, and interconnection [26]. Increasingly, government authorities in most regions have been influenced by the continuing American commitment to service the entire domestic population. Problems of interconnection and efficient networking remain continuing concerns, as does price stability. Licensing, particularly in terms of electromagnetic spectrum held in perpetuity by wireless firms, endures as a contentious and sensitive issue. The key to consummating these agreements and overcoming government resistance lies in placating regulators tied to these criteria. At stake is a multinational corporate competition that will guide the immediate future of global telecommunications infrastructure.

While mergers, in the initial stage of telecommunications, seem to be occurring on the infrastructure and networking side, there is evidence that the fusion of content and distribution is also unfolding [27]. As the telecommunications industry restructures on a global basis, business opportunities surface throughout domestic economies. These consequences can precipitate new strategic partnerships and strategic alliances.

9.3.2 International Strategic Partnerships

Chapter 8 introduced a working definition of strategic partnerships: agreements, often temporary in nature, that typically, though not exclusively, join service and content providers in telecommunications. For purposes of this definition, a partnership is generally shorter term in design than in an alliance. International partnerships are therefore forged with the concrete understanding that the agreement suffices only as long as the interests of both parties are served. Unlike mergers or acquisitions, the partnership is dedicated to preserving the independence of both parties while encouraging focus on a specific task or end. Partnerships have the virtue of encouraging organizational flexibility and adaptability in foreign markets. Additionally, in a communications world made unpredictable by technological advance and obsolescence, the partnership remains an expedient tool in minimizing strategic risk. The alternative risk, of course, is that spectacular success in a partnered enterprise does not assure long-term growth and opportunity; either party may venture in a new direction with a vested proprietary interest. For this reason, contracts written with cross-cultural clarity and understanding are essential to global collaboration. As suggested in Chapter 8, the word "nimble" might be applied to the best of strategic partnerships. The speed with which they are consummated and effectively introduced to market should be the gauge by which their success is measured.

9.3.3 International Strategic Alliances

As previously discussed, the terms mergers, acquisitions, alliances, and partnerships are applied with increasing frequency in telecommunications. The multiplicity of contexts in which these business expressions are identified has led to confusion in the minds of many people. For purposes of our discussion, the distinction between partnerships and alliances lies in the motive supporting the agreement and the time frame projected for that collaboration. An alliance may be defined as a long-term commitment between two or more enterprises that find that the merger/acquisition route may not be feasible or desirable. As suggested in Chapter 8, symbiotic relationships can be formed through this endeavor without necessitating legal and regulatory delays prompted by foreign governments. Thus, a strategic approach to circumventing the drawn-out procedure associated with judicial approval is the international strategic alliance.

Alliances, if formed to the long-term satisfaction of all, also represent a method of integrating two or more partners in a venture that could not otherwise be financed. In other words, the costs of capital, labor, and technology can be reduced proportionate to the number of partners coordinating the alliance. In the United States alliances formed in interactive television in 1994 and 1995, for example, have typically resulted in three or more partners consummating an agreement at diminished financial exposure. The same occurred in December 1996, when five RBOCs agreed on adopting software created by an Internet provider. Increasingly, the high costs of capital and time expended to bring new products to market dictate that international firms carefully weigh the potential advantage of alliances.

In cases where long-term commitment to new markets is a necessity, where firms may not possess all the in-house technical expertise they require, where cultural synergies lead to greater insight about marketing to global consumers, the alliance is a viable route. These factors presuppose that the organizational cultures of international firms can be made to align efficiently and that collaboration—not confrontation—becomes the engine of change.

9.4   Venture Capital and International Telecommunications

The venture capital industry supplies, as noted in Chapter 8, a critical financial resource for emerging telecommunications companies. This is no less true in the international environment, but the full impact of venture capital will not be felt in the global telecommunications environment for several years [28]. Venture capitalists throughout the world are poised to finance start-up companies, particularly in the fields of wireless communications and Internet content and delivery [28]. International risk capital is often controlled by fund managers who assess market potential based on political stability, regulatory restrictions, and rising consumer consumption. As is the case with American venture capitalists, these fund managers seek out telecommunications firms that are projected to grow in access of 25% during their initial five-year period of development. Predictably, there is great variation in the dispensation of start-up funds based on geographic territory and density of consumer markets.

International venture capital markets may be categorized as "visible" or "invisible" [29]. The visible venture capital sector consists of between one and two thousand firms worldwide; half this number reside in the United States. Perhaps something on the order of $70 billion is managed by these firms, the majority of which is funneled to high-technology firms. The key to securing international venture capital, as is the case in the American market, lies in the development of an innovative, structured business plan that offers credible projections and identifiable markets. Obviously, in regions where political risk is substantial, venture capitalists will demand higher rates of return relative to their American counterparts. In this sense, new ventures born in nations other than America or Britain stand a higher probability of procuring capital if an American partner can be found. In the short term, we may therefore see the emergence of international telecommunications start-ups in which a high fraction of equity interest is controlled by an American.

The "invisible" venture capital market is comprised of several million individuals whose net worth exceeds $1 million [30]. These persons are scattered around the globe, act as independent entrepreneurs, and collectively invest perhaps $30 billion annually. Sometimes referred to as "business angels," these entrepreneurs channel funds to start-ups at levels of risk substantially higher than venture capital fund managers traditionally assume. As a result, business angels—so named because they network in often mysterious, intangible ways—provide a foundation of capital emerging as important to international ventures. Because the invisible venture capital market remains nebulous, a critical issue for international telecommunications is the initial point of solicitation. Often, a group of six to eight angels will coordinate their investment effort after verifying the credibility of a venture. Referrals to business angels are occasionally facilitated by venture capital firms, but the development of the Technology Capital Network (TCN) is likely to be a significant vehicle for international telecommunications entrepreneurs in the future [31]. TCN is committed to bringing angels and entrepreneurs together, and recent international ventures have included sponsoring Americans.

9.5 International Telecommunications Business Opportunity Paradigm

A BOP for telecommunications was introduced in Section 8.3. The data assembled suggest that new technologies, when achieving 10% market penetration, inevitably invite imitators. These imitators form the first layer of competition sparked by each successive innovation in telecommunications. In many cases, the instinct of established telecommunications firms will be to absorb such innovators; whether innovators respond to these overtures or simply capitulate in the face of competition, the interaction of veteran and entrepreneurial companies define a BOP that generates new opportunities throughout the economy.

As previously described, tertiary (third-ordered) economic consequences are the products of telecommunications innovation. The BOP for domestic telecommunications represents a synthesis of three themes: (1) constant innovation leads to product substitution; (2) large firms seek to preempt small firms through advantages of capital and economies of scale; and (3) support industries will emerge to sustain the originating technology. The synthesis of these three themes provides a model by which some prognostication of future telecommunications development can be applied. The significance of these interconnecting themes is no less significant for international entrepreneurs than it is for their American counterparts.

Chapter 8 recounted the questions that telecommunications entrepreneurs must ask and answer for themselves in defining strategic plans.

- If we introduce Product X, how will domestic and global competitors respond?

- In designing this product, have we created the demand for supporting sales/maintenance infrastructure?

- Are we vulnerable to predatory instincts of larger firms? Can our product and strategy be quickly imitated?

Figure 8.1 addressed these questions, formulated a model, and linked *customer, competitor,* and *industry responsiveness* so as to estimate future business opportunities in the telecommunications and ancillary industries. Figure 9.1 postulates an international model of future business opportunities promulgated by the continued advance of telecommunications technology and convergence of the telephony, television, and computer industries. This model also implies that some of the most significant economic opportunities emerging from telecommunications innovation are embedded in the tertiary consequences prompted by that creativity. In other words, telecommunications firms are increasingly compelled to evaluate and forecast unfolding business and consumer transformation in order to define the firm's own future strategic path. This phenomenon dictates that in the next century global telecommunications providers must execute a strategy of *creator, manager,* or *distributor* of information rather than maintaining a profile as a discrete member of the telephony, television, or computer industries.

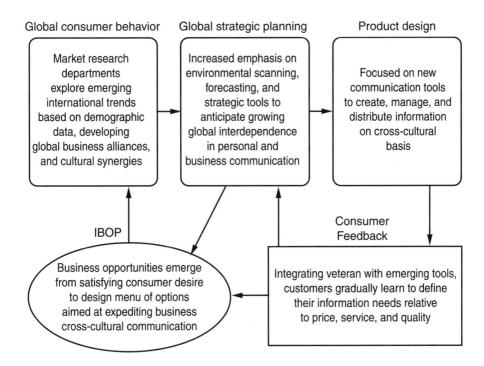

Global consumer behavior   Global strategic planning   Product design

Figure 9.1 International telecommunications business opportunity paradigm.*

9.6 Deregulation and International Telecommunications Marketing

The early stage of global deregulation, given inertia by the American and British experiments, has inspired a definite trend in the marketing of telecommunications products. Marketing strategy has already sped new products to market,

---

*Premise: Future international telecommunications product/service innovation is a function of customer-led customization, stimulated by deregulation in the United States, United Kingdom, and other nations experimenting with regulatory relaxation.

Assumptions: (1) The use of advanced, global telecommunications services becomes a necessity to conduct international business. (2) The adoption of new services compels associates to acquire comparable tools for purposes for communications. (3) The generation, storage, transmission, and retrieval of information represent "inexhaustible" markets; that is, the growth of information precipitates continuous global demand for improvements in telecommunications management.

Inference: Modern telecommunications corporations of various sizes gradually define themselves as creators, managers, or distributors of information. Business opportunities emerge from cooperative and synergistic ventures in which firms of national origins participate to build new markets.

even in those countries most resistant to regulatory relaxation. The economics of international telecommunications marketing thus represent a new experience for multinational corporations and international entrepreneurs.

Chapter 7 distinguished two important concepts for new and veteran firms approaching deregulated markets: market research, the process by which prospective customers are identified, monitored, and served; and competitive research, those techniques involved in tracking and predicting competitors' strategic plans. The interaction of these two critical tools, both of which are tied to determining appropriate strategic planning, are in evidence in the early development of deregulated telecommunications markets. We note the following early priorities in the economics of international telecommunications marketing [32]:

- Emphasis on the mechanics of *market segmentation*, particularly in Europe, whereby companies stratify their consumer and business markets and adjust ongoing marketing accordingly;
- *Cherry picking*, a predictable tool of large firms who seek at the beginning to carve out a presence among high-end users who seek value-added services;
- *Government lag*, in which regulators are slow to respond to licensing, tariff, interconnection, universal service, and price control concerns raised by corporations.

Superimposed upon these three concerns is the dynamic of telecommunications innovation. The speed with which firms introduce new technologies, especially with regard to cultivation of value-added Internet services, has overwhelmed international regulators. Communications technology advances at an exponential pace, while legislators and regulators promote deregulation at an arithmetic rate. As suggested, the intrinsic inability of government to adapt its policies to the needs of domestic providers puts these firms at competitive international risk.

Despite these continuing concerns, a critical mass of customer need and industrial demand could well provoke a replication of the American experiment throughout the world as the next decade passes. In that sense, the models described in Chapter 10, focusing on alternative future scenarios for the American telecommunications industry, may anticipate global communications policy.

9.7 The Emergence of NII/III Cooperation

This chapter discussed efforts by the GATT and WTO to encourage the telecommunications industry, chiefly through the liberalization of tariff schedules.

The International Telecommunication Union continues to promote common technical standards, which are critical in encouraging cooperation among international firms. One of the more intriguing political issues surrounding global communications development, however, concerns the potential linkage between the NII and the III. Some governments in Western Europe, most notably France and Germany, are concerned that the absence of cooperative public efforts to encourage growth in new telecommunications services—that is, explicit government effort to strategize an environment in which multinational corporations collaborate in building out networks—might leave American firms with a permanent competitive advantage. American deregulation is not only transforming the domestic industry, it has set in motion newly emerging alliances, partnerships, and mergers between U.S. and other international firms. Americans are pursuing the most attractive value-added markets available and are generally first to market with these services.

Despite the fear that this early stage of deregulation will only enhance the global standing of American telecommunications companies, some governments, including our own, have concluded that complementary NII/III policy might be in everyone's best interest. In thwarting the advance of international agreements, governments only punish their own economy, with no guarantee that protectionist measures will achieve their intended objective. Cooperative standards in licensing and other technical measures may thus simultaneously boost every nation's economy. The movement in this direction has been gradual but is nevertheless discernible in recent trade policy. The North American Free Trade Agreement (NAFTA), passed by Congress in 1994, has established precedent for closer collaboration among governments in matters of trade. A clue to possible cooperation that might unfold can be ascertained by evaluating the impact of free trade on growth rates in Mexico, Canada, and the United States, signatories to NAFTA. Success at this level might inspire other nations to follow suit and, in so doing, further promote the international telecommunications market. The fastest growing sector of the global economy may thus produce an accelerated standard of living throughout all regions and nations.

References

[1] Elliott, T. L., III, and D. Torkko, "World-Class Outsourcing Strategies," *Telecommunications*™, April 1996, pp. 47–49.

[2] "Global Alliances Must Fight for Growth," Reuters News (London Newsroom), Nov. 3, 1996.

[3] Taoka, G. M., and D. R. Beeman, *International Business: Environments, Institutions and Operations,* New York: Harper Collins Publishers, 1991.

[4]   Kennedy, C. H., and M. V. Pastor, *An Introduction to International Telecommunications Law*, Norwood, MA: Artech House, 1996, p. xvii.

[5]   Taoka and Beeman, *op. cit.*, pp. 111–112.

[6]   "Investments in Engineering Telecommunications," *PRNewswire*, Nov. 19, 1996.

[7]   "Getting a Line on Europe's Telecom Free-for-All," *Business Week*, June 19, 1995.

[8]   "Europe: The Age of Mergers," *Business Week*, April 15, 1996.

[9]   Kennedy and Pastor, *op. cit.*, pp. 206–220.

[10] "Tearing Up Today's Organization Chart," *Business Week*, Nov. 18, 1994.

[11] "New Global Communications Survey," *Business Wire*, Sept. 19, 1996.

[12] "NTTC May Not Be Broken Up," *Reuters News*, Dec. 6, 1996.

[13] "The New Asian Manager," *Business Week*, Sept. 2, 1996.

[14] "A Free Phone Market in Japan," *Business Week (International Edition)*, March 11, 1996.

[15] Kennedy and Pastor, *op. cit.*, pp. 237–239.

[16] Appleyard, D. R., and A. J. Field, Jr., *International Economics*, Homewood, IL: Irwin Press, 1992, pp. 152–153.

[17] Kennedy and Pastor, *op. cit.*, pp. 170–171.

[18] "Chile: The Strangest Telecom War Yet," *Business Week*, Jan. 15, 1996.

[19] "Wireless on a Global Scale," *Wireless*, Oct. 1996, Vol. 5, No. 9, pp. 18–20.

[20] "A $15 Trillion Private Piggy Bank," *Business Week*, June 24, 1996.

[21] Taoka and Beeman, *op. cit.*, pp. 202–206.

[22] "Israel . . . Or Go Forth to the U.S. and Prosper," *Business Week*, Feb. 26, 1996; "Telecommunications: The Developing Leap," *Wall Street Journal*, Feb. 11, 1994, p. R15.

[23] "Global Competition: The Second Act," *Tele.com*, Nov. 15, 1996, pp. 11–33.

[24] "How Merger Mania Has Redefined the Communications Landscape," *Telecommunications™*, Aug. 1996.

[25] "Europe: The Age of Mergers," *Business Week*, April 15, 1996; "British Telecommunications/MCI Announce Definitive Merger," *PRNewswire*, Nov. 3, 1996.

[26] Kennedy and Pastor, *op. cit.*, pp. 147–164.

[27] "Answer the Call of the Future," *PRNewswire*, Oct. 22, 1996.

[28] "World Telecommunications Is Undergoing a Revolution," *Reuters News*, Nov. 3, 1996.

[29] Bygrave, W. D., *The Portable MBA in Entrepreneurship*, New York: John Wiley & Sons, 1994, pp. 172–194.

[30] *Ibid.*, p. 174.

[31] *Ibid.*, pp. 175–176.

[32] "Deregulation and a Segmented European Market," *Telecommunications™*, July 1996, pp. 64–66.

10

Telecommunications Deregulation and Market Planning: A 10-Year Forecast

Forecasting the fate of an industry is dangerous conjecture. It is especially dangerous when one contemplates the telecommunications industry. The sheer size of this industry, as earlier demonstrated, translates into a myriad of intangible opportunities, hazards, and pitfalls. The pace of technological change spurred by the communications revolution and its resultant impact upon peripheral enterprises are effectively transforming all aspects of human endeavor. Any serious attempt to forecast the transformation of this industry represents, by implication, an effort to indirectly outline the future structure and productivity of the national, and indeed international, economies.

Despite this daunting task, we have little choice but to elaborate and refine the probable alternative future scenarios for the industry. Strategic planners, marketers, and corporate heads will demand it, public institutions will compel it, and investors will require it. We cannot plan unless we have a sense of where the telecommunications industry is headed. In defining a set of scenarios, the author does not presume that one scenario is intrinsically more probable than another; he merely contends that forces presently in place have impelled both the industry and economy to move along a path that isolates the following outcomes as most likely. In the end, the status of the industry in 2010 will be a function of the assimilated legal, regulatory, political, technological, and organizational dynamics presently installed. It is on this basis that we are permitted the opportunity to project the future.

The initial chapter of this study explored the pending transformation of the telecommunications industry in light of the Telecommunications Act of 1996. Mounting evidence suggests that during this decade the interaction of the telecommunications, computer, and entertainment industries will constitute the single largest segment of the American economy. Annual industrial growth among established firms is likely to be the greatest in this field. Rapid growth will be a catalyst for other industries: From services to agriculture, from mining to manufacturing, advances in telecommunications will facilitate productivity gains never before attained. Thus, before we identify the potential scenarios likely for communications during the coming decade, let us review a "macroeconomic context" for the American and international economies within the same time frame.

10.1   A Macroeconomic Profile of Telecommunications Services

10.1.1   The Macroeconomics of Telecommunications

Chapter 1 established an empirical framework for the American economy. Each sector of the economy contributes the following share to the national economy [1]:

- Services: 32%;
- Trade (wholesale and retail): 27%;
- Manufacturing: 19%;
- Finance (banking, insurance, real estate): 7%;
- Communications (includes utilities): 6%[1];
- Construction: 5%;
- Agriculture (includes mining, fishing, forestry): 4%.

If we project a 3% average annualized growth rate (adjusted for inflation) in the national economy, a figure consistent with the historic national average sustained throughout most of the post–World War II period, then American GDP could attain a level in excess of $15 trillion by 2010 [3]. If telecommunications grows at a rate double the output of projected GDP during the same period, then the industry could contribute one-fifth of all national wealth, a

---

1. Does not include the peripheral contributions of the entertainment and computer industries; these numbers, if transposed from the service sector, raise the "telecommunications contribution" to national wealth to approximately 10%. Because telecommunications permeates all other sectors, some observers estimate that as much as 15% of our economic output is directly or indirectly generated by the industry [2].

staggering annual sum of $3 trillion! Significantly, some analysts have suggested that growth rates for certain segments of the industry, particularly in the field of wireless communications, could substantially exceed these projections [4].

10.1.2 The Economics of Tertiary Effects

As suggested earlier, the significance of telecommunications in today's economy lies in the fact that it influences the productivity of all other forms of enterprise. In the 1950s and 1960s steel was thought to be the linchpin of commerce: The supply and demand of this material indirectly influenced the production and price of all other commodities. In the 1970s and 1980s the bellwether of consumer and business behavior was petroleum: The prevailing price of oil charted the outline of economic growth, inflation, and interest rates. It is now increasingly evident that telecommunications—and the pivotal role this industry plays in *generating, transmitting, retrieving,* and *storing* information—represents the critical link in sustaining competitive performance. Success in business, domestically and internationally, is now directly related to state-of-the-art telecommunications performance [5].

Economists refer to the "tertiary effect" in describing the commercial impact of emerging telecommunications technology. Tertiary consequences, and the opportunities they promote, are largely unanticipated by producers at the time a new product or service is introduced. Chapter 7 elaborated a set of forecasting tools, the object of which was to estimate future competitive behavior. Those empirical methods, when applied to the telecommunications industry, dictate a model (Figure 10.1) elaborating the tertiary effect.

Figure 10.1 identifies the unfolding market effects of telecommunications innovation. When a new product is introduced, serving its constituency, we immediately note a primary tertiary effect unintended at the time of introduction. For example, the development of network television effectively created first-ordered tertiary effects: New sales and maintenance professions, along with support services required to sustain them, were immediately necessitated by the innovation. A second-ordered tertiary consequence (all consequences precipitated by the primary effect are categorized as secondary) was the impact felt by the motion picture industry: The audience for film was reduced by a factor of two-thirds in less than a decade [6]. The commercial success of television and its resultant tertiary impact on enterprises immediately outside its own domain, represented an unintended, and thus unpredicted, consequence. A third-ordered effect prompted by the rapid expansion of television was the restructuring of the advertising industry, a phenomenon that reorganized corporate planning and promotion in all industries. By implication, we also infer fourth-ordered, fifth-ordered, and sixth-ordered tertiary consequences and so

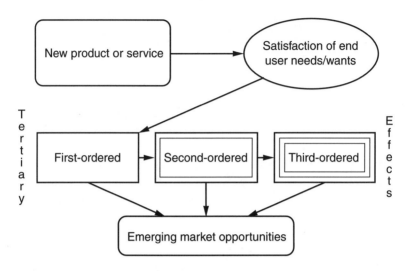

Figure 10.1 The telecommunications tertiary effect.

on [7]. Technological innovation influences every layer of organizational life, public and private.

The importance of tertiary effect analysis, as applied to the telecommunications industry, lies in its identification of prospective market change. With the release of every successful communications product or service, we note the emergence of new market opportunities. Significantly, telecommunications has become the dominant engine of change in an economy marked by an expanding service sector. The generation, exchange, and evaluation of information will increasingly define success, individual and organizational; the evolution of telecommunications thus provides the framework for future economic change. It will become increasingly apparent to individuals and organizations that they will be compelled to forecast the long-term tertiary effects of telecommunications innovation. New markets will gravitate to those able to predict the tertiary consequences embedded in today's organizational decision making.

10.1.3 The Demographics of Telecommunications

Increasingly, market planners in the communications industry are cultivating an incisive view of their prospective market base. The key determinant of future macroeconomic performance, and the critical variable underpinning the direction of telecommunications services, is the unfolding demographic character of the American consumer. There are several emerging demographic statistics that are essential to understanding the significance of the market paradigms discussed in this chapter.

A standard market research tool designed to estimate future demand for goods and services is the *consumer S-curve*. The S-curve, applied to prospective household and business behavior, identifies the stages of developing demand for new products. An extension of the logic implicit in Figure 10.1, the consumer S-curve holds that "early adopters," responsible for initial household penetration rates of 10%, dictate the fate of new technologies. A critical demographic link in the success of newly introduced products is the influence of early adopters. Early adopters are those who, by nature, are experimental, curious, and trend setting. We note that the unfolding pattern of a consumer S-curve is the behavior of these early adopters; no other consumer will purchase new products in the absence of the influence of those whose lifestyle and income permit such experimentation (for further information on S-curve applications, see Chapter 7). Using the consumer S-curve, substantiated by cyclical economic evidence, we affirm the following model (Figure 10.2) with regard to household and organizational adoption patterns.

From the viewpoint of the firm, the maintenance phase becomes highly unpredictable as the struggle to maintain market share precipitates great uncertainty (hence, the indicated question marks).

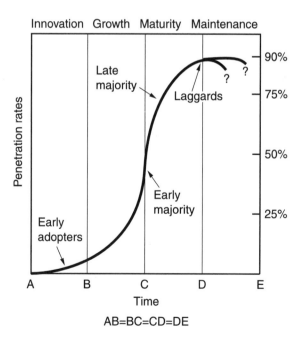

Figure 10.2 Consumer adoption rates. During the late phase of the growth stage, competitors seek to develop alternatives to newly emerging technologies. At the beginning of the maturity phase, we note the commercial marketing of alternative technologies.

During the late phase of the growth stage, competitors seek to develop alternatives to newly emerging technologies; at the beginning of the maturity phase, we note the commercial marketing of alternative technologies.

The principal demographic variables driving the market acceptance of new technologies include income, occupation, regional origin, family structure, age, and education. Literally hundreds of other nominal and ordinal variables are of relevance, of course, but these data correlate very highly with new product adoption. In considering the array of data, and their collective influence on marketing during the next 10 years, bear in mind the following [8].

- Approximately 80 million Americans will be entering their peak spending years. Americans are becoming increasingly discerning and demanding with regard to product quality and durability.

- The strategic value of time, that is, the efficiency with which mundane tasks are accomplished, is gaining a higher priority with respect to emerging consumer taste.

- The appetite for information is expanding and accelerating rapidly, but the time necessary to assimilate that knowledge is concurrently shrinking for precisely the same reason.

- Customization of product line, for all industries, is the collective result of the three forces identified above.

In short, the data repeatedly indicate that Americans are prepared to pay for information that uniquely serves their needs, enhances their productivity, and allows them to spend time on matters of "felt need." These felt needs may be as simple as spending more time with family or as concrete as improving professional efficiency. Forty years ago marketers geared their strategy to the *segmentation* of consumers by virtue of generalized demographic analysis. In the 1970s and 1980s the strategy shifted to *niche* marketing, in which those same demographics were measured against emerging consumer tastes via an expanded array of segments. Today, as we assess new product development in all markets, it has become clear that niche marketing has given way to *particle* marketing; the object of market research has become the satisfaction of consumer tastes on an *individual* basis. Each consumer thus represents a new market. These markets can change suddenly, dramatically, and unpredictably.[2]

---

2. The shift to particle marketing suggests an increased emphasis on a study of the psychological and social psychology influences on individual consumer marketing. The transitions from segmentation to niche to particle marketing imply dynamic customization of telecommunications services.

The implications of these broad demographic movements and their impact on prospective marketing are profound and are crucial to the success of telecommunications firms in the age of deregulation. You will note differences in how the five basic paradigms to be outlined respond to the changing character of American society, but every model factors into its imprint the great change implicit in these trends. During this decade, business success will be defined by those firms who can make the difficult and delicate transition to particle (or one-to-one) marketing.

A few final macroeconomic and demographic facts should be weighed before we proceed to an evaluation of emerging scenarios for the industry. First, as the Baby Boom generation ages (the Baby Boom is broadly defined as that generation born between 1946 and 1964), produces children, endures an increased debt burden, and seeks relief from the time constraints of daily living, the most successful telecommunications firms are likely to be those able to "bundle" their services [9]. In essence, firms that can merge telephony with television, Internet, and other services will secure a rising fraction of the market. Evidence suggests that the "Boomers," the single largest generation in American history, may seek a simplified menu as competitive rates.

Second, a contentious but nevertheless credible scenario for the American economy suggests a period of rising prosperity between 2000 and 2010. Consider, for example, the following facts [10].

- Every generation attains peak spending (income + savings + debt = spending) on average between the ages of 47 and 49. The Baby Boomers, comprising nearly two-thirds of the labor force, will produce repeating waves of spending within this age group through 2012.
- While consumer debt will increase between now and 2007, resulting in rising demand for all goods and services, public debt should continue to abate during the same period. The result is that we may well be headed for a sustained period of stable prices with concurrent modest interest rates. We noted in Chapter 1 the basic equation used to estimate annual national wealth:

$$C + I + G + Ex - Im = \text{GDP}$$

where C = consumers, I = business investment, G = government, Ex = exports, and Im = imports. GDP is thus a function of consumer spending combined with business investment and capital spending; government spending, when added to exports (and the overseas wealth they generate), but adjusted for imports (income and savings departing the country in the name consumption), dictates national wealth. Viewed in their totality, the macroeconomic data suggest an expanding capacity for future savings and investment, particularly as Boomers approach retirement.

- The Baby Boom generation could well be the wealthiest generation in history, for two reasons. (1) It is the best educated generation in the nation's history, fueling its capacity to adapt to economic change at levels of rising income; and (2) it is about to inherit the vast wealth accumulated by its parents.

These facts, however interpreted, suggest a more positive general economic environment than often outlined by economists. We must approach any serious discussion of economic modeling, however, with a set of assumptions, and those assumptions must be consistent with the empirical evidence that directly bear on consumption for telecommunications services. Armed with these facts and assumptions, we can outline five discrete scenarios for the telecommunications industry through 2010. These scenarios include:

- *Service Explosion Model:* in which the dream of deregulation, framed by the authors of the TRA, achieves a full flowering. It is a scenario in which rising demand for telecommunications products and services is met with an ever-increasing number of providers able to deliver state-of-the-art services.

- *Corporate Consolidation Model:* a scenario greatly feared by proponents of the TRA, in which initial deregulation inevitably promotes oligopolical restructuring in a manner reminiscent of airline deregulation since 1978.

- *Customer-Led Customization Model:* a scenario suggesting that while the initial years of the TRA will lead to an array of products and services dictated by corporate research and development, it will be the consumer that eventually determines the emerging shape of innovations to come.

- *Price Implosion Model:* a nightmare for telecommunications firms in which prices collapse, a function of constant technological innovation and rapid substitution of veteran product lines. Under terms of this vision, (1) the profits are so diminished that no firm has any incentive to maintain existing infrastructure, and (2) new product development is stifled because of inadequate wireline maintenance.

- *Short-Run Chaos (SRA)/Long-Run Stability (LRS):* in which turmoil for the remainder of this decade inevitably invites economic stability during the first decade of the next century. This model presumes that public utilities will eventually assume infrastructure responsibilities that were formerly the domain of telephone monopolies.

If one assumes that the future of telecommunications during the coming decade will be a function of deregulatory evolution reinforced by accelerating technological change, we can project these five alternative visions. It is essential to bear in mind that a hybrid of two or more of these models could emerge in response to dynamic competitive forces. In the years immediately following deregulation, we have experienced a rapid proliferation of mergers and acquisitions; this trend, accompanied by one or more of the forces embedded in these illustrations, will guide the reconfiguration of the industry.

10.2   Future Demand for Telecommunications Services

10.2.1   The Service Explosion Model

As noted in Chapter 2, the TRA was given impetus by the convergence of a variety of interest groups: telecommunications providers, consumer groups, consultants, and economists. In the years preceding 1992 these groups were largely divided as to economic self-interest and competitive philosophy. Those divisions were eventually submerged as providers embraced the opportunity to enter *all* markets. Opportunity thus superseded security as the emerging industry paradigm [11].

If the framers of the 1996 Act are correct in their assumptions about present competitive forces and the innovations likely to come from firms seeking to enter diversified markets, we can conclude that within the next several years we will experience a rapid increase in the supply of telecommunications goods and services. This is an industry in which there is little point in dichotomizing product from service.

Services cannot exist in the absence of infrastructure to support them; product development will not be initiated in the absence of projected consumer demand for such services. With this reality in mind, the TRA authors envisioned a simple scenario in which the forces of supply and demand would consistently reduce the price of emerging telecommunications goods and services [12], as illustrated by Figure 10.3(a).

In a utopian scenario, supply would constantly expand relative to demand, precipitating gradual, and in some cases, profound reductions in price. A majority of legislators contended that constant innovation in the telecommunications industry would precipitate rapid substitution of veteran product lines in favor of new, cheaper substitutes. Indeed, these innovations would be spurred by entrants from peripheral industries; some government officials and

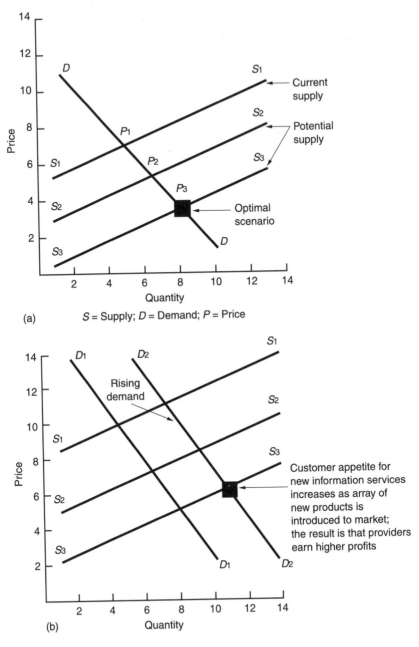

Figure 10.3 Service Explosion Model: (a) what legislators foresaw (price continues to fall as new supply is introduced to market; P_3 would constitute optimal pricing for new products) and (b) what providers foresaw.

industry players concluded that, because the legislation permitted "entry" by any competitor, expansion of supply in the short run might drive down prices by a factor of 50% or more.[3] Eventually, stability in market supply, demand, and price would be attained, but in the interim, consumers would benefit from expansive choice and diminishing cost.

The simplified account of the impending competitive battle in the industry was not shared by all proponents of the legislation. Some observers of this phenomenon perceive that the industry will develop in a manner fundamentally desirable for both consumers and providers. Viewed from their perspective, and applying the same reasoning, we might witness the scenario illustrated in Figure 10.3(b).

Both proponents and critics of the TRA recognize that the industry will experience dynamic shifts in the lines of supply and demand illustrated in Figure 10.3. However, those who envision that the market will efficiently decide winners and losers also assume that any organizational turmoil will be minimal within several years following passage of the legislation. Eventually, they contend, this model will define market equilibrium to the advantage of both providers and consumers. As we explore alternative modeling, we will note that this traditional conjecture about market competition constitutes the cutting edge of interpretative dispute among forecasters. In short, will competition successfully manifest itself in the telecommunications industry as it traditionally has in other sectors of the economy? Or are there economic characteristics uniquely associated with "information businesses" that alter competitive behavior over the long run?

Authors of legislative reform, and the lobbyists who encouraged them, thus presume that providers will seek to optimize market share in the short run by maximizing supply. Reductions in pricing would ensure nominal market share at the beginning of this competitive wave; thereafter, providers would be faced with the daunting task of expanding market share in the face of sophisticated players pursuing the same goal [13]. Significantly, it is generally estimated that the cost of recruiting a new customer approximates five times the cost of retaining one [14].

10.2.2 The Economics of Corporate Consolidation

Some observers believe that "historical analogy" represents the most effective method of predicting the fate of the industry over the space of the next decade. For these individuals, it is the *pattern* of historical evidence that provides the

---

3. Increasingly, telecommunications providers must regard themselves as having competitors whose interests lie outside the service domain itself, but whose resources inevitably lead to the creation of partnerships with other firms destined to enter their industry.

best guide to future business performance. Experimentation with deregulation in the aviation and trucking industries since 1978, they contend, therefore presents the best clue as to the fate of telecommunications.

In Chapters 1 and 2 we identified the political and economic significances of deregulatory practices. One may distill the historic evolution of regulation and deregulation, adjusting for time relative to the industry examined, as illustrated in Figure 10.4.

Historically, perceived excesses in free markets precipitated government regulation in the name of protecting the consumer and preserving competition, as identified in Figure 10.4. Such perceptions gave way eventually to attempts at deregulating those same markets as regulation stifled innovation. In the early stages of deregulation we note the proliferation of competitors and resultant declining price. However, as competitors seek to maximize market share (often through vertical and horizontal integration), oligopolies emerge, thus reinventing government involvement to redress problems of diminished competition. In the case of the airline industry, for example, the second and third stages of this process have taken 15 years to fully unfold. Now, Congress is besieged by complaints from consumers who believe that diminished competition has impaired quality and raised price.

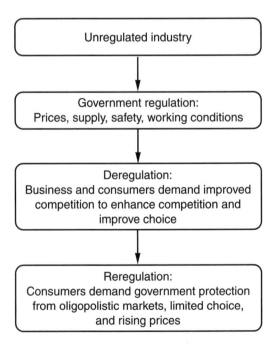

Figure 10.4 Regulatory evolutions.

Applying this historical parallel, the telecommunications industry might operate at optimal competitive levels through 2010. Thereafter, an oligopolical pricing structure would emerge from the interaction of a reduced number of players who simply absorbed their competitors. Such horizontal integration would consolidate perhaps 80% of telecommunications market share among three, four, or five providers.

In the absence of testing this hypothesis, one cannot be sure of the outcome, of course. But this conjecture is given some credence by similar experiments in deregulation attempted by the federal government since 1978. To argue that historical analogy is a reliable forecasting tool, however, one must consider the unique characteristics of the telecommunications industry.

Several factors distinguish telecommunications from other industries. As noted in Chapter 4, certain fields within the industry presently operate on duopolistic, monopolistic, or effective oligopolistic bases. The cellular industry, for instance, consists of two providers per metropolitan area. While emerging competition is coming on-line in the form of SMR and PCS technologies, cellular incumbents nevertheless enjoy intrinsic competitive advantages. The same dilemma confronts long-distance providers who seek to enter local loop markets. These patterns are repeated at competitive variance depending on wireline or wireless services.

Additionally, unlike other industries in which products or services are comparatively tangible, the information generated via telecommunications is *intangible* and *portable*. Prospectively, every telecommunications service— voice, data, and image—presently offered through wireline can be replicated through wireless access. The result is that future commercial transactions are likely to be consummated at any location at any time: We thus make the transition to "anytime, anywhere information." The significance of this development, relative to industry consolidation, lies in the necessity of every wireline provider to augment its product in the form of wireless service. Thus, a substantial fraction of capital investment otherwise exclusively devoted to wireline maintenance is then diverted to wireless infrastructure.

Consider Figure 10.5 in light of the expressed fears of those who believe that deregulation, whatever its short-term benefits, will inevitably impel industry oligopolies. The telecommunications industry, dominated by AT&T through the mid-1980s, enters a period of intense competition during the first five years following passage of the 1996 Act. Beginning early in the twenty-first century, the industry experiences substantial competition through multiple provider access, but mergers and acquisitions gradually consolidate the market. By 2010 the industry essentially pivots on the market behavior of four or five providers, collectively controlling a minimum of 60% of market share. Approximately a decade will pass before phase 3 is concretely evident.

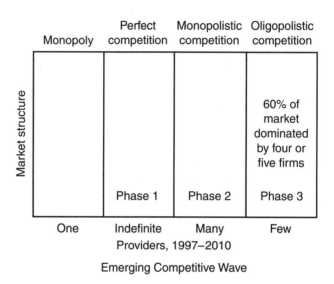

Figure 10.5 Corporate consolidation: a 13-year projection.

For purposes of the corporate consolidation model, phase 1 acts as merely a theoretical construct. In practice, it is highly improbable that perfect competition will immediately precede monopolistic (many competitors dividing market share on the basis of branding) competition; however, proponents of this model would contend that the greatest opportunities for entrepreneurs will reside within this stage of competitive development.

Given the need of incumbents to diversify services, particularly as wireless products approach maturation in the next century, there exists a natural propensity to purchase firms established in this arena [15]. Quite apart from the need to expand market share, industry consolidation is prompted by the desire of large enterprises to acquire the marketing expertise of young, successful firms. A knowledge of consumer behavior and forecasting will be at a premium for the long-distance giants as well as the Baby Bells [8]. The impetus toward horizontal and vertical integration is thus assured.

10.2.3   Customer-Led Customization Model

An underlying assumption embedded in the models thus far elaborated has been that providers will dictate the outline of future telecommunications products and services. This premise, underscored by the historic development of the industry, is now challenged by an alternative forecast stressing the growing significance of particle marketing. Proponents of this view argue that business will now follow the lead of the consumer in designing and promoting services intended to meet the specific needs of the customer [16]. One notes the historic evolution in marketing in Figure 10.6.

The importance of the recent transition to particle marketing is revealed in Figure 10.6. In prior eras, telecommunications providers originated a new product and then sought to find or create a market for it. Under the umbrella of particle marketing, providers must first understand the unique needs of *individual* customers and then attempt to develop services that satisfy those multifaceted needs. Marketing therefore becomes highly complex and fluid. If this development truly manifests itself in the near future, then the industry will be completely reordered—mass marketing will give way to customized market research, and the success or failure of a firm will hinge on its ability to meet consumer needs on an ongoing basis. In other words, consumers—both households and organizations—will dictate the terms of the service they receive. "Total marketing" therefore emerges as a model equipped to forecast changing consumer tastes.

If this model represents a valid paradigm for the telecommunications industry over the space of the next decade, then we can infer that providers will attempt to maintain continual, positive, and personal relationships with their customers. As a result, organizational capital once directed to basic re-

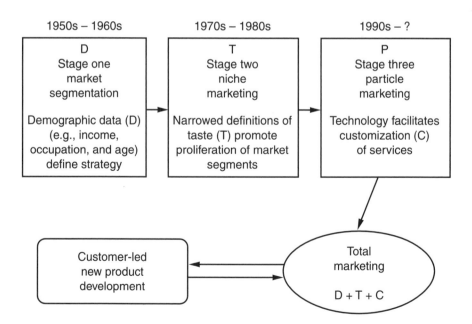

Figure 10.6 Customer-led customization model.

search will be diverted increasingly to applied research. A rising fraction of these funds will also be directed to marketing and strategic planning departments, where the firm's primary objective involves gaining a keener understanding into the dynamics of consumer behavior. In short, the customer will tell the industry what to produce and when. If a firm fails to meet that standard, the model implies that we can expect to see a wave of new market entrants, thus generating competition over the long run.

10.2.4   Price Implosion Model

As the TRA recedes into the fabric of the marketplace, one can detect a faint whisper of genuine concern not expressed at the time this legislation was enacted. That concern is the theoretical possibility of a price implosion: descending prices for telecommunications goods and services with paralleled declines in profit margins. In other words, the unique characteristics of this capital-intensive industry—increasingly governed by innovation in software applications—might lead to substitutions in technology so rapid that financial disincentives will divert providers from maintaining or improving existing infrastructure. Simply, if telecommunications providers cannot be certain of recapturing the costs of new product development, why would they make them in the first place? And, if providers do not make these investments, will emerging competitors be thwarted in their efforts to provide new services?

To gauge the credibility of this scenario, which is uniquely pertinent to the telecommunications industry, consider the recent evolution of the Internet. The Internet, originally created to transmit information between computing networks, was not intended to be a commercial enterprise. In recent years, however, households and businesses are increasingly using this tool for commercial applications. Among those applications are recent innovations in software that permit voice traffic. The result is that telephony providers may be denied the profit incentive to maintain the infrastructure that others are exploiting—in effect, "free telephony" will eradicate the incentive to maintain current infrastructure. Five years ago this prospect was unthinkable; now it is attainable.

The same phenomenon holds for other telecommunications providers as well. In 1994 the FCC granted IVDS licenses to investors via auctions. The IVDS industry, as well as the interactive television industry generally, have been stifled in their attempts to bring product to market because of perceived commercial threats as implied by Internet development [17]. Private enterprise and the investment community are discouraged from generating new product lines in many of these fields because of the emergence of this new platform. Prices of new product lines might contract as innovators learn to substitute the ubiquitous Internet for prior applications. The economics of substitution create, in effect, "free" services for those previously proprietary in nature.

This scenario oversimplifies, of course, the pending development of the telecommunications industry. The fact is that providers in all industries continually seek to exploit the economics of substitution as new markets unfold; this is a traditional tactic of enhancing profit margin. However, TRA is engineering a "free market environment" at a time when major advances in telecommunications, particularly those sensitive to software applications, make it highly uncertain as to what new ventures are likely to be successful. The risk of excess supply, based on innovations that cannot be forecasted, exists at any phase of the production/sales chain. If risk can no longer be calibrated and if prices might implode based on sudden overcapacity, have we put the industry at great long-term risk? That is the message implicit in the model illustrated in Figure 10.7.

The figure implies that competitive threats surfacing from outside the industry, spurred by technological innovation, influence the outline of future profit margins. Typically, a company will concern itself only with those innovations introduced by competitors from within the industry itself. In the case of telecommunications, however, a new supply by traditional competitors is reinforced by additional services that are not susceptible to traditional forecasting. The result is that the firm that assumes the cost of new product development is sometimes denied the profits that would otherwise accrue. Prices fall or collapse, reducing profit margins proportionately.

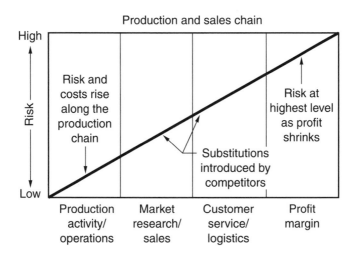

Figure 10.7   Price implosion model.

10.2.5   Short-Run Chaos/Long-Run Stability Model

Some observers contend that market confusion and chaos will evitably follow implementation of the TRA. It is their contention, however, that it is just as inevitable that market stability will be restored after 10 to 15 years of experimentation. A highly capital-intensive industry whose progress was guided by government regulation, they believe, must educate itself in market dynamics never before experienced. The absence of competitive knowledge and experience means that industrial giants must learn during a period of trial and error that can only be described as chaotic. Thereafter, acquired knowledge by current players and the assumption of risk by new entrants establish a framework of growing maturity and stability. One might express this evolution through the model illustrated in Figure 10.8.

We note in the figure the instability associated with short-term innovations in telecommunications. The proliferation of new entrants, accompanied by uncertainty and risk in infrastructure maintenance, translate into an industry shakeout before 2005. (Empirical evidence suggesting this time frame is supplied by the experience of airline deregulation in the years immediately

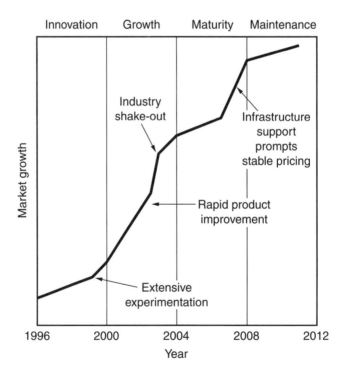

Figure 10.8   SRC/LRS model.

following 1978.) The model holds that long-term stability will manifest itself by 2008, however. It is during this period that risk diminishes as new infrastructure providers finally stabilize the economic environment. Put another way, the proliferation of service providers in the early years of deregulation is followed by the intervention of capital-intensive wireline maintenance companies. The logical candidate for wireline support during this period is the electric utility: These providers, already facing emerging deregulatory efforts by the states, will seek out new markets where their knowledge of state-of-the-art infrastructure can be readily exploited. The model implies that, by 2008, a vacuum in infrastructure maintenance will invite electric utilities to enter the competition.

Once dismissed as pure fantasy, a scenario in which public utilities enter and stabilize the telecommunications industry is now embraced by some observers. By entering the field of infrastructure support, electric utilities effectively permit other telecommunications companies to pursue diversified strategies and newly emerging markets, both domestic and international [18].

10.2.6   Evaluating the Models

With these five alternative visions in place, the challenge remains, of course, to identify that paradigm which most closely approximates the fate of the industry by 2007. Although elements of each model are based upon empirical evidence, the interpretative emphasis embedded in each imparts contentiousness. Instead of relying upon a single model to guide planning and strategy, one might consider testing the efficacy of each model as the early stage of competition unfolds. In other words, the outcome defined by each paradigm will be a function of the underlying assumptions postulated early in the framework of that model. We can thus evaluate the validity of these paradigms as each stage of competition unfolds. As suggested in Chapter 2, approximately three to six years are required for a regulated industry to fully exhibit the benefits of an unregulated market system. As a result, we are about to enter an important transitional period as we advance toward a salient market model.

It should also be noted that models serve three discrete functions:

1. To *describe* and categorize future economic development;
2. To *prescribe* appropriate strategy in response to those developments;
3. To *project* future outcomes based on continuing changes in the marketplace.

Examined from this perspective, each model should be seen as a tool to measure the advance of competition and its impact on the telecommunications industry. We are left with a menu of potential outcomes for the industry rather than a certitude about its direction. We must bear in mind that the experiment of deregulation, as it now applies to the formation of unfolding telecommunications products and services, is fundamentally unique.

10.3   The Foundation of Future Telecommunications Services

10.3.1   Beyond Economic Models: Telecommunications and Continuous Markets

Empirical evidence established in Chapters 2 and 7 of this volume confirmed that the shift from industrial to post-industrial societies would create knowledge-based economies in the future. Fundamentally, the transmission, storage, and use of information would define individual and organizational success and failure; winners and losers in the economy would be determined by the incisive use of information to differentiate and select alternative courses of action. Creativity, speed, adaptability, and flexibility would be the determinants of reward in an era so dependent on the dexterous requirements of information management [19].

Whatever the efficacy of the models herein described, as we extend our vision to the next decade it is becoming increasingly evident that telecommunications deregulation will accelerate change. Social, political, cultural, technological, and economic change will be greatly aided by the facile application of information to guide personal and organizational decision making. Mounting evidence suggests that long-term profitability in the industry will pivot on the development of "continuous markets" to sustain future growth.

Telecommunications continuous markets are ones in which recurring need, and therefore profitability, are generated by the use of communications technology. Primary, secondary, and tertiary markets for telecommunications in health care, education, transportation, environment, entertainment, agriculture, and manufacturing are formed and reinvented when users fully exploit the value of information to perform their tasks. As a result, we may designate *learning*—personal and organizational—as key to long-term viability. For both service and content providers, to facilitate learning is to expand market and market share.

Learning, the pursuit and acquisition of knowledge, constitutes the great long-term domestic and international market for telecommunications firms. Whether one dichotomizes or synthesizes a vision of the telephony, cable, or broadcasting markets, the recurring need of the individual to explore, apply, or assimilate knowledge—to learn continuously—is likely to become the heart of future aggregate demand for communications services. The future telecommunications market may be structured through competitive proliferation or consolidation, but treating each consumer as a market unto him- or herself constitutes the greatest opportunity.

To tap the continuous market—to uncover the recurring revenue of continuous personal and professional need for information—we must understand the prospective market response to the learning process. We note two driving forces of learning now manifested in the knowledge-based economy: education and training. Education involves the transmission of knowledge through formal instruction and results in a body of knowledge whose economic value does not erode. Language, history, mathematics, and the sciences are examples of education in a formalized sense. Training, in contrast, represents practical instruction aimed at accomplishing specific tasks; because innovation drives training, inevitably the value of prior training recedes proportionate to the level of improvement introduced. We thus encounter a situation in which education and training must be continuous, a circumstance necessitating perpetual certification and recertification. The markets for education and training will grow commensurate with the level of knowledge and skill demanded by the private sector. Individuals will be compelled to provide tangible evidence of state-of-the-art knowledge in their professions. Communications technology thus drives economic change, which drives education and training, which in turn invites further improvements in communications, thereby accelerating the demand for further training. It will be a common occurrence for professionals, technicians, and administrators—indeed, people in all areas of endeavor—to return to schools at two or three critical junctures in their lives. These schools, physical or virtual, will focus on customized learning, tailored to the precise present and future objectives of these individuals—and will unite vocation with avocation [20]. Telecommunications providers, whether focused on service or content, will rise to meet this enormous prospective demand. In the largest sense, learning—in all its manifestations—is the most attractive market for this industry.

The continuous learning market is likely to consist of three discrete segments and will evolve in ways as yet unimaginable. These three concrete markets may be examined from the perspective of Table 10.1.

The potential economic value of these opportunities cannot yet be measured with precision. Already, these markets are beginning to establish a foothold in the communications industry. Major providers, geared to the short-term strategic implications of deregulation, have yet to develop an arsenal of tactics to customize information on a "particle" basis. However, technological innovation fused with continuous learning make this vehicle the most attractive long-term market in the age of deregulation.

Table 10.1
Continuous Learning Markets

Market	Consumer Focus	Consumer Goals
Certification	State-of-the-art knowledge/ refined training; upgraded skill base and enhanced productivity	Degree; professional designation
Collaboration	Cross-departmental or occupational understanding and communication	Partnerships; alliances; joint projects
Community	Shared learning experiences based on similar philosophy and value base	Permanent learning networks (vocational or avocational)

10.3.2   Personal Information Management: The Individual During the Era of Telecommunications Deregulation

If we fuse the central thesis of this book—that deregulation will precipitate communication innovation at lower price—with the pending development of continuous learning markets, then we may infer that it is in the industry's best interest to promote personal information management. In other words, the convergence of the computer, telephony, and broadcasting industries may eventually lead to the assimilation and manipulation of information by individuals in a way that today is only achieved by sophisticated organizations. Telecommunications deregulation, in its ultimate manifestation, may usher in an era of personal information management replicating today's business demand for data warehousing, data mining, and multimedia command.

Should this scenario take hold, deregulation's greatest influence may be indirect but nevertheless profound: Telecommunications may facilitate cooperative efforts throughout the world never before attainable.

References

[1] Colander, David, *Economics*, Burr Ridge, IL: Irwin Press, 1995, pp. 84–89.

[2] *Ibid.* See also Alexander, Peter, "What Hath Telecom Reform Wrought," *Telephony*, June 24, 1996.

[3] Colander, *op. cit.*

[4] "Technology 1996: Analysis and Forecast Issue," *IEEE Spectrum*, Jan. 1996.

[5] "Choosing the Right Path," *Convergence*, May 1996. For a discussion of historical developments leading to these business opportunities, also see *Traveling the Information Superhigh-*

way, Feb. 1994, monograph produced by Columbia University's Freedom Forum Media Studies Center.

[6] Asthana, Praveen, "Jumping the Technology S-Curve," *IEEE Spectrum*, June 1995.

[7] For a discussion of sequential ordering of tertiary consequences, refer to Cornish, Edward, *The Study of the Future*, World Future Society, Washington, D.C., 1977.

[8] Dent, Harry S., *The Great Boom Ahead*, New York: Hyperion Press, pp. 21–46.

[9] Dent, *op. cit.*

[10] Dent, Harry S., *Job Shock*, New York: St. Martin's Press, 1995, pp. 53–57.

[11] Gross, Neil, Peter Coy, and Otis Port, "The Technology Paradox," *Business Week*, March 6, 1995.

[12] Thyfault, Mary, "Deregulation Has IS Support Wired," *Information Week*, Sept. 9, 1996.

[13] See President Clinton, *Technology for America's Economic Growth, A New Direction to Build Economic Strength*, NTIA Publications, Washington, D.C., Feb. 22, 1993.

[14] Cawley, Joel, "The Era of Intelligent Communications," *Telecommunications™*, April 1995, p. 4.

[15] Baig, Edward, "Read, Set—Go On-Line," *Business Week*, July 12, 1994, p. 26.

[16] For additional information on emerging customization, refer to Dent's "Incremental/Radical Innovation" paradigm, elaborated in *Job Shock, op. cit.*, pp. 241–243.

[17] See Galatowitsch, Sheila, "Internet: Telephony's New Era," *Convergence*, Sept. 1996, for a discussion of emerging issues surrounding development of Internet telephony as it prospectively will influence wireline local and long-distance competitiveness.

[18] Rivkin, Steven, "If Competition Won't Build the Nil, Utility Partnerships Will," *New Telecom Quarterly*, Aug. 1996, pp. 19–23.

[19] Firat, A. Fuat, "Literacy in the Age of New Information Technologies," *New Infotainment Technologies in the Home: Demand-Side Perspectives* (R. Dholakia, N. Mundorf, and N. Dholakia, Eds.), Mahwah, NJ: Larwrence Erlbaum Associates, pp. 173–198.

[20] Gilbert, Steven, "How to Think About How to Learn," *ACB Trusteeship*, Special 1996 Issue, Washington, D.C.: American Association for Higher Education.

11

Contours of the Information Economy

Corporate and entrepreneurial strategy cannot be successfully formulated in the absence of a macroeconomic analysis of the Information Economy. Macroeconomics must be effectively applied to the marketplace in order to ascertain whether a firm's strategy is congruent with the competitive environment. The author does not assert herein a certitude or rectitude about strategic direction in light of deregulation. There is growing consensus, however, regarding the questions that must be addressed to prepare for impending volatility and uncertainty.

Among the questions critical in unlocking competitive advantage in this "new economy" are these:

- Does empirical evidence support the contention that there is indeed an "Information Economy," and if so, what does this development portend for the communications industry?
- What are the contours—the forces—shaping this economy, and how does a firm respond to and exploit such characteristics?
- To what extent does the business cycle influence the demand for telecommunications products and services?
- Does the prospective proliferation of new competitors in the global marketplace stimulate or suppress the supply and demand for U.S. communications services?
- To what degree will competition intensify during the next decade, and can these forces be accurately projected?
- What dynamic forces will drive the competitive and oligopolical market structures for each sector of the industry?

- What variables will guide the economics of competitive interaction between the telephony, broadcasting, and computer industries?

- How will changes in telecommunications influence other sectors of the economy, and how will this resultant adaptation guide future demand for communications products and services?

- How should communications firms evaluate their own competitive positioning during the next decade?

- What communications products will likely dominate the future competitive marketplace, and what strategies will guide their distribution to consumers?

- Which synergies will define future telecommunications competition?

- Will traditional strategic management techniques be useful in the newly emerging marketplace, and if not, how should they be redesigned to respond to the unique needs of this industry?

It is the dynamic complexity of these questions that increasingly disorients communications firms as they devise strategy [1]. This phenomenon is no less apparent in other sectors of the domestic and global economies, but the rapidity of technological change that typifies the telecommunications enterprise amplifies the priority for strategic clarity. The author does not attempt to answer these questions in specificity through this work, but has nonetheless assembled a body of knowledge increasingly acknowledged by practitioners as a useful first step toward strategic illumination.

11.1   Contours of the U.S. Macroeconomy, 2000–2010

The U.S. economy has grown substantially during the past decade. Following the recession of 1990–1991 (the most recent of nine recessions punctuating the post–World War II period), the U.S. economy has experienced declining rates in unemployment, inflation, and interest rates while concurrently enjoying higher corporate profitability, rising stock valuations, and spectacular increases in both consumer and business investment. This era has represented a remarkable period in U.S. economic development. During the late 1990s, Americans became wealthier while the world concurrently experienced recession and depression [2]. Government deficits have evaporated, businesses have become more productive, and manufacturing has become substantially more competitive. U.S. firms have increased output while restraining price inflation and reducing unit costs of production [3].

If these patterns of economic development have characterized the 1990s American economy, what reasonable macroeconomic projections might be made with respect to the coming decade? As is often the case, economists have developed no absolute consensus about our probable long-term economic future. We may therefore speak only in terms of probability as we assess emerging business opportunities during the next 10 years. We do, however, have at our disposal a set of macroeconomic determinants that assist in providing an educated forecast (see Appendix E).

Macroeconomic data suggest that American economic expansion, though always subject to temporary downturns, may well extend through 2010. The resurgence of productivity, positive demographic factors, reduction of business and government debt, and the restoration of price stability imply favorable business conditions for all industries. The challenge confronting telecommunications planners is a deceptively straightforward one: to what extent will the communications industry share in this prosperity, and how do these forces define the future of the sector's competitive landscape? With which firms should veteran telecommunications players align themselves in this economic environment? Does a period of sustained national economic growth necessarily expand the demand curve for communications services? If it does, is there any assurance that the established telephony, broadcasting, and computer sectors will all benefit from rising consumer consumption of these services? Are newly emerging firms poised to seize traditional market share from veteran companies?

A positive economic forecast based on the variables indicated above obviously does not guarantee a favorable path for either established or entrepreneurial firms. However, the collection of data—reflecting enhanced productivity, heightened entrepreneurialism, expansive consumption, and strong demographics—does suggest the following:

- Inflation is likely to remain very low, reducing the risk of capital acquisition and investment. Thus, business demand for communications products and services is likely to expand significantly during the coming decade, particularly as the world economy emerges from recession after 2000.

- Low inflation dictates low interest rates over the long term. Again, this factor increases the probability that both business and consumer demand for communications services will rise significantly: The purchase of these products defines positive future demand for telecommunications services, particularly for the telephony industry in the short term. The dramatic reduction in and elimination of government deficits during the late 1990s strengthens this forecast, as government at all levels—federal, state, and local—no longer absorbs capital from the financial markets.

- Unemployment is likely to remain low during the next 10 years. This development is a function principally of the decline in new workers entering the workforce after 2000. With a decline in the aggregate numbers of Americans seeking work, wages are likely to rise in real terms (wage increases minus inflation equal increases in real income). The capacity of Americans to consume will rise, and it is not unreasonable to infer that demand for advanced communications will accelerate, particularly in regard to those services to which consumers become "dependent." These might include Internet usage, wireless data and voice service, and other products hardly yet conceived. It is instructive to recall that basic telephone and cable television service were both regarded as luxuries at the time of their inception. The strategic conversion of "luxury" items to "staples" has now become a principal priority of the modern communications firm.

- The telecommunications industry increasingly displays a propensity toward inelastic demand; that is, individuals, households, and businesses must use communications services irrespective of downturns in the business cycle. This development suggests that consumers require such services independent of cost. Such a predilection does not mean the elimination of price discrimination among consumers—that users will not seek out competitive pricing—but implies that, once adopted, consumers regard such services as necessities. If this tendency asserts and sustains itself during the coming decade, then communications firms may become impervious to fluctuations associated with the business cycle; demand for veteran products and services may not be directly tied to changes in income, productivity, wealth, or national economic policy.

- The advance of electronic commerce (Internet-based transactions) represents a new conduit for business and is fueling recent global economic growth. This activity started in the United States in 1994 and has accelerated at a spectacular rate. By some estimates "e-commerce" will generate $1 trillion in gross revenues by 2002, and perhaps double again before 2010 [4]. One cannot overstate the significance of this transitional era in U.S economic development. The federal government acknowledged the criticality of this phenomenon in a report issued in 1999 [5]:

The digital revolution is also changing the respective roles of government and the private sector. In the 19th and for much of the 20th centuries, governments played a key role in helping build or actively regulate much of the country's infrastructure. The federal government made extensive land grants to encourage private capital to expand the nation's rail network. Government subsidies were used to stimulate the development of an airline industry. Federal and state dollars combined to build and maintain the interstate highway system. In communications, the government granted a virtual monopoly to a single company and regulated the industry after its breakup. Most power companies have been regulated monopolies at the state or federal level.

The federal government funded and developed early versions of the Internet for national security and research purposes. It will continue to provide funding for research and development on future Internet and high-performance computing technologies. However, most of the capital to build the computing and telecommunications infrastructure is being provided by the private sector.

The pace of technological development and the borderless environment created by the Internet drives a new paradigm for government and private sector responsibilities. Creating the optimal conditions for the new digital economy to flourish requires a new, much less restrictive approach to the setting of rules.

Governments must allow electronic commerce to grow up in an environment driven by markets, not burdened with extensive regulation, taxation, or censorship. While government actions will not stop the growth of electronic commerce, if they are too intrusive, progress can be substantially impeded.

Where possible, rules for the Internet and electronic commerce should result from private . . . action, not government regulation.

Governments do have a role to play in supporting the creation of a predictable legal environment globally for doing business on the Internet, but must exercise this role in a nonbureaucratic fashion.

. . . competition in telecommunications and broadcast industries should be encouraged so that high-bandwidth services are brought to homes and offices around the world and so that the new converged market place of broadcast, telephony, and the Internet operate based on laws of competition and consumer choice rather than those of government regulation.

. . . there should be no discriminatory taxation against Internet commerce.

The Internet should function as a seamless global marketplace with no artificial barriers erected by governments. As with any major societal transformation, the digital economy will foster change and some upheaval.

The Industrial Revolution brought great economic and social benefit, but it also brought about massive dislocations of people, increased industrial pollution, unhealthy child labor, and unsafe work environments. Societies were often slow in responding to these negative side effects. Similarly, the digital economy may bring potential invasions of privacy, easier access by children to pornographic and violent materials and hate speech, more sophisticated and far-reaching criminal activity, and a host of other as yet unknown problems.

The private sector and government, working together, must address these problems in ways that make the Internet a safe environment while not impeding its commercial development.

These conclusions by the federal government are highly revealing. Many public officials now acknowledge not merely their inability to "manage" electronic commerce, but implicitly concede that national economic growth and competitiveness are vitally tied to the course of a digital economy. The migration from a conventional to digital economy creates winners and losers, as well as new efficiencies, but also discontinuities and dysfunctions. These effects are inevitable when consumer choice fuses with deregulation [6]. The communications industry thus instigates precipitate threats and opportunities for all sectors of the macroeconomy.

Analysis of domestic economic data leads us to evaluate the following objective criteria and to project these variables forward. Our concern here is to learn more about the impact of the telecommunications industry on the national economy during this era and, in turn, to assess unfolding macroeconomic trends on the industry itself. It is now clear, and largely beyond dispute among academics and practitioners alike, that we have opened an economic period essentially unprecedented in its ramifications for the future. We can calibrate the contours of the information economy by dissecting trends in unemployment, productivity, corporate profitability, stock valuations, gross domestic product, inflation, business investment, consumer spending, and government budgeting. Refer to the data in Appendixes E and F when reviewing the following criteria.

11.1.1 Unemployment and the Information Economy

Unemployment in the United States during the last two years of the twentieth century hovered at 4% or less, a figure that economists cite as "full." When we connect demographic trends with rising consumption, we can deduce why unemployment has fallen sharply during the past decade. With the economy growing at a healthy rate, and the number of workers entering the economy for the first time (first-time workers) at subdued levels, we can reasonably project exceptionally low unemployment throughout this coming decade. Indeed, there is some evidence to indicate that if the economy grows significantly during this era, we may be facing prospective labor shortages in every sector of the economy. In this regard, recruiting and retaining skilled workers may become a critical strategic issue for communications firms, particularly before 2004. The lack of skilled workers is a likely scenario confronting the telecommunications industry in the years ahead, especially with regard to those firms compelled to customize their services for their customers.

11.1.2   Productivity and the Information Economy

There is now hard evidence that the American economy is accelerating its productivity gains. In the 20 years following World War II, average annual productivity exceeded 3%, a figure unprecedented in the nation's history. In the 30 years following the mid-1960s, annual productivity fell by a factor of 50%. We can now confirm, in reviewing the data, that productivity has been hovering near historic levels since 1997. In 2000, productivity climbed as high as 5%, a truly extraordinary development.

Productivity is pivotal to national competitiveness because it signals the efficiency with which products are generated and distributed to market. A firm that maximizes its efficiency is a company that will capture expanding market share by pricing its product favorably; enhanced productivity achieves this objective by containing the costs of inputs. National competitiveness hinges on the collective ability of industry to exploit these productivity advances. As indicated in the "competitiveness scorecard" found in Appendix D, the U.S. economy has never been more competitive and productive, relative to its global counterparts.

Recent studies confirm that productivity has been spurred in recent years through advances in supply-chain management, inventory control, outsourcing, and just-in-time deliveries. We have convincing evidence that the costs of running an enterprise in all operational phases—transactions, production, and distribution—have been declining since at least 1996. These declining cost curves verify that productivity is measurably improving. Most revealing is the fact that these productivity improvements are now cutting through every sector of the economy.

11.1.3   Corporate Profitability and Stock Valuations and the Information Economy

Corporate profitability improved substantially after the nation's last recession in 1990. In fact, corporate profits rose at a rate significantly higher since 1995 than any comparable period during the past 15 years. In part, higher profits have fueled improvements in productivity and cost containment. In part, rising incomes, which in turn are traceable to higher productivity, have instigated higher revenue streams and profits. And, in part, business investment in high-technology products and services (among these, networking, communications equipment, advanced telecommunications infrastructure) has stimulated new demand. Additionally, new markets have opened to U.S. firms in the wake of the collapse of command-control economies in the former Soviet Union, Eastern Europe, and Asia.

Higher stock valuations normally precede profitability, and the U.S. stock market has enjoyed spectacular gains during the past decade. In 1992, the Dow Jones average sustained a rate slightly above 3,000; as the decade closed, this figure stood at nearly 11,000. Significantly, these advances in stock valuations have channeled new investment to domestic communications companies. Among those firms that have benefited most from these new sources of capital have been telecommunications companies whose products have facilitated the buildout of networked commerce.

11.1.4   Gross Domestic Product and the Information Economy

As the data assembled in this volume confirm, U.S. gross domestic product (GDP) has risen sharply during the past decade. As the last decade opened, U.S. GDP stood at approximately $4 trillion; as the century closed, nominal GDP had climbed to $10 trillion. National wealth, even after adjusting for inflation, has been expanding substantially in recent years. Improvements in productivity, positive demographics (that is, a significant fraction of Americans are now in their peak years of consumption), and rising output have combined to accelerate this extraordinary development. As GDP has risen, so too has the ability of Americans to procure the latest in communications technology. In effect, Americans use a substantial fraction of their newly acquired wealth to purchase these communications products and services, which, in turn, allow them to create new markets and new wealth.

The Department of Commerce estimates that 35% or more of recent increases in GDP are broadly linked to information technology, communications equipment, and, most notably, networked commerce. It is impossible to gauge with precision the impact of the Internet on the national economy, but it is now clear that much of the improvement in the nation's standard of living has been stimulated by this technology. Recent forecasts hold that business-to-business and business-to-consumer networked commerce may account for as much as 20% of U.S. GDP by 2004.

11.1.5 Inflation and the Information Economy

Inflation in the United States has remained at a tepid 2% during the past three years. Average inflation rates have remained well below 3% for virtually the entire past decade. Price stability is linked to efficiencies in business. As indicated herein, cost efficiencies in transactions, production, and distribution account, in part, for this era of contained inflation. Additional factors include rising global competition and deregulation. In excess of 90% of the world's economies—some two hundred in number—are now nominally capitalist and participative in the global environment. Before 1990 perhaps 40% of the world's nations operated in a command-control environment, not interactive in the other half of the old bipolar world. It is only a slight exaggeration to assert that the United States has doubled the number of producers and consumers in the global economy during the past decade [7].

It should also be underscored that major advances in inventory control, particularly with regard to manufacturing, have meant that firms have been able to restrain inventories while fully satisfying their customer base. Essentially, companies have learned to shave costs of inventory and insurance through just-in-time management techniques.

11.1.6 Business Investment and Consumer Spending and the Information Economy

Among the least appreciated developments during the past decade has been the extraordinary boom in business investment, particularly since 1994. With the exception of World War II, we have not witnessed an investment spree of comparable dimensions. Average annual business investment in new plant and equipment, telecommunications infrastructure, networking components, and information technology has been rising by a factor of three times during this period. Firms now recognize that they cannot successfully compete without deploying state-of-the-art equipment and services in each of these areas. Such business investment is tied directly to reducing per-unit costs of production, which is critical to competing in a world of intensified competition. More appropriately, we might refer to this era as hypercompetitive, a period impelling such investments [8].

Similarly, consumer spending has been rising extraordinarily during this era. The empirical evidence suggests that 1995 marked the beginning of sharply rising consumption for virtually all goods and services during this span. The telecommunications industry has benefited mightily from this development. With capital available at reasonable interest rates, and Americans using the bulk of their disposable incomes to fuel consumption, demand has been rising for all goods and services. Projections for consumer spending appear strong throughout the U.S. economy for most of the coming decade [9].

11.1.7 Government Budgets and the Information Economy

Perhaps the most surprising development in this era has been the evaporation of government deficits, and the unfolding, prospective string of surpluses. In 1992, the federal government incurred a deficit of $290 billion. By 1998, the federal government entered a surplus position on the budget. Two years later the annual budget surplus swelled to $240 billion. Most forecasts now place accumulated surpluses in excess of $3 trillion by 2012. State and local governments are similarly well capitalized.

The conversion of government deficits into surpluses represents something more than an arcane intellectual exercise. Assuming these surpluses are sustained, the federal government will not be competing in the macroeconomy with producers and consumers for available capital. As a result, interest rates are likely to remain moderate during this period, again facilitating debt accumulation to finance consumption. For the communications sector, this development augers well for the years ahead. Firms will be able to borrow money (or run corporate bond issues) at reasonable rates, assuring a continued positive investment climate.

11.2 Telecommunications and the World Economy, 2000–2010

There is no concrete method to chart the course of the entire world economy, and the role of telecommunications in influencing it, for the coming decade. We can, however, make reasonable inferences about the prospective supply and demand of communications services. Assuming that the pattern of post–World War II growth rates is sustained through the coming decade, we can presume global economic growth of 3–8% on an average annualized basis [10]. If the world economy grows only moderately during this period, we might see growth at 4%; a high growth rate (beyond 6%) is conceivable if global trade and productivity accelerate.

Two hundred nations comprise the world economy. Each one finds itself in a slightly different stage of economic development. For telecommunications firms seeking optimal strategy, these varying stages of development connote a pecking order with regard to investment and marketing. It is probable, and, indeed, empirical evidence presently confirms, that multinational communications corporations will pursue investment initially in the most advanced industrialized societies [11]. It is in these regions that profitability will be maximized in the short term. We may also infer that these firms will pursue business-to-business marketing strategies, while corporate customers are more likely to purchase high-end services, thus promoting economies of scale for the most aggressive carriers. Thereafter, households with high disposable incomes and established patterns of communications consumption will represent the most attractive customer base. Such consumers, given the uneven distribution of global wealth, are scattered in all nations and signify viable markets only when they can be successfully aggregated. Concentrated groups of business and household users, sharing these patterns of behavior, will lure communications investment at the front end of the next decade [12].

Nominally, it may be argued that the pattern of U.S. communications investments during the 1990s—in mergers, acquisitions, affiliations, partnerships, alliances, and start-ups—will be replicated elsewhere in the decade to come. There are several reasons for this phenomenon. Experiments in deregulated monopolies, coupled with privatization, are seen with increasing frequency throughout the world as the new decade begins. The collapse of command-control economies—principally those in the former Soviet Union and Eastern Europe—has greatly aided this trend, of course. By some estimates, free-market economies in this decade have consequently absorbed an increment of more than 1 billion consumers for prospective communications products and services. Elementally, it is not feasible to participate in the world economic system without conforming to the rules adopted by the largest players.

Second, the encroachment of information technology in all areas of human endeavor necessitates speedier, more adaptable organizational structures that might consolidate market clusters [13]. The most expedient of way of efficiently distributing services to global markets lies, in the short term, in exploiting the advantages of large, multinational corporate forms: large players can leverage vast amounts of capital so as to quickly build required infrastructure [14]. This step must naturally precede the development of those service enterprises, large and small, which must rely on such infrastructure to attract their customers.

Finally, the surge in networked economies—linking common interests across the globe without regard to geographic boundaries—has impelled governments to accede to intrusive corporate interests. Any government of national, regional, or local origin effectively weighs a trade-off in responding to this development: an attempt to forestall the vanguard of multinational corporate consolidation runs the risk of inciting consumer and business opposition. This opposition is a function of local interests, which sustain competitive disadvantage if not permitted access to state-of-the-art communication services. Global governments, for better or worse, now find themselves drawn into the broader web of multinational corporate investment typifying consolidation in the late 1990s. The pattern of corporate investment and consolidation is explored further in Chapter 8.

11.3   The Emerging Economics of Telecommunications

Economists and market researchers have discerned the emergence of new industry structures governing the creation and distribution of telecommunications products and services. In effect, innovation in communications products has precipitated economic changes in other industrial and consumer sectors of the economy; this resultant economic transformation has influenced subsequent innovations in market planning. The telecommunications industry is fostering an explosion of product proliferation, which has introduced a metamorphosis in economic and strategic analysis. The planning of strategy must now consider these emergent economic developments:

- *Aggregation:* the method by which firms seek to expand and retain a customer base;

- *Complementation:* the synergy of two or more firms cooperating to develop new products or cultivate new markets;

- *Consolidation:* the procedure in which competitors merge, acquire, or otherwise transform their company structures to achieve aggregative or complementary strategies;

- *Disintermediation:* the unfolding loss of traditional distribution markets in which firms and their employees lose access to their customer base via the encroachment of new firms that directly link producers with consumers;

- *Convergence:* the circumstance in which a traditional sector of communications is either invaded by, or chooses to invade, an unrelated sector as a result of technological innovation.

The following sections isolate the economic determinants of aggregation, complementation, consolidation, disintermediation, tertiary analysis, and convergence. One must bear in mind that the economic and strategic significance of these tools remains in flux, and are influenced by variables as yet not fully understood. We can empirically verify the unfolding presence of these economic phenomena but not arrive at certitude about their long-term meaning for strategic planners.

11.4 The Economics of Aggregation

To aggregate—to attract and retain—customers has become an overriding concern for communications firms during the past decade. The strategy of market aggregation is, in principle, no more significant for telecommunications than for other sectors of the economy. However, new methods of estimating and projecting the economic value of communications customers have attained heightened status in recent years. There are two primary reasons for this development: first, in the face of deregulation, customers—households and businesses—are becoming adroit at pursuing alternative providers of both products and services; second, the continued expansion of Internet services has promulgated many new firms whose own valuations pivot on the dynamics of aggregation [15]. Simply, communications firms are groping for a method of estimating their own true prospective value in a changing marketplace. Traditional methods (among them price-earnings ratios, subscriber revenues, cash-flow management) of company valuations are often inadequate in and of themselves to provide such a determination.

How does a telecommunications firm assess its own value amid mounting competition stimulated by deregulation and the convergence of the telephony, broadcasting, and computer industries? A common basis of valuation is required, and such a quantitative method must conform to the new contours of technological innovation. In this environment, we are slowly witnessing a consensus among market forecasters: The economics of *reach* and *loyalty* define present company value [16]. "Reach" is a quantifiable variable indicative of the degree to which a firm's message penetrates markets. In other words, that fraction of the market to which a firm's message and image are exposed asserts an essential element of underlying value. "Loyalty" refers to that percentage of a customer base that may be regarded as continuing customers [17]. The interaction of these two variables increasingly correlates with the perception of a firm's value; the stock and venture capital markets in recent years reaffirm the growing significance of economic aggregation. A firm that can demonstrate (see Figure 3.1) growing valuations via these two variables is rewarded for such success, whether publicly or privately held. Furthermore, the capacity to float bonds or borrow from the capital markets—often necessary for expansion—is considerably eased through valuations that rise in response to strategies that exploit loyalty and reach. Costs of running a business are reduced through lower interest charges, and profitability is increased through the attainment of this strategic objective.

Economic aggregation, the measurement of loyalty and reach, is likely to also be a critical factor in instigating mergers and acquisitions, as well as business alliances and partnerships. A firm adept at effectively designing strategy applied to both loyalty and reach is often rewarded for its long-term prospects, even if short-term profitability is nonexistent. The capacity to perpetually engage the customer's attention represents extraordinary long-term value to communications carriers, content providers, and other distributors of information [18]. Investors presume that perpetual customer exposure to a firm's message signifies primary, secondary, and tertiary value for the long term. These issues are discussed further in Chapter 4.

11.5 The Economics of Complementation

Complementation is the process by which two or more firms cooperate by exploiting their counterpart's competitive advantage. In dynamic industries, where change is impelled by dramatic transformation of social, cultural, political, or economic systems, or through technological innovation, we can confirm the rapid reconfiguration of business enterprises [19]. Such reconfigurations—new organizational schemes—are intended to penetrate new markets, expand market share, and overcome prior obstacles to market entry. In the case of telecommunications, political, economic, and technological changes have coalesced to fuel such synergies.

It is no longer feasible for a single telecommunications firm to dominate all avenues of infrastructure, just as no content provider can direct or influence the course of broadcasting. The disintegration of AT&T's monopoly, the elimination of arbitrary dichotomies between the principal communications sectors, and the competition—domestic and global—promulgated by deregulation, have combined to necessitate cooperation of competitors. This is not to say that these firms do not regard themselves as industry rivals or long-term opponents. To the contrary, the shifting competitive landscape can alter the very reconfigurations of those business forms that surfaced in response to such change.

It remains salient, however, that a broadcaster may itself be in need of software support only supplied by the computer sector; the telephony provider may be compelled to absorb a cable firm with access to markets it cannot presently secure; the computer firm may wish to facilitate the objectives of leading firms in other sectors to assure a ready demand for its product. The multiplicity of scenarios is expansive and ever-changing, and demands farsighted strategic management. These matters of coopetitive dynamics are explored in greater detail in Chapters 7 and 8, but, as elaborated in Chapter 1, there exists a menu of criteria isolating the economic dynamics at work [20].

The economics of complementation hinge on these simple criteria [21]:

- A competitor is a complementor if consumers value a company's product more when value absorbed from that competition makes that product more desirable than all others available at that time. If your cooperation with another firm leads to an expanded consumer base—whether or not that competitor resides in your industry sector—it is foolish to ignore a prospective alliance, partnership, or affiliation.

- A competitor represents a prospective complementor, after an alliance is forged, if suppliers are more likely to provide a continuity of supply at favorable prices than if the same firm operated in the absence of such an alliance.

- A competitor is a complementor if this is the only mechanism available to facilitate immediate improvement in price, quality, service, or post-purchase satisfaction.

- Complementation is a viable strategic priority if these criteria are met and acceptable standards of benchmarking—methods of accurately evaluating competitive performance—can be established. These benchmarking methods can include productivity performance, consumer satisfaction, market share percentages, profitability, cost reduction, and similar measures.

Adam Brandenburger and Barry Nalebuff, the principal authors of "complementation," have simplified their elaborate theory in this way: ". . . there are win-win and win-lose relationships with *customers, suppliers, complementors, competitors* . . . ," and these relationships continually shift [22]. Identification of the time to forge or disband complementary relationships is partly intuitive, but also partly empirical science. Employing these criteria to define optimal moments for engineering such relationships refines strategic alternatives.

11.6 The Economics of Consolidation

The extraordinary proliferation of telecommunications mergers and acquisitions following the passage of regulatory reform in 1996 is noteworthy in terms of size, scope, and complexity. Consolidation, the product of merger and acquisition, represents a strategic attempt to combine the resources of symbiotic partners while reducing unit costs of production.

Consolidation is inevitable in the early stage of deregulation in virtually every industry. In trucking, aviation, petroleum, and natural gas during the late 1970s and 1980s, and in banking, financial services, and electric utilities during the 1990s, we can verify this repetitive experience. Economists debate the value and purpose of consolidation from a social and political perspective. The motive for such a strategy, viewed exclusively in business terms, remains unmistakable: lower production costs coupled with greater market share yields higher profitability and competitive advantage.

Integration, *vertical* and *horizontal,* is the tool of such consolidation, and both procedures have been seen with increasing frequency in the communications industry since 1996. Indeed, this trend has been manifested in both domestic and global mergers in the past several years. Mergers are defined as a combination of two or more firms; an acquisition represents the purchase of one firm by another. Vertical integration is defined as the ownership of networked production or distribution of a product or service. A firm engaged in vertical integration chooses to purchase or merge with those enterprises that support its delivery of products or services. Vertical integration is expressed as "forward" or "backward": if a cable television firm purchases a manufacturer of set-top boxes for consumer selection of programming, its strategy is supportive and therefore backward. If a motion picture studio purchases a television network so as to enter a new market for distributing its product, its strategy may be termed "forward." It is forward rather than backward integration that incites antitrust investigation and enforcement.

Horizontal integration occurs when one firm purchases another so as to expand market share. Viewed benignly, such integration can enhance economies of scale; if reduced production costs result in lower prices, both producer and consumer may benefit. If reduction in competition yields higher prices and less innovation, which are sometimes the results of such strategy, prospective improvements in production and distribution are stifled.

These standard definitions of horizontal and vertical integration do not fully reveal the economic complexities of consolidation. The telecommunications industry, given the converging telephony, broadcasting, and computers sectors, is stimulated today by shifting domains of content and infrastructure. In this respect, increasingly sophisticated attempts to deliver multidimensional content to consumers dictate advanced infrastructure; innovative infrastructure, in turn, enables content providers to lure more consumers to high value-added services. The interests of content providers and producers synergistically blend in each of these three sectors. Thus, the wave of mergers and acquisitions *within* each sector will evolve over the longer term to a comparable trend *between* these sectors. Consolidation may paradoxically provoke intense competition between and among these converging sectors, although this scenario remains pure conjecture at present. Table 3.2 offers a hypothetical projection of horizontal and vertical integration, as well as the objectives and motives of such strategy, through 2010 for the communications industry.

11.7 The Economics of Disintermediation

U.S. business has now entered the first stage of what might be termed "economic disintermediation." Disintermediation is the process by which mid-level management and distribution processes are effectively contracted or eliminated. The advent of an Internet-based economy has facilitated, for the first time, direct transactions between producers and end users. Consumers, through electronic commerce, are now able to circumvent formal distribution channels to command lower price, simultaneously accelerating rates of transaction. The result is that mid-tier managers and other workers—traditional "middlemen"—are vulnerable to job loss. Those firms who choose not to participate in the digitized economy lose competitive advantage and market share, in addition to prospective clientele. Economic disintermediation is becoming a fact of contemporary business life and cannot be avoided or evaded in light of advances in communications technology [23]. While advances in telecommunications technologies have imposed this restructuring of modern commerce, so too have they introduced similar opportunities and restrictions in this industry.

In the transition from an intermediated to a networked economy, a firm now faces the challenges of *managed disintermediation*:

- Applying the economics of intermediation, disintermediation, and re-intermediation;
- Identifying shifting customer tastes amid proliferating choices in the networked economy;
- Training sales and marketing staff to specify emerging market opportunities;
- Specifying prospective opportunities for disintermediation in the distribution chain;
- Planning sector strategy during the networked transition;
- Elaborating new methods of reintermediating value in the distribution chain;
- Isolating appropriate forecasting methods for networked market objectives;

- Leveraging technology to maximize market exposure at minimal cost;
- Differentiating product value in networked distribution chains;
- Pricing products and services during transitional reintermediation.

In this reconfigured networked economy, both individuals and organizations must learn the techniques of reintermediation [24]. Consider the following strategic scenarios:

1. How does a computer reseller insert itself back in a distribution chain when other firms have seized its market through direct producer-to-buyer sales?

2. How does a television broadcaster respond to the proliferation of niche cable channels, resulting in displacement of traditional mass audiences?

3. How do telephony carriers retain traditional revenues in the face of Internet telephony?

4. How do veteran traditional telecommunications providers in telephony and broadcasting respond to dramatic innovations in software development that replicate or simulate standard services formerly the domain of large infrastructure competitors?

These contingencies are among recent developments now seen in the communications industry. Scarcely a month has passed since the passage of regulatory reform that dramatic evidence of disintermediation has not been seen in the U.S. economy. The motive for disintermediation lies in cost reduction; declines in production costs lead to more competitive pricing; competitive pricing, for those firms able to exploit such advantages, leads, of course, to rising market share. As noted in subsequent chapters, however, those firms that successfully practice the art and science of disintermediation may later be subjected to the same tactic. Those who disintermediate can themselves be disintermediated. Strategic design aimed at exploiting and responding to disintermediation is discussed in Chapters 7 and 8, along with the issues described above.

The century-long trend in adding intermediaries—intermediation—is now giving way to the technological turmoil and opportunity of disintermediation. Conceiving tools of reintermediation now beckon all firms, as well as the professionals who manage them. In this regard, both the domestic and global economies shift to a new paradigm for those businesses seeking the benefits of strategic disintermediation [25]:

The key managerial challenge of the emerging global infostructure will be to reconfigure organizations to create and maintain high profits in a low transaction cost environment. These costs include those of searching for and identifying products, drafting, negotiating and safeguarding the terms of a sale or purchase, payment and settlement, and the costs incurred to enforce contracts or to correct and resolve contract disagreements. As communication becomes inexpensive, and software agents, technology and transaction templates are better defined, the transaction costs incurred by firms and individuals will fall. This in turn will make markets more efficient.

Firms will have to reconfigure their strategies and organization to respond to inexpensive communications, lower transaction costs and reductions in profit opportunities from market inefficiencies in the emerging global infostructure. As management and economics researchers have noted, successful firms focus their strategies along a few salient dimensions for competitive advantage.

For communications firms, the heightened priority of cost reduction, and the organizational development requirement to instigate and sustain it, is likely to remain predominant during the coming decade. Savings produced via cost reduction can be reinvested back into value-added services necessary to strengthen customer retention. The proliferation of new business formations seen after the passage of the Telecommunications Act of 1996 certifies the inevitability of this transition. Chapter 8 outlines the multiplicity of motives surrounding these creative business restructurings and the strategic path major competitors have now carved for the immediate future.

11.8 The Economics of Convergence

The decision of a communications firm to penetrate an unrelated sector of the industry—that is, launch new product lines or cultivate emerging markets outside its own domain—can be construed as sector convergence. Ordinarily, the expression "convergence" is used as a euphemism for the dynamic interaction among the broadcasting, telephony, and computer industries [26]. This recent phenomenon is directly attributable to deregulation, which effectively diminished or eliminated the artificial boundaries dividing these sectors.

Technological innovation demands that new economic tools be applied to estimate the business effects of convergence. The strategic significance of these overlapping sectors, and the economic changes they promulgate, are as yet only vaguely understood by economists and business strategists [27]. While little methodological progress has been made in forecasting the dynamics of economic convergence, the parameters of the problem are becoming apparent [28]:

- Technological innovation has accelerated the process of product substitution in recent years, thereby dramatically contracting development schedules and expediting product release dates. Firms are thus compelled to "obsolete" their own product lines at ever-increasing rates of acceleration.

- There is increasing evidence that it may be more important to be "first to market acceptance" rather than be "first to market" relative to new product lines: A firm must anticipate, more than ever, the risks associated in investing in *product development versus market research*. Amid dynamic convergence between these three sectors, does a firm prosper more if it is the originator of new technology . . . or should the same firm permit (or even encourage) its competitors to make such an investment while it strategizes a procedure to capture market share after replicating that product?

- To what degree does synergistic interaction between these sectors affect prospective demand elasticity for communications services? In other words, as these sectors both compete and cooperate, which products and services represent future platforms for the industry? Will the computer industry dominate telecommunications services in the future, or will consumer demand be linked primarily to veteran firms whose focus has been an improvement in traditional infrastructure? Will content development drive advances in infrastructure, or does infrastructure define eventual communications usage and distribution?

The following chapter explores economic tools that can be applied by strategic planners to estimate future consumer demand, the elasticities of that demand, and its probable determinants. One is cautioned to recall that the principles and models of macroeconomics and microeconomics are being reshaped in the post-Act era. Information usage—and the products and services that support it—follows a logic of its own. Unlike all other economic resources, information—and the infrastructure that undergirds it—proportionately increases relative to its consumption. The more we exploit knowledge—the key economic resource of the modern economic period—the more expansive our need for data storage, distribution, and interpretation [29]. Knowledge and information cannot be depleted; this is an inescapable and overriding tenet of the information era.

References

[1]   Hope, J., and T. Hope, *Competing in the Third Wave*, Boston: Harvard Business School Press, 1997, pp. 109–128.

[2]   Dent, H. S., Jr., *The Roaring 2000s*, New York: Simon and Schuster, 1998, pp. 14–27.

[3]   Business Intelligence Unit, "Report on the U.S. Economy," New York: DeLoitte, Touche, in cooperation with DeLoitte Consulting, Jan. 15, 1999.

[4]   Kambil, A., "Electronic Commerce: Implications of the Internet for Business Practice and Strategy," *Journal of Business Economics*, Oct. 1, 1995, pp. 27(7).

[5]   Department of Commerce report, *The Emerging Digital Economy*, Washington, D.C.: U.S. Government Printing Office, 1999.

[6]   Shaw, J., "Future Scenarios for the Telecom Industry: A Ten-Year Forecast," *New Telecom Quarterly*, Austin, TX: Technology Futures Inc., Oct. 1996, pp. 25–33.

[7]   Schiller, B., *The Economy Today*, New York: McGraw-Hill, 2001, pp. 370–377.

[8]   Shapiro, C., and H. Varian, *Information Rules*, Boston: Harvard Business School Press, 1999, pp. 173–182.

[9]   Dent, *op. cit.*

[10] Schiller, *op. cit.,* pp. 134–135. Also review Lindert, P., and T. A. Pugel, *International Economics*, New York: McGraw-Hill, 1998, pp. 251–253.

[11] Buckley, P., and M. Casson, "Models of the Multinational Enterprise," *Journal of International Business Studies*, March 22, 1998, pp. 21(24).

[12] Baron, D. P., "The Nonmarket Strategy System," *Sloan Management Review*, Vol. 37, Sept. 22, 1995, pp. 73(13).

[13] Frost & Sullivan Report, "Telecomms Deregulation to Act as a Spur for Growth in Intelligent Networks Market," M2 PressWIRE, Oct. 1, 1998 (press announcement).

[14] Buckley and Casson, *op. cit.*

[15] Martin, C., "The Strategist," *New Media*, March 1999, p. 20.

[16] Wagner, C. G., "Digital Engine Powers New Economy," *The Futurist*, Bethesda, MD: World Future Society Press, April 1999, p. 20.

[17] For information on new methodologies designed to estimate the value of Internet-based marketing penetration, refer to data culled by industrystandard.com. Also see Golum, R., "E-Commerce Crazy Economics," *Investor's Business Daily*, March 26, 1999, p. 12.

[18] Schwartz, E. I., "Brands Aren't Everything," *The Standard*, May 10, 1999, p. 41.

[19] Brandenburger, A., and B. Nalebuff, *Co-opetition*, New York: Doubleday Press, 1996, pp. 36–39.

[20] *Ibid.*

[21] *Ibid.*

[22] Scott, D. L., "Disintermediation," *Wall Street Words*, Jan. 1, 1998, p. 22.

[23] Wilder, C., "Special Report: Emerging E-Commerce," *Information Week*, Oct. 20, 1997, p. 94.

[24] Weiner, E., and A. Brown, "The New Marketplace," *The Futurist*, Vol. 29, May 1, 1995, pp. 23–29.

[25] Case, K. E., and R. Fair, *Principles of Economics*, fourth edition, Englewood Cliffs, NJ: Prentice-Hall, 1998, pp. 444–458.

[26] Shaw, J. K., "The Emerging Business Economics of Telecommunications," *New Telecom Quarterly*, April 1997, pp. 9–19.

[27] Rosenbush, S., *Telecom Business Opportunities*, Newport, RI: Aegis Press, 1998, pp. 35–68.

[28] Asthana, P., "Jumping the Technology S-Curve," *IEEE Spectrum*, June 1995, pp. 59–54.

[29] For an analysis of the unique characteristics of information as a resource input for production, refer to: Davis, S., and C. Meyer, *Blur: The Speed of Change in the Connected Economy*, Reading, MA: Addison-Wesley Press, 1998, and Economic Intelligence Unit, *Designing Tomorrow's Organization* (special report), DeLoitte, Touche Consulting, 1998. Also review: Tapscott, *Blueprint to the Digital Economy*, New York: McGraw-Hill, 1998, pp. 96–110. Also recommended is: Eisenach, J., T. Lenard, and S. McGonegal, *The Digital Factbook*, Washington, D.C.: Progress & Freedom Foundation, 2000.

12

Electronic Commerce and the Telecommunications Enterprise

The emergence of electronic commerce as a platform for the delivery of communications services signals the reorganization of the modern telecommunications enterprise. Such a reconfiguration is destined to reshape the underlying strategy of the firm, its competitive positioning in markets, and the volatility of the competitive landscape in which it must function. These dynamics, in their collective interaction, will also result in a structural metamorphosis of the firm, triggering innovative attempts to empower employees on the front line of customer relationships. There is some evidence that the most assertive firms have already positioned themselves to exploit the emerging network environment, but there remains little consensus about the most appropriate method of formulating and implementing strategic priorities in this virgin territory.

This chapter discusses the unfolding environment of networked commerce and its impact on strategic planning and the prospective reorganization of most corporate structures. We have learned, through recent studies, that certain preconditions are necessary for effective strategic design. It is also clear that the most successful firms have restructured themselves to respond to market turmoil with nimbleness and flexibility, key ingredients for mounting hypercompetition.

12.1   Networked Commerce, Competition, and Strategic Priorities

Networked commerce—euphemistically defined as e-commerce, e-business, e-retailing—has imposed upon producers a new set of environmental forces. Among these environmental factors are [1]:

- *Interactivity:* effective enterprise nurtures customer relationships with continuous attention and responsiveness;
- *Continuous improvement:* incremental improvement in product innovation is no longer acceptable, as the number of new competitors rises in response to the globalization of competition;
- *"Always-on" competition:* in networked commerce, the sun neither rises nor sets on consumer demand or competitive rivalry, and it is greatly aided by sophisticated multimedia delivery coupled with high-speed architecture;
- *Information maximization:* information proliferates, not depletes, relative to its use—while information is maximized for consumers seeking the optimal choice in this environment, the same is also true for industry rivals tracking the course of competitors;
- *Frictionless transactions:* research confirms that "time" is the most important resource for consumers in this environment—telecommunications firms must therefore create services designed to reduce the time invested in buying and selling (such reductions result in higher productivity and diminishing consumer friction);
- *Ascendancy of IT:* the ability to differentiate product line, enhance customer satisfaction, heighten productivity and time management, reduce friction, and launch future multimedia platforms is increasingly linked to the quality and creativity of the company's information infrastructure and supporting staff.

These broad environmental factors, increasingly transparent, are impelled by the strategies of the most successful firms, and investors are rewarding those enterprises with higher valuations in the capital markets [2]. We are only on the cutting edge of what promises to be a global business revolution in electronic commerce, but these forces are clearly linked to customer loyalty, the ultimate testament to strategic effectiveness.[1]

---

1. The methodology applied in this chapter is regarded by the author as a contentious treatment, but nevertheless represents a starting point for evaluating strategic efficacy. The author operates under the presumption that effective strategic plans result in higher stock valuations over the long term as investors endorse or repudiate a firm's policies.

12.2 Networked Commerce and Strategic Success

A straightforward method of identifying the common attributes of strategic success is to isolate criteria generally accepted as denoting success in the marketplace. There is little question that these criteria include [3]:

- *Sales:* measured in aggregate;
- *Profitability:* that sum generated by a firm after costs have been subtracted from revenues;
- *Productivity:* deduced in monthly, quarterly, yearly increments, and evaluated with respect to organizational and employee output;
- *Customer satisfaction:* ultimately manifested in the willingness of consumers to return to the same company for subsequent purchases;
- *Market share:* that fraction of the market retained by the firm;
- *Sales growth rate:* the percentage a firm's sales are rising, and tied directly to market share;
- *Competitive performance:* a collection of measures intended to benchmark the performance of the company to the firm's most aggressive competitors.

These criteria are the ones typically applied by a firm, and embedded in the planning process, to adjudge its strategic effectiveness. Whether viewed individually or collectively, and regardless of the strategic model adopted, these canons afford a reasonable gauge of strategic congruence with the marketplace. Methodology intended to apply these criteria to the marketplace are discussed below, but we pause to observe an important caveat.

12.3 Telecommunications Strategy: Measuring Corporate Performance in a Networked Environment

Our objective in weighing the criteria outlined above is to determine whether the strategy imposed by the firm is: (1) maximizing the full potential of the company, measured in sales, profitability, and long-term competitiveness; (2) sustaining long-term growth potential; and (3) satisfying the demands of customers so as to attain objectives 1 and 2. Note, however, that these criteria and their methodological evaluation are applied by management itself. The conclusions drawn from this procedure, clearly necessary and essential, are nevertheless judgments formed by strategists—they are not determinations made by the market. For critics of traditional strategic management methods, the ultimate test of strategic efficacy lies in the appraisal of the investor. Do investors think enough of a company (and its accompanying strategy) to invest in it? The answer, argue these observers, lies in a simple assessment of valuation [4].

Market valuation—or capitalization—is computed by multiplying a firm's outstanding shares by their current market (or stock) price. For investors narrowly focused on this criterion to ascertain strategic effectiveness, it is a simple matter to deduce strategic success. A firm's strategy is successful if it results in raising a company's valuation; if its market price falls, the presumption is that strategists have failed in their task. There is good reason to consider this method of evaluating a firm's success. Sales and profitability are clearly linked to an enterprise's value; additionally, this simple method represents a gauge of strategic effectiveness that everyone (inside and outside the firm) can understand. There is nothing arcane or nebulous about the method. There is, too, the larger significance of market valuation: If a firm's stock rises in value, its capacity to raise capital eases, thus resulting in lower costs associated with investment in plant and equipment [5].

Despite the acknowledged simplicity of market valuation as a tool in discerning strategic efficacy, the myopia of this tool limits its usefulness. Strategy is clearly one component in building a firm's market price—it is, however, merely one element. Other factors incite stock valuations: the company's relationship to fluctuations in the business cycle; the movement of funds into and out of capital markets, which can be temporarily unrelated to a firm's capacity to attract customers; the dynamics of new product development outside a firm's own control, but which nevertheless may shift the industry's pool of investment capital; the capriciousness of political, social, or cultural factors that can momentarily influence a stock's market value [6]. In short, a stock's price can be "disconnected" at any given time from a firm's true value—and strategic impact.

For these reasons, applying stock valuation as a measure of strategic success is legitimate only if evaluated over the course of time, relative to the long term. Over the space of several years, however, a firm's strategy should result in higher valuations if it was well conceived and implemented. It is not unreasonable, in this context, to chart a company's stock valuation over the long term so as to determine its effectiveness. The only exception to this judgment is in those instances where a firm explicitly states that its strategic plan is designed to create, attract, or divert market share over a defined duration exceeding 36 months.[2]

---

2. Contemporary strategic planning has experienced a time-horizon contraction in recent years. Twenty years ago, strategists generally regarded a five-year time horizon as long-term strategic

We therefore have two credible, measurable methods of ascertaining strategic success—a benchmarking of criteria as outlined above, accompanied by a serious attempt to weigh the self-interested appraisal of investors. These two approaches to strategic evaluation should be perceived as complementary, but individually inadequate to the task. Given the state of telecommunications strategic management since 1996—in the aftermath of regulatory transformation—it is reasonable now to apply these interlocking methodologies to identify those attributes of firms successful in the competitive landscape.

The empirical evidence leads us to several significant inferences about strategic performance in a networked environment. In recent studies performed by the Federal Reserve Board, the Bureau of Labor Statistics, and the Progress and Freedom Foundation, it has been confirmed that three critical costs of managing the networked enterprise are falling in real terms:

- *Transaction:* the costs of processing orders is diminishing as automated software is employed, typically by customers themselves in networked interaction.

- *Production:* the fusion of automated software, sophisticated information architectures, and robotics are lowering unit and inventory expenses.

- *Distribution:* wireless wide area networks instantaneously determine transmit customer orders, thus facilitating both transactions and production schedules.

Review data assembled in Appendixes C and D to identify these emerging trends in electronic commerce. Most revealing is the remarkable diminution in inventory-to-sales ratios, signifying remarkable new efficiencies in just-in-time (JIT) inventory and labor management. It is now salient that the networked environment alters traditional strategic management by establishing new benchmarked standards in transaction, production, and distribution processes. The most successful firms are exploiting these competitive advantages to the detriment of their slowly adapting rivals. This competitive landscape is unforgiving for those firms unable to meet the standards set by the industry's most aggressive innovators.

---

planning. Short-term strategic planning was generally defined as consisting of a 36-month lifecycle; mid-term planning was widely defined as running from years 3 to 5. With product lifecycles shortened by vast improvements in innovation—some high-technology products have lifespans today of 60 days or less—communications strategic planning has been reformulated to regard three-year planning cycles as "long run" in nature.

12.4   Telecommunications Strategic Management Effectiveness: A Post-Act Assessment

We have defined the contours of contemporary telecommunications competition. Empirical evidence affirms:

- Proliferation of domestic and global competition, with new entrants to market instigated, in part, by advancements in the computer and information technologies industries;

- An increase in new business formations (alliances, affiliations, partnerships, intrapreneurships) promulgated by deregulation and privatization, domestic and global;

- Major improvements in new technologies, impelling prospective substitution of veteran product lines by newly designed alternatives;

- Business reengineering, the reconfiguration of organizational structures to accommodate technological advances, and concurrently reshaping the future of telecommunications consumption;

- Customer fastidiousness regarding pricing, service, and quality, and its resultant effect on product lines and the strategies designed to introduce them.

The debate regarding the effect of the Telecommunications Act of 1996 has been provided in Chapter 2. Whether the impact, positive or negative, has been substantial or minimal, it is clear that the Act has had an influence on business formations and strategic management during the past several years. We have witnessed a heightened expansion in the number of mergers and acquisitions, partnerships, affiliations, intrapreneurships, and alliances following adoption of the statute. Whatever else one might deduce about the law's strategic impact, it has clearly set in motion greater interaction between the telephony, broadcasting, and computer industries. This deduction is not conjecture; it is empirical.

In light of these developments, one may begin a serious consideration about strategic success by gauging the stock performance of companies before and after the imposition of deregulation. There are restrictions on this form of analysis, as discussed below, but this approach represents a first step toward the measurement of strategic efficacy. The following case studies are presented in an attempt to clarify strategic performance and evaluation.

12.5   Telecommunications Strategy and the Five-Force Model

Michael Porter authored the ubiquitous Five-Force Model a quarter-century ago. The model isolates the determinants of industry profitability, and in so doing, facilitates optimal positioning of the firm within its competitive environment. Drawing upon the fields of strategic planning, industrial economics, political economy, and market research, Porter has verified that five variables determine ongoing profitability within an industry (see Figure 12.1): threat of new entrants; industry rivalry; threat of substitutes; buyer power; and supplier power. Market planners throughout the world accept Porter's thesis and prescription for optimal strategic performance.

To the extent that these forces are positive for a firm—that is, that barriers to entry are high, supplier and buyer power are minimal, rivalry is limited, and the substitution effect nonexistent—industry profitability will be high. Thus, potential profitability for a firm rests on these intersecting five determinants. In contrast, to the extent that these forces are negative for an industry (particu-

Competitive Industry Analysis

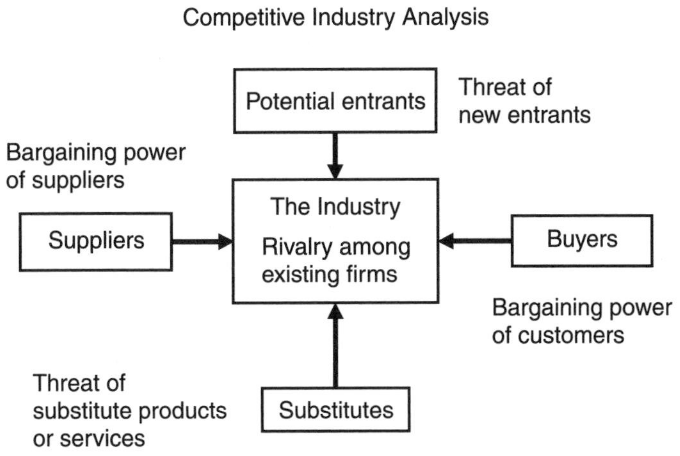

Figure 12.1 The Five-Force Model.
Note: To the extent that these forces are positive for a firm, that is, that barriers to entry are high, supplier and buyer powers are minimal, rivalry is limited, and the substitution effect is nonexistent, industry profitability will be high. Thus, potential profitability for a firm rests on these interactive five forces. In contrast, to the extent that these forces are negative for an industry (particularly with respect to minimized barriers to entry), consumers may expect expansive choice with little pricing power exercised by the firm. Prices, in short, will be constrained by competition. In this sense, "negative" five forces impel firms to compete largely on the basis of price. Such firms must learn to contain costs so as to offer products or services at competitive prices.

For the telecommunications sector, deregulation has meant an expansion of new firms as barriers to entry have fallen. Sector structure and capital availability dictate the varying influence of the other four forces on the constituent elements (broadcasting, telephony, and computer) of the communications industry.

larly with respect to minimized barriers to entry), consumers may expect expansive choice with little pricing power exercised by the firm. Prices, in short, will be constrained by competition. In this sense, "negative" five forces impel firms to compete largely on the basis of price; this implies that such firms must learn to contain costs so as to offer products/services at competitive prices.

We now juxtapose to the Five-Force Model the notion of "competitive positioning." Competitive positioning is defined as the placement of a firm within a 2 × 2 matrix (note Figure 12.1), thus defining its market profile. According to Porter, a firm must define its image in the marketplace and has four available options to do so.

A firm may elect, for example, to define itself in the market as a *broad cost* enterprise. It seeks access to the entire market within its industry and competes largely on the basis of price. The *broad differentiation* position pursues the entire market but seeks to charge a higher price by offering greater value to the consumer (better quality or superior service). The *focus cost* company aims at a particular segment of the market and attempts to compete on the basis of price; to do this, it must restrain costs. Finally, *focus differentiation* positioning holds that a single segment should be served through superior quality or service; such a firm establishes a higher price for this higher level of service. Porter supplies illustrations of successful enterprises in all four cells of this matrix. A firm can be successful in any of these cells, but, according to the author, it must clearly identify its positioning and must not obscure its image by seeking to appeal to all elements of all four consumer constituencies. There are two reasons for this: (1) customers require clarity when confronted with a multiplicity of choices; and (2) a company's labor force becomes confused if the company "fuzzes" its strategic mission and goals. Thus, for purposes of consumer clarity and company focus, a firm must avoid the strategic error of obfuscation—the failure to select one of the four paths to industry success constitutes the most common strategic error of the modern enterprise.

While Porter's model may appear contentious, it is important to note that no leading thinker or critic has successfully refuted the author's assertions. Over and over again, we find mounting empirical evidence that these five forces, and the positioning options that follow, are reaffirmed in competitive strategy. The most serious challenge to Porter's argument lies in the author's assertion that a firm should not alter its competitive positioning more frequently than once every four or five years. In the Internet age, many firms feel compelled to modify strategy as frequently as every year. Porter warns against this but has not provided an addendum to his model to accommodate the hypercompetitive networked environment. Porter contends that his model remains the most effective device in industry analysis, and in a 1999 interview he indicated, ". . . there is no evidence that the Five-Force Model is no longer relevant or effective [for Internet commerce] . . . these five determinants still work."

One's success in applying Porter's model hinges on the ability to grasp his primary two-step model. You must feel comfortable with auditing dynamic influences impinging on each determinant and must also understand the procedure by which the five forces are linked to competitive positioning. We may summarize Porter's key strategic findings:

1. It is possible to be successful (profitable) in any one of the four cells of the competitive positioning model, but a firm must guard against an attempt to simultaneously appeal to all four consumer constituencies.

2. Successful strategies are ones that are consistently implemented over long periods of time.

3. Strategy must change if the industry structure or competitive landscape fundamentally shifts.

4. It takes a long time horizon to formulate, implement, and evaluate effective strategic plans, often in excess of four years.

5. A low-cost strategy must begin with a good product, though it may lack features encased in other competitors' products.

6. Successful "cost competitors" find ways to make a virtue out of the product features that lead to low cost.

7. Differentiated firms (broad or focus) always provide superior service. If they are unable to sustain high quality, they must remove themselves from either of these cells and shift to "cost" positions (either broad or cost).

8. Firms have a right to shift their position when market or internal forces dictate, but cannot do this more frequently than once every four to five years.

"Old Economy" firms continue to adhere to Porter's model, but incumbent and emerging enterprises on the Web find his theory, in many cases, to be less practical vis-à-vis the "New Economy." The issue here is not whether Porter is relevant; there is no hard empirical evidence to discount the credibility of the Five-Force Model. In hypercompetitive markets, however, can the telecommunications firm—large or small—afford to adopt a strategic planning cycle in which fundamental alterations to strategy occur along a four-year planning horizon? Is such "strategic patience" congruent with the volatile communications landscape? There is yet no consensus about this critical developing concern in telecommunications strategic management and the uncertainties promulgated by the networked competitive landscape.

12.6 Telecommunications Strategic Management Performance: The Telephony Sector

Thousands of companies, large and small, comprise the telephony segment of the telecommunications industry. Divisions of this sector are generally grouped into long-distance communications, local or residential communications, wireless communications, telecommunications equipment, professional services, data networking, data services, home networking, internetworking, data conversion, networking markets and management, public data and voice networks, and wireless networking. Many observers also include communications information technology as an additional component of the group. The multiplicity of subdivisions within these groups is growing rapidly, as multimedia, Internet telephony, video services, videoconferencing, and many other technologies are brought to market.

A modest initial attempt to contend with the question of strategic management effectiveness might include a financial survey of publicly held firms whose charge it has been to introduce their products and services in the post–Telecommunications Act of 1996 era. (The reader might wish to examine the Web sites of those major telephony enterprises featured in Appendix C to evaluate firms of professional interest. Customized charts, similar to the ones indicated below, can be designed by the reader to track the strategic performance of any public corporation; refer to the economic sites identified in Appendix C.) Six case studies are examined (two in each sector of the industry—telephony, computer, broadcasting—as defined Chapter 1) in the following sections.

Two case studies are encased in this portion of the chapter. As is the situation with the other case studies examined in this chapter, the author does not assert the examples cited here are emblematic or even necessarily representative of the sector of which they are a part. Instead, the author has identified large capitalized firms that possessed the necessary funds to respond to the strategic imperative of deregulation during the past several years. The reader will note that the stock valuation charts, and attached case descriptions, focus on the period beginning in 1994; in this way, we might compare the strategic performance of each company immediately before and after the imposition of deregulation in February 1996. Four elements combine to form the following strategic evaluation: (1) company summary, elaborating the essential organizational activities and mission of the firm; (2) strategic profile, specifying past and present strategic emphases; (3) financial profile, charting the stock's performance since 1994 (framing a five-year analysis of stock valuation relative to the Dow Jones Industrial Average, and integrating its historical volume of trading); and (4) strategic findings, extracting the lessons of the firm's strategic plans relative to the future direction of the industry [7].

12.6.1 SBC Communications

12.6.1.1 Company Summary

SBC is a local telephone company created as a result of AT&T's breakup during the 1980s (the other divested local telephone companies similarly constructed were Pacific Bell, Bell South, Ameritech, Bell Atlantic, NYNEX, and US WEST). SBC is particularly intriguing because they represent the most aggressive of those firms practicing horizontal integration. SBC engineered the acquisition of Pacific Bell in 1996, extending strategy with a bid for Ameritech in 1998. One can deduce the general stance of this firm with regard to strategy: lock in key acquisitions early during deregulation and build up assets to form the foundation of collateral for comparable future acquisitions. The addition of Ameritech makes SBC the largest local telephony provider in the United States. The firm has not limited this acquisitive philosophy to traditional wireline business, however; it operates wireless services in 10 nations. SBC eschews cultivation of elaborate new communications services in favor of solid management of traditional products. It is ambitious in its market planning, but is austere with regard to its palate of services. SBC is more concerned about being "first to market acceptance" as opposed to "first to market." It is a company that avoids incurring the costs associated with new product development, instead preferring to learn from the experience of other providers before incorporating new services.

12.6.1.2 Strategic Profile

SBC is a provider of local telephony services to nearly 40 million customers throughout California, Texas, Arkansas, Connecticut, Kansas, Missouri, Nevada, and Oklahoma. It also serves some 7 million wireless consumers in more than 15 states, complementing voice communications with paging services. SBC operates in 10 nations and is planning to penetrate additional international markets in the near future. Its expansionary strategy was established from the first moment telecommunications deregulation was inaugurated: An offer to acquire Pacific Bell was accepted within weeks following passage of the statute. This was the first major acquisition of a major telecommunications provider by a direct competitor, and prompted the FCC to carefully review the pattern of mergers and acquisitions to follow. Indeed, the Pacific Bell negotiations signaled a simple but aggressive effort to expand market share through competitor absorption rather than innovative product introduction. In this way, SBC has been able to leverage operational costs. After acquiring its competitors, SBC embarked immediately on an internal strategy to reduce administrative costs by stripping personnel redundancies and organizational overlap. It has also carefully cultivated contacts in state and local governments, lobbying effectively to strengthen its image in communities. It has generally been successful in this endeavor.

12.6.1.3 Financial Profile

SBC's aggressive strategy, measured in market terms, compares most favorably with its peer competitors. Unlike many of the Baby Bells, as well as AT&T, SBC's stock performance has been reasonably congruent with broader market averages since 1998. By avoiding the sometimes excessive costs associated with "first to market" technologies and adhering strictly to proven market lines while expanding its customer base, SBC has goaded investors into an affirmation of its strategy. SBC is not glamorous, but it is widely regarded as a dependable provider, and fierce competitor, of standard telephony services. Figure 12.2 shows the financial profile of SBC Corporation from 1995–2000.

12.6.1.4 Strategic Findings

We have in SBC a rugged competitor and acquisitive strategist. This is a firm that will seize upon any opportunity to expand its customer base and market share through horizontal integration. It is also corporation managed in a rather traditional organizational ambiance; centralized decision making at the top of its pyramid is principally responsible for its strategic positioning. If the price is justifiable, and a potential competitor can be absorbed through effective leveraging, one can anticipate its strategy. The specter of a firm braced for competition through the application of simple business principles—a path that some dismiss given the risks attendant in a dazzling, dynamic industry—invites serious contemplation regarding its strategic design. Whatever its effect on the competition, SBC in recent years affirms that traditional business strategy has not yet faded from view.

12.6.2 AT&T Corporation

12.6.2.1 Company Summary

AT&T is the nation's largest market capitalized firm. In the aftermath of divestiture after 1984, the firm groped for an appropriate market strategy, ini-

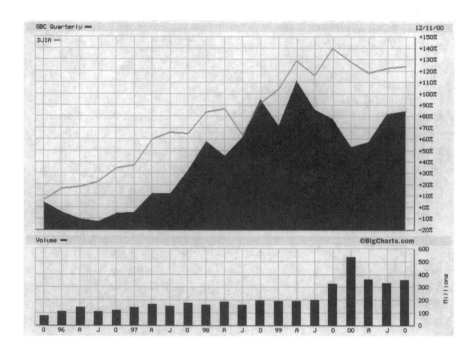

Figure 12.2 SBC Corporation, 1995–2000. (*Source:* BigCharts.com, December 2000.)

tially emphasizing company spin-offs fused with diversified penetration in a multiplicity of communications markets. During the years immediately following deregulation, the corporation pursued a policy of acquiring telecommunications firms whose assets maximized perceived potential synergies. In essence, the company has embarked on an aggressive acquisition strategy aimed at getting access to as many customers as possible. The company acquired TCI in 1997; this purchase, with its recent acquisition of MediaOne, makes AT&T the nation's leading cable operator. We have in this firm, therefore, the largest long-distance provider with a second line to entering the homes of approximately 100 million consumers. No other firm in the industry can make such a claim. An interesting irony, or perhaps paradox, in the strategic development of AT&T concerns the fact that it can now theoretically penetrate the home or business of virtually every American with a greater multiplicity of products than during its most prominent period as a regulated monopoly.

12.6.2.2 Strategic Profile

AT&T had lacked a clear, cohesive market strategy prior to 1996, although it had committed to experimenting with entering new markets. Its management of strategy since 1996 is salient and unambiguous: It aims to secure the largest conceivable market base in the short term—based on wired telephony and cable—then introduce value-added services based on bundled packages. This is a costly investment, as the corporation has had to take on extensive debt to accumulate the firms necessary to seize desired market share. One might refer to this strategy as one based on horizontal integration, the obvious motive an attack on as many high-end value users (observers refer to such practices as "cherry picking" the market) as feasible in the short term. In short, they seek access to as many customers prepared to pay high fees for variety of services— telephony, cable, Internet, wireless—as possible. Customer retention will be especially vital for this firm, as interest costs on debt will be significant for years to come. It is clear that AT&T regarded passage of the 1996 statute as a clarion call for aggressive acquisition strategy.

12.6.2.3 Financial Profile

Since 1996 the market has adopted a "wait-and-see" attitude with regard to AT&T's recent assertive strategy. One clear concern of investors has been the wisdom of pursuing so expensive a strategy; the return-on-investment (ROI) will have to be substantial to absorb interest costs. If ROI is not significant, the company would be limited in pursuing other strategies. In effect, the company is banking its future existence on a single strategy, albeit the most impressive wireline connection to consumers in industry history. Figure 12.3 reveals the market assessment of AT&T's strategy, and it reflects an inconclusive reaction. AT&T's stock price lagged broader market averages, both within and outside the telecommunications sector. This is understandable, in part, because AT&T's strategy is clearly focused on the long term. If its strategic priorities achieve the intended result, profitability will not be enhanced for some years to come, and in a mature corporation such a strategy is not rewarded with investor confidence until such profits are recorded. Nevertheless, the lag in stock profitability suggests that the market questions the wisdom of a one-dimensional strategy that could be rendered ineffective by technological change capable to diverting consumers to wireless substitutes in telephony, broadcasting, and Internet usage.

12.6.2.4 Strategic Findings

AT&T's long-term strategic focus on the acquisition of customers through traditional wireline access signifies that category of communicators providers which rely on horizontal integration to "grow" the business. A cutting-edge debate among telecommunications strategists today concerns strategic positioning and its efficacy. We note a gradual shift in strategic emphasis following deregulation: How does a firm place itself relative to others with regard to the philosophy of accumulating customers? If it is acquisitive, it asserts financial

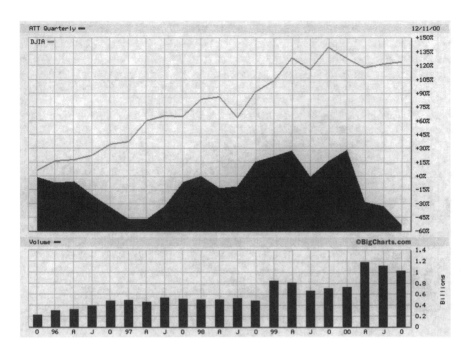

Figure 12.3 AT&T Corporation, 1995–2000. (*Source:* BigCharts.com, December 2000.)

authority to acquire other firms—adapting horizontal integration—to mini-mize competition while growing market share; if it is interested in securing predictable supplies or reducing the cost of production, it might be inclined to pursue a vertical integration strategy. In the case of AT&T, it is presumed by the company that if it maximizes its available customer base through acquisitions, it can sustain and increase market share. It presumes, indeed, that customer loyalty can be built based on quality and reputation. For AT&T, this strategy remains an untested hypothesis, and will require some years to affirm or reject.

12.7 Telecommunications Strategic Management Performance: The Computer Sector

Some observers might contest the appropriateness of including the computer industry as a subcomponent of the communications industry. We have outlined the principal reasons behind this inclusion in Chapters 1 and 2. The computer industry has progressively involved itself in telecommunications during the past decade because it has evolved from an information tool to a communications tool. The multiplicity of features now standard in home computers, for example, has induced households to increase their consumption of telephony and broadcasting services. The elimination of cross-ownership restrictions has invited telephony and broadcasting firms to negotiate alliances and partnerships with firms in computer hardware and software. The proliferation of content providers in the Internet domain has encouraged interaction among all three groups. The list of factors instigating interaction between telephony, broadcasting, and computer firms is lengthy and growing rapidly [8].

There is wide agreement today that an important field of specialization within the telecommunications industry is information technology [9]. Advances in software are driving improvements in communications products and services; they are also speeding organizational changes within other sectors of the economy, thereby encouraging vast prospective demand for sophisticated communications. We might divide information technologies along these lines to gauge its future impact on telephony and broadcasting for both businesses and households:

- Advanced intelligent networks (AINs);
- Data warehousing and mining;
- Geographic information systems;
- Network management hardware and software;
- Object-oriented technology;
- Billing network systems;
- Fault tolerance and clustering;
- Transaction processing software;
- Data networking components;
- Digital signal processing (DSPs);
- Internet telephony;
- Fiber optic transmission technologies;
- Wireless ASICs;
- Broadband multimedia support systems;
- Web hosting services;
- Extranet support systems;
- Intranet support systems.

The first half of this list directly impinges on the management of telecommunications firms themselves; the latter half pertains currently or prospectively to the Internet and the growing significance of electronic commerce. The influence of electronic commerce on strategic management is explored in the following chapter, but it is now salient that the computer industry is inspiring much turbulence in the telephony and broadcasting sectors. Indeed, advances in hardware and software—industries untouched by government regulation—are responsible for transforming the method by which services are developed in their regulated counterparts. By extension, this simple logic means the computer sector is transforming strategic management in telephony and broadcasting. The computer is indeed transforming the nature, design, implementation, and evaluation of strategic plans in all industries. We explore now strategic management practices in two leading computer firms, and their role in inciting changes elsewhere in the communications industry.

12.7.1 Intel Corporation

12.7.1.1 Company Summary

Intel is the premiere microprocessor maker in the world, controlling in excess of 80% of the market. The firm is noted for emphasizing continuous product improvement and rewarding its employees in furthering innovation. During the past two decades, the firm has pursued development of microcontrollers, networking products, and flash memories. Its products are regarded as state of the art and sustain targeted market share. Although Intel's earnings have recently been uneven, largely because of the decline in the global economy during the late 1990s, the company nevertheless outperforms its competitors in sales, market share, productivity, and innovation. The company has embarked on a recent strategy to encourage entrepreneurial ventures in every area of the private sector; it believes this approach will assist in sustaining demand for its products in both small businesses and households. Key buyers of Intel chips are Dell Computer and Compaq; the fate of these firms is tied directly to Intel's self-interest. Early in the 1990s, Intel had made a number of projections pertaining to productivity. Intel had contended that the economy would experience a rising tide of productivity (labor efficiency) as a result of the full flowering of the computer industry. As noted in Chapter 2, recent empirical evidence has begun to support this lofty claim. During the past three years, the corporation has explored new ways to spur development of electronic commerce, both domestically and internationally.

12.7.1.2 Strategic Profile

As a firm committed to continuously growing its market through product improvement, Intel has few peers. The historic presumption has been that its open organizational style, system of rewards and incentives, and aggressive multinational development have been principally responsible for this growth. These factors may constitute significant variables in its development, but the firm's mission is clearly congruent with the destiny of an economy that pivots on information, knowledge, and organizational intelligence. Twenty years ago, Intel could hardly conceive that its product would be as important in driving the communications industry as the indigenous innovations promulgated by the sector itself. It would not be literally accurate to infer that Intel is a communications provider, but its contributions in information technology have altered telecommunications strategy. There is little doubt that Intel will play a significant part in advancing Internet, multimedia, e-commerce, and broadband services, and the company will be pivotal in defining those products in computer software upon which telecommunications depends. The long-term prognosis for Intel is very favorable, provided its strategy is congruent with the future contours of the global economy. If its products are characterized by continuous value-added quality, it seems unlikely that its global competitors will substantially penetrate its overwhelming market share.

12.7.1.3 Financial Profile

In reviewing Intel's chart (Figure 12.4), we note that it has experienced exaggerated swings in demand for its product. These movements in stock price reflect the link between the company's product and prevailing economic demand. In other words, this is a firm that must seriously study fluctuations in the business cycle so as to refine its production scheduling and inventory management. Note in Figure 12.4 that the stock consistently outperforms the broader market average—a hallmark of a consistently successful firm. Implicit in these data, however, is volatility in volume and valuation, which signifies the need to develop strategy consistent with prevailing market demand. How does Intel handle the challenge of strategic fit—ensuring that its strategy exploits market potential without resulting in excessive or unneeded supply of its product? They have adopted remarkably effective forecasting methods to estimate variations in the economy, consumer demand, and competitor behavior. Their methodological framework for the design of strategy, in this regard, has resulted in one of the great high-technology success stories in the post–World War II era. Note, too, that the firm has been even more successful in the years following deregulation than in the era that preceded adoption of the statute. Deregulation of the communications industry has benefited Intel by encouraging dynamic interplay between telephony, broadcasting, and computer and information technology.

12.7.1.4 Strategic Findings

The evidence suggests that Intel has excelled in three critical endeavors: (1) framing strategy consistent with generating products congruent with market

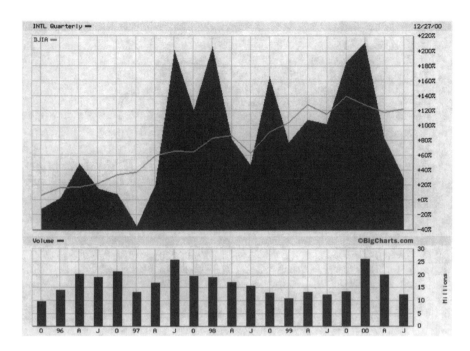

Figure 12.4 Intel Corporation, 1995–2000. (*Source:* BigCharts.com, December 2000.)

demand; (2) successfully gearing production scheduling to effective forecasts; and (3) encouraging consumers and businesses to buy its products by emboldening them—through counsel, technical assistance, and financial support—to pursue entrepreneurial endeavors. This is a proactive, assertive firm that relies chiefly on its internal resources to generate growth rather than pursuing an acquisitive policy. Its strategy is to pursue high value-added products that simply cannot be replicated by its competition at any given moment in the production cycle.

12.7.2   Microsoft Corporation

12.7.2.1   Company Summary

Microsoft's meteoric success as the global leader in software products has been reinforced recently by a strategy focused on diversification and entry to new markets. With a market capitalization that now makes it the world's wealthiest company, the firm has placed itself in the position of product developer and acquisitive operator. The corporation has pursued new markets in broadcasting, Internet and multimedia services, cable television, and a host of other communications enterprises. It has actively pursued strategic partnerships, affiliations, and alliances with cable television firms, broadcasters, content providers, and a wide array of other firms. It seizes market opportunities in virtually every communications domain, apparently concluding that it cannot pick winners and losers with precision, but can, nevertheless, participate in every growing ancillary market. Equally apparent is the hostility it is now drawing from firms both within and outside its own domain. For this reason, a constant strategic concern involves antitrust action by the Justice Department (a contingency that lobbyists for other firms pursue with some vigor). Critics charge that Microsoft is less an innovator than a copier of competitor products. The fact is that Microsoft is pragmatic, pursuing every opportunity to penetrate new markets. As a result, some observers express concern that the company may lose its strategic focus—that the firm may abandon its original mission as software developer. If an observer makes a serious effort to distill the company's mission today, Microsoft is interested in pursuing any market aimed at creating, transmitting, interpreting, applying, processing, or storing information. This broader strategic focus—beyond that contemplated by any other software manufacturer—makes Microsoft a communications provider in the fullest sense of that term. One should not dismiss Microsoft's potential pursuit of any market within the telephony, broadcasting, or computer sectors.

12.7.2.2   Strategic Profile

Microsoft's ventures in Internet content and provision, cable television, satellite communications, and a host of related industries define Microsoft as a modern telecommunications corporation. There is no example of business formation—acquisition, alliance, partnership, affiliation, intrapreneurship—that it has not exploited during the past decade. Business synergies, creating cross-ownership opportunities in telephony, broadcasting, and cable television—ignited after 1996—have been pursued with vigor. What is remarkable about the proliferation of business alliances that the firm has entertained is the fact that it could have pursued many more market opportunities in recent years. The company could have penetrated banking, electronic commerce, publishing, and a wide array of other industries if not intimidated by prospective antitrust administration. Bill Gates, the firm's CEO, has had to defend his corporate strategy before numerous congressional inquiries in 1997 and 1998.[3] This is

---

3. At the time of this writing, the Justice Department's suit against Microsoft has not been resolved. Viewed in terms of its implications for the entire telecommunications industry, the final dispensation of this suit extends beyond Microsoft's relationship with its immediate competitors. An unfavorable judicial outcome for Microsoft is likely to act as a precedent for prospective antitrust administration in those cases of industry consolidation perceived as restraining potential competition. A favorable outcome for Microsoft may signal major acqui-

a firm whose acquisitive instinct has been tempered, for better or worse, by government prosecution. The merits of the government's antitrust case are not entertained here, but there is an obvious strategic footnote that all communications firms and related high-technology industries should contemplate: the effect of antitrust application on the competitive landscape. Should a firm factor such government activity into its strategic formulation? Should a firm renounce potential market opportunities in an effort to placate government regulators? These questions cannot be addressed in the absence of assembling empirical evidence in the years to come. It is instructive that firms of comparable characteristics in other parts of the globe feel no such reticence in the current climate.

12.7.2.3 Financial Profile

Microsoft's market success has been unrivaled during the past 20 years. Given its number one market capitalization position, its control of in excess of 90% of its market, its global dominance, and its tertiary influence on all other industries, it is not unreasonable to contend that this is the most important corporation in the world. As we begin the new millennium, it is conceivable that Microsoft could double in market capitalization every three to four years. Note in Figure 12.5 that Microsoft's stock valuation has greatly exceeded market averages in the DJIA and NASDAQ. Its pattern of trading volume also certifies that it attracts willing investors on a perpetual basis. It stock price is high but investors continually reward the firm with a plentiful supply of new funds; this is a firm able to leverage its market capitalization to secure funds at minimal costs. In short, Microsoft can pursue its multilevel strategy of acquisition and diversification without fear of enduring inadequate future capital.

12.7.2.4 Strategic Findings

Microsoft is a multilayered communications and information-technology company. It is in the most enviable position of all communications entities, save the prospective intervention of government sanctions. Its acquisitive strategy is likely to produce market winners and losers with respect to new product introduction, but the firm has also assured itself that there is no attractive telecommunications market of which it will not be a part. Such is the fortune of the most significant global corporation in 2001. If antitrust policies intervene to influence strategy, Microsoft can take comfort that the vigor of such enforcement will vary from administration to administration. Microsoft emerges in this era with optimal strategies stressing "first to market" and "first

---

sition efforts by this and other well-capitalized computer firms seeking entry to the broadcasting and telephony sectors.

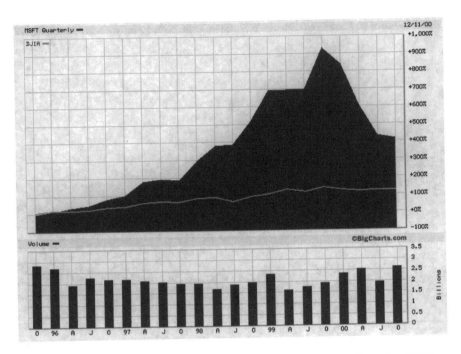

Figure 12.5 Microsoft Corporation, 1995–2000. (*Source:* BigCharts.com, December 2000.)

to market acceptance" philosophy depending on the vagaries of each market opportunity.

12.8 Telecommunications Strategic Management Performance: The Broadcasting Sector

The broadcasting industry has been reconstituted in recent years. It is no longer occupied by competitors exclusively focused on television and radio, or even on wireless platforms. The encroachment of the Internet, and its expansive array of new multimedia services, promises industry change on two key fronts: (1) content development, with increased emphasis on customized information and knowledge designed for the unique needs of individuals, households, and businesses; (2) network delivery, with a striking multiplicity in the number of alternative methods to organize, distribute, archive, and retrieve specialized programming.

If technology facilitates a growing range of alternative providers, this development surely portends transformation of content development. Again, the computer industry has been decisive in spurring this metamorphosis. Innovation in information technologies has led to value-added improvement in the quality of deliverable knowledge; the thirst for such knowledge—in education, health, entertainment, business, and government—further incites consumer demand. The distillation of knowledge—information that can be applied to improving the quality of life or material well-being of customers—confers upon the user comparative advantage. Those who possess state-of-the-art information hold the upper hand in all forms of endeavor. This factor ensures that the broadcasting industry will move in the direction of customization, value-added programming for both domestic and global consumers. Content development will reconfigure network delivery in radio, television, satellite, Internet, and other emerging forms of delivery; improvements in network delivery will inevitably enlarge both the number of providers and prospective demand for such services. It is in this environment that we explore the dramatic strategic transformation of the broadcasting sector. Two examples of contemporary strategic management in the industry are now explored.

12.8.1 Disney Corporation

12.8.1.1 Company Summary

In market capitalization, the Walt Disney Corporation is the second leading media firm in the world. The company is involved in television and film production, Internet commerce, the publishing business, amusement parks, transportation, and professional sports. It is at once both a multinational corporation and growing conglomerate. Its ventures in communications, now pronounced with the recent acquisition of ABC/Capital Cities, makes it a broadcaster in the most profound sense of the world. The Disney brand has been cultivated and refined over many decades, and the firm welcomes any project that builds and enhances its global prominence. Disney is both innovative and acquisitive, originating new broadcasting product but also active in seeking out prospective alliances and affiliations that leverage its brand name to greatest effect. The result is that the company has generally exceeded market averages in valuation throughout its history, save recent turbulence in stock trading. The firm maintains a diversified product and service line but is sensitive to movements in the business cycle, particularly with regard to global fluctuations in consumer income.

12.8.1.2 Strategic Profile

Strategically, Disney, as a broadcaster, has attempted to penetrate television and Internet industries by developing its own content while simultaneously purchasing networks. It has seized upon its forecasts of changes in consumer preference by diversifying its content; the firm is not reactive to change but is definitely vigorous in its pursuit of new markets. The firm immediately responded to the deregulation of the communications industry, for example, by intervening to purchase ABC Network before other competitors could prepare bids. Disney has little difficulty raising capital in new stock offerings or entering the financial markets for an infusion of new funds. Some competitors in the broadcasting sector therefore view Disney as a predator in the marketplace.

12.8.1.3 Financial Profile

During periods of general economic prosperity, Disney excels in outperforming market averages. As a member of the Dow Jones Industrial Average (a select group of 30 of the nation's largest firms), Disney's performance is directly factored into, and responsible for, significant variation in the DJIA. Note in Figure 12.6 that Disney's acquisition of ABC in the mid-1990s was well received on Wall Street, and widely regarded as a harbinger of similar broadcasting acquisitions in the years to come. Such acquisitions provide the network platform for its growing content interests. Recently, the firm has had to digest a series of lawsuits, a decline in attendance at its global amusement parks, along with concurrent decline in the global economy. This last factor reaffirms the importance of the business cycle relative to consumer demand for Disney products. Within the United States, however, Disney's production, broadcasting, and Internet-based businesses have performed well. The decline of Disney's stock valuation in 1998 and 1999 reflects these recent problems, but its long-term strategy remains intact. Most observers regard current financial difficulties as transient, short-term phenomena.

12.8.1.4 Strategic Findings

Disney's success in the broadcasting sector has been replicated by few rivals. As was the case with Microsoft in its sector of communications, Disney trades on a powerful brand name, diversified product line, and acquisitive tactics. It has succeeded in penetrating every potential market targeted for experimentation and maintains the financial leverage to do so at reasonable scale economies. In short, it is expansive without being inefficient. Disney's experience in broadcasting signifies an emerging trend in communications, for those fortunate enough to emulate or adapt the appropriate strategy: when content is developed, designate new networks necessary to bring it to market. Ignore no alliance, affiliation, partnership, or acquisition that advances this objective before the competition can react.

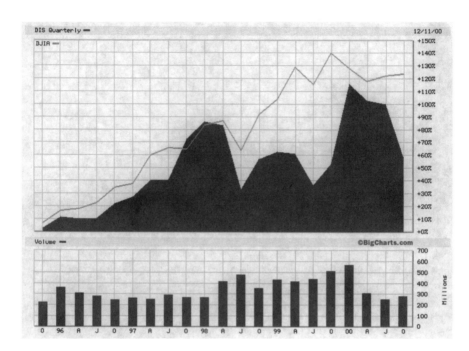

Figure 12.6 Disney Corporation, 1995–2000. (*Source:* BigCharts.com, December 2000.)

12.8.2 America Online

12.8.2.1 Company Summary

This Internet provider is also a broadcaster. America Online (AOL) has moved from its position as a link to the "information superhighway" to a full-fledged innovator of content. It is the organization, design, and reformulation of content that attracted customers to its service. In an industry of numerous providers formed at the beginning of the World Wide Web, America Online has managed to differentiate its service, build a brand name, and generate a subscriber base approaching 20 million. No company in this sector compares favorably with this performance.

A key element of the company's strategy has been the relentless pursuit of business affiliations and alliances intended to generate value-added content for its subscribers, and minimal cash is diverted to this task. In other words, AOL has successfully built economies of scale—the cost of producing content has fallen over the long term—resulting in an expansive array of services that no competitor can match. AOL has complemented this strategy with a sustained emphasis on vertical and horizontal acquisition. The firm has purchased its direct competitors (Compuserve is a recent acquisition), while acquiring network and browser operators (Netscape was acquired in 1999, for example). Continuous improvements in infrastructure management, fused with content development, are the synergies defining the early evolution of this firm.

12.8.2.2 Strategic Profile

AOL is one of the few major Internet firms presently recording profit—an extraordinary development in light of the fact that the company must necessarily plow all available profit back into network management and content development. In the early stage of a young company operating in a seminal industry, it is unreasonable to anticipate profitability. Yet, AOL's strategy of using other individuals and firms to generate content under its brand name—thus assuring value-added performance at minimal cost—has proved remarkably prescient. We have seen in this firm the development of a new business model, a model discussed at greater length in the final chapter of this text. This business model establishes a precedent that few firms have begun to understand in the advent of electronic commerce, let alone imitate. The model embraces four interactive tactics:

- *Brand name development:* substantially cultivated through business alliances forged to exploit complementary attributes of other firms;
- *Economies of scale:* reducing costs of managing the business concurrent with improved product, and resulting in rising profitability;
- *Proprietary technologies:* exploiting state-of-the-art hardware and software, differentiating itself from the competition and restraining competitor encroachment on its market share;
- *Value-added focus:* continuously improving the applicability and customization of services available to consumers.

Whether America Online conceived of the criticality of these tactics early in its development can be debated. What cannot be refuted or dismissed is the success that has been realized through these strategic priorities. AOL's growth has not been limited to the United States. The firm has been zealous in its penetration of international markets, as well.

12.8.2.3 Financial Profile

AOL's financial performance has been extraordinary since its inception in 1990. The firm has grown by a factor of 15,000% in less than a decade, a feat unequaled by its competitors. No company in the telecommunications industry during the past decade has come close to AOL with respect to generating customers (reach) and sustaining that base (loyalty). It has thus outperformed every market average, every direct and potential competitor, and is thus a broadcaster in the literal sense of that term. Figure 12.7 shows the financial profile of America Online Corporation from 1995–2000.

12.8.2.4 Strategic Findings

AOL literally broadcasts content in every form—data, audio, and video—and continually reformulates information and knowledge to expand its customer base. Its financial profile reflects the dynamism of a firm that has not merely stumbled upon, but exploited, the emerging economics of the telecommunications industry. The four tactics defined above, when applied to sustaining customer reach and loyalty, result in a winning financial formula. The business model designed by America Online will draw many future imitators in the telecommunications industry.

12.9　Networked Commerce and Strategic Assessment

Six case studies have been assembled in this material, each representative of a company coping effectively with extraordinary strategic risks and uncertainties.

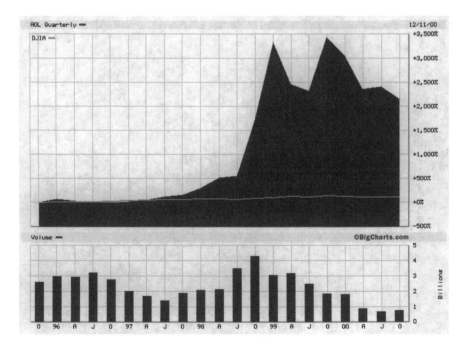

Figure 12.7 America Online Corporation, 1995–2000. (*Source:* BigCharts.com, December 2000.)

We discern in these profiles important lessons about the design, monitoring, and evaluation of successful strategy. Although these cases reflect different strategic experiences in different sectors of the industry, we can reasonably assert these common attributes of modern strategic management:

- Successful firms in their respective sectors do not react to change; they embrace it and seek new avenues to add value to their product.

- Successful firms in the newly defined competitive landscape have learned to negotiate affiliations, alliances, and partnerships that either improve their product or create new distribution routes to consumers.

- Successful firms exploit the Internet as a new distribution channel to consumers; they do not attempt to supplant this competitive threat through traditional strategic design.

- Successful firms make every effort to reinforce brand image and management, now more important than ever.

- Successful firms have learned to exploit scale economies, critical in strengthening profitability.

- Successful firms develop proprietary technologies or products, where feasible, such that their competitors' efforts at replication are forestalled.

- Successful firms adopt either one of these strategic philosophies: (1) they define themselves as "first-to-market" product developers; or (2) they emphasize a "first-to-market-acceptance" image clearly recognizable to employees, customers, and investors.

- Successful firms "obsolete" their own product before the competition does it for them.

- Successful firms have learned the most important lesson of contemporary telecommunications economics: Reach (customer base) loyalty (customer allegiance) rewards providers and content developers, and lays the foundation for future growth.

- Successful firms exploit organizational synergies to enhance value and growth rates by cooperating with any firm, including competitors, that assures this contingency.

These strategic commonalities signify a serious attempt to contend with the uncertainties and vicissitudes of networked environments and pending changes in consumer preferences.

References

[1] Turban, E., J. Lee, D. King, and H. M. Chung, *Electronic Commerce: A Managerial Perspective*, Upper Saddle River, NJ: Prentice-Hall, 2000, pp. 25–39.

[2] For additional background in current issues confronting strategic planners, refer to Wright, P., M. Kroll, and J. Parnell, *Strategic Management: Concepts and Cases*, Upper Saddle River, NJ: Prentice-Hall, 1998. Other recommended titles include: Tapscott, D., *Digital Capital*, Boston: Harvard Business School Press, 2000; Thompson, A., and A. J. Strickland, *Strategic Management*, Upper Saddle River, NJ: Irwin Press, 2000.

[3] These benchmarked standards are traditionally applied in measuring the firm's performance. For an introduction to the advantages and pitfalls of applying this tool, refer to Porter, M., *Competitive Strategy*, New York: The Free Press, 1998, pp. 126–155. Also refer to the Web sites listed under management-specific functions in Appendix C.

[4] Fahey and Randall, *op. cit.*, pp. 171–194.

[5] Bernstein, P. L., (ed.), *The Portable MBA in Investment*, New York: John Wiley & Sons, 1995, pp. 111–197.

[6] *Ibid.*, pp. 308–342.

[7] The author has culled primary data on financial and industrial developments from these sources: IndustryStandard.com; CompanySleuth.com; Hoovers.com; and BigCharts.com. See Appendix C for related Web sites on charting the performance of stocks.

[8] Shaw, J., *Telecommunications Deregulation*, Norwood, MA: Artech House, 1998, pp. 38–48. See also Heldman, P., R. Heldman, and T. Bystrzycki, *Competitive Telecommunications*, New York: McGraw-Hill, 1997, for additional information the competitive implications of converging technologies. Also refer to: Matsuura, J. H., *A Manager's Guide to the Law and Economics of Data Networks*, Norwood, MA: Artech House, 2000.

[9] Shaw, *op. cit.*

13

Deregulation and Hypercompetitive Markets: Managing Customer Reach and Loyalty

Communications strategists confront a baffling array of strategic priorities in the emerging competitive landscape. Superimposed over the daunting goal of designing competitive strategy is a watershed era of economic, social, political, and technological volatility. The macroeconomy is reconfiguring to accommodate the technological innovations introduced by the telecommunications industry. This chapter elaborates on those developments that are unfolding as decisive in differentiating future competitive performance, and it also projects methods of managing a customer base amid the era of hypercompetition. Deregulation has been a prime, but not exclusive, variable in triggering a hypercompetitive environment.

13.1 The Advent of the Hypercompetitive Market

Some consideration of economic analysis, particularly with respect to the coming decade, is essential to defining a firm's strategic options during this period. A recapitulation of the forecasts outlined earlier in this volume asserts the following (review Appendixes D and E):

- The U.S. economy is likely to grow substantially between 2000 and 2010, based on technological innovation fused with positive demographics, and resulting in mounting productivity. The world economy is now poised for significant economic growth, beginning in 2001 and extending through the same period. The engine of growth will be principally technological, with people only now learning to exploit the intrinsic value of information. This positive economic environment reflects the growing significance of communications and information technologies. Information cannot be depleted; unlike all other resources, it increases proportionately, indeed geometrically, relative to its use.

- The economy is readjusting to reward those who learn to exploit and leverage information and communications technologies. Competitive advantage is conferred upon those who speed products to market without sacrificing customer responsiveness or sensitivity.

- Mergers, acquisitions, affiliations, alliances, partnerships, and intrapreneurships will proliferate and cross national boundaries. No firm can control its own fate, except in those instances of proprietary technologies. More importantly, no company retains full control over infrastructure, distribution, or content. Public monopolies are about to end in the form to which we were accustomed, and will be replaced by a mixture of competitive, oligopolistic, and monopolistic competitive industrial market structures.

- Increasingly, electronic commerce or "e-business" will be the platform by which firms interact with suppliers, distributors, customers, even industry rivals. Telecommunications companies that cannot improve the value-added component of their services at a rate comparable to their most aggressive competitors will wither in this competitive landscape. Value-added improvements in supply-chain management, production, transaction, inventory, and customer service processes will be the initial beneficiaries of this transformation.

- The telephony, broadcasting, and computer industries will continue to intersect, with principal strategic synergies driven by software development. The computer sector will come to predominate in influence over its telephony and broadcasting cousins.

- New business formations in other nations will intensify in frequency, complexity, and reconfiguration. The wave of deregulation and privatization typified in the United States after 1996 is likely to be replicated elsewhere, with cross-national collaboration strengthening as a result of implementation of key provisions of the 1997 World Trade Organization agreement.

- Demand for U.S. telecommunications equipment exports (equipment and services) will rise substantially after 2001, as the global economy recovers and it becomes increasingly apparent that competitive advantage is conferred upon those able to achieve first-mover advantage.

- Productivity will rise significantly after 2001, resulting in higher wages and profitability, and will instigate continuing demand for communication services. Inadequate skilled labor in the communications industry will simultaneously serve as a stimulus to higher wages within this sector; for communications firms, a critical managerial issue will become the recruitment, retention, and continuous retraining of skilled personnel.

- Regional trade (multinational) agreements will combine with bilateral (country-to-country) arrangements to forge dramatic demand for communications products in both developed and developing nations. Command-control nations, economies in which government outlines the goals and means of production, will not benefit from these improvements in technological innovation. The empirical evidence suggests that those nations committed to relaxing trade restrictions or sanctions (tariffs, quotas, and regulations) will profit from most of the coming innovations introduced by both domestic and global telecommunications enterprises. Liberalized trade will define the future of telecommunications imports and exports.

- The demand exerted elsewhere for U.S. communications services of all kinds will rise faster than the demand for physical products, resulting in a boom for telecommunications management consulting. A substantial fraction of telecommunications equipment will be produced in other nations. U.S. communications services will be provided almost exclusively by Americans themselves. All U.S. and European telecommunications firms should prime themselves for emerging demand for these value-added services, particularly those centered on customer service.

These are the broad contextual developments about to assert their influence. Their blast effect will be felt very shortly. We refer to the collective impact of these forces as hypercompetition, a landscape characterized by intensified rivalry, resulting in the ascendancy of the consumer in guiding industry production and distribution. New products and services are sped to market in this environment, spurred by globalized, networked interaction.

We discern common attributes in the most successful communications firms currently designing strategy. Successful firms:

- Speed product to market; listen and respond to customers;
- Benchmark their performance against the best global—not domestic—standards;
- Elect and adhere to one of two product tactics: first-to-market or first-to-market-acceptance;
- Embrace continuous improvement in product design, distribution, and post-purchase satisfaction;
- Design and introduce customized product menus;
- Learn and apply the precepts of tertiary analysis and their role in projecting future demand;
- Exploit relational marketing as a component of strategic management.

13.2 Telecommunications and E-Business

It is now agreed that the commercial application of networked information and knowledge will profoundly redefine the domestic and international economies. Those sufficiently prescient to anticipate the exponential growth of electronic commerce have already profited from the transition. The data in Appendixes B and C reveal what is increasingly obvious in this regard. What is less apparent is the importance of e-commerce in reshaping relationships between employers and employees; between consumers and producers; between suppliers, distributors, wholesalers, and retailers; and between governments and citizens. We have already asserted empirically the unfolding phenomenon known as disintermediation, in which individuals and firms are already being removed from the distribution chain.

Viewed strictly in competitive terms, we are beginning to divine essential strategic principles governing effective commercial use of the Internet:

- Most firms will be compelled to cultivate narrow market niches in this competitive environment.
- Raw information or data offers little value to consumers in this environment: it must be distilled in a form that can be applied to productive decision-making. Strategy in firms will be redirected to attain this objective.
- Aggregation of a customer base becomes the greatest strategic, or competitive, advantage in networked environments.
- Firms gravitating to Internet-based services must educate customers as to the intrinsic value of the information they provide and must per-

suade them that their firm is superior in delivering information and knowledge.

- Target marketing, illuminating precisely the information that improves consumer decision making (in professional or personal endeavors), will confer competitive advantage to the firms that learn to exploit this tactic.

When viewed in their totality, these interacting elements, empirically verifiable, redefine the method by which we assess the strategic efficacy and market valuations of communications firms. Evidence was provided in Chapter 11 that clearly demonstrated the relationship between market reach and market loyalty. Increasingly, portfolio managers and venture capitalists apply qualitative and quantitative methodology to define, differentiate, and project the fate of the communications firm in the deregulatory environment. *Market reach* represents the size of the customer base; *market loyalty* represents the degree to which a customer remains with, or returns to, a firm. The intersection of these two variables signifies the market valuation of the firm in this competitive landscape.

13.3 Managing Reach and Loyalty: Customer-Led Customization

Formerly, telecommunications providers developed a new product line and then sought to create a market for it. Providers must now verify the peculiar requirements of *individual* customers and then innovate services that satisfy personalized, multifaceted needs. Market research becomes a highly complex task in this environment. The industry may well be reshaped—mass marketing will give way to customized market research, and success or failure will pivot on the ability to meet the dynamic and fastidious requirements of consumer tastes. Consumers will dictate the terms and distribution of the services they require, dramatically reconfiguring organizations. Competitive advantage will fall to those firms that are first off the mark in customizing their products. "Total marketing" therefore emerges as a model equipped to forecast changing consumer tastes and establishes a framework for strategic management in this environment.

Assuming this is a credible paradigm for the delivery of communications services, then we may conclude that service and content providers will treat customers as the principal focus of strategy. An increasing fraction of a firm's earnings will be redirected to marketing and strategic planning departments, where the firm is closer and most sensitive to customer behavior. Customers, in a customer-led customization environment, dictate what is to be produced. A firm that chooses to ignore this development will be superseded and supplanted by companies prepared to serve this emerging market.

13.4 Managing Reach and Loyalty: The Emergence of Continuous Markets

Telecommunications continuous markets are ones in which demand for a variety of services is perpetually manifested through communications technology. Primary, secondary, and tertiary markets for telecommunications in health care, education, transportation, environment, entertainment, agriculture, engineering, science, construction, and manufacturing are created and reconfigured when consumers learn to exploit information and knowledge to perform daily undertakings. In this environment, there is little difference between work and learning; work *is* learning, and it is tied to directly to the production of value-added services necessary to maintain competitive advantage. For communications providers, a strategy stressing learning, in the largest sense of that term, may be vital in securing new markets and sustaining market share.

Learning is the pursuit and acquisition of knowledge. Whether one dichotomizes or synthesizes a vision of the telephony, cable, or broadcasting markets, the recurring need of the individual to assimilate information and knowledge may become the core of future demand for communications services. To tap the continuous learning market, one must understand the prospective market response to the nature of learning itself. Two developments are reconfiguring knowledge-based, post-industrial societies. Education involves the transmission of knowledge through formal instruction and results in a body of knowledge whose economic value does not erode. Language, history, mathematics, and the sciences, for example, represent bodies of knowledge not subject to obsolescence. Training, however, is practical instruction focused on accomplishing specific professional or avocational goals. Innovation spurs training and requires periodic updating. Education and training must be updated and expanded on a regular basis, as the useful half-life of knowledge is contracting. There is little dispute that markets for education and training will expand proportionate to the level of knowledge and skill compensated by the private sector.

Communications technology increasingly directs economic change. It is already common for professionals to return to school at two or three junctures in their lives. These institutions are very likely to focus on customized learning, geared to professional objectives. Telecommunications providers, whether focused on service or content, will inevitably respond to this emerging demand. Learning, personal and vocational, represents the most attractive long-term market for this industry, and strategists are just now awakening to this market reality.

13.5 Managing Reach and Loyalty: Developments in Relational Marketing

Using the customer as a tool to project, define, and launch new product line is the heart of *relational marketing*. Euphemisms for relational marketing include *one-to-one marketing* or *particle marketing*. The undergirding strategic implications of relational interaction with consumers (discussed in Chapter 8) impel a firm to eschew unprofitable customers in favor of profitable ones. Relational strategists presume that customers whose demand for services is limited—or unprofitable—can be deflected to unsuspecting firms. Because companies traditionally invest heavily in recruiting all available customers, the cost of recruitment per customer is higher than that incurred by firms focused exclusively on acquiring the most profitable consumers.

The incisive insight of relational strategists, however, concerns a shift in customer strategy. Historically, a firm creates or improves a product, then finds a market for it. This has been the typical pattern of new product development in telecommunications and high technology. Relational strategy holds that customers should be managed, not products. In short, strategic effectiveness and profitability result from managing a lifetime customer. Because the cost of retaining a veteran customer is cheaper than recruiting a new one, the firm that achieves customer loyalty among its most profitable consumers is likely to be the most successful. There exists variance by product and service, but it is generally accepted in the telecommunications industry that solicitation of new customers exceeds the cost of retention by a factor of five to eight times.

If one adopts this line of reasoning, the economics of customer loyalty will come to dominate strategic design in the most successful communications firms. Hence, we return to the theme that strategists might consider managing customers rather than products. We would determine what tactics would be necessary to manage a customer's demand over the lifetime of that consumption. Some analysts have noted that in telecommunications, the top quarter of the customer base is responsible for as much as 100% of the firm's profitability. If this statistic holds in the immediate future, we will witness a reinvestment of profits in market research aimed at retaining a firm's best customers. Customer prospecting will thus give way to reinvestment in the firm's most loyal and dedicated consumers.

13.6 Managing Reach and Loyalty: Envisioning Strategic Choice

We have noted through empirical confirmation the determinants of industry change and thus the emerging requirements of strategic management. These determinants are both contextual and content based. The array of regulatory, technological, economic, and demographic variables that comprise these two categories are growing increasingly difficult to forecast with precision. These difficulties will be further compounded by the intensification of global interaction in the years ahead. In the face of these challenges, we require new methods to project the pattern of determinants destined to define industry change. We have at our disposal a set of methods designed to anticipate such change, thus aiding in effective design and implementation of strategy. The application of these methods again pivots on the intersecting forces of market reach and loyalty, spurred by the dynamics of competition unleashed by deregulation.

Historical analogy is an economic methodology that asserts that all industries endure a series of evolutionary phases, in which a strategist may define the competitive environment of a firm. Understanding those forces governing a firm's competitive landscape enables one to position a firm and estimate probable strategic interactions with industry rivals. The most compelling feature of the contemporary communications environment (as noted in Chapter 2) is the prospective unfolding of competition associated with deregulation. A mature industry, steeped in regulatory compliance and incumbent control of market share, is now headed for intense competition and technological convergence.

We have some understanding of the dynamics of deregulation and the role it plays in reordering industry structure while innovating product line. In aviation, trucking, petroleum, natural gas, financial institutions, among others—industries all subjected to deregulation since 1978—we can outline a similar pattern of competitive development.

Some observers have contended that market confusion and chaos will inevitably follow implementation of the Telecommunications Act of 1996. It is their contention that it is just as inevitable that market stability will be restored after 10–15 years of experimentation. A highly capital-intensive industry whose progress was guided by government regulation, they believe, must educate itself in market dynamics never before experienced. The absence of competitive knowledge and experience means that industrial giants must learn during a period of chaotic experimentation. Competitive exposure and the assumption of risk by new entrants establish thereafter a framework of growing maturity and stability.

The proliferation of new entrants, accompanied by uncertainty and risk in infrastructure maintenance, translates into an industry shakeout before

2005. The model holds that long-term stability will manifest itself before 2008. It is during this period that risk diminishes as new infrastructure providers finally stabilize the economic environment. Market dynamism eventually begets consolidation, stability, and product maintenance. This model must be adjusted for each sector of the industry. Computer and broadcasting market structures, largely unburdened by regulatory litigation, are characteristically far more dynamic in the early stages of deregulation. The telephony sector, presently embroiled in a series of lawsuits designated to forestall regulatory sanctions, has effectively slowed the process of competitive metamorphosis.

Scenario-writing, applied to strategic management, itemizes prospective avenues for industry development and company positioning. A scenario is a model built on a potential determinant or catalyst for change. These determinants may be economic, social, regulatory, political, or technological in nature. Every scenario carries a set of assumptions about how industry rivals interact in response to these determinants. Scenarios are aimed at one or more of these objectives:

1. *Describing* and thus categorizing future competition;
2. *Prescribing* appropriate strategy in response to industry developments;
3. *Projecting* future outcomes based on continuing changes in the marketplace.

Scenario-writers bring both intuitive insight as well as empirical analysis to their models. As a result, scenarios vary considerably in terms of model-building and drawn inferences. The essential concern here is to bring order and clarity to strategic management when quantitative methodology cannot be applied effectively. Scenarists have designed a multitude of theoretical constructs to their design of models and have borrowed extensively from similar methodology in the public sector.

Another method of forecasting the fate of the communications industry involves a key lesson culled from industrial economics and the stage of consolidation known as the Rule of Three. The Rule of Three states that oligopolical forces—market structures characteristically consisting of three (or sometimes four) primary providers—emerge in capital-intensive industries.

Such a contingency is likely in most areas of telecommunications. In such a competitive landscape, three firms would generally acquire 80% or more market share, would offer a full array of bundled services, and would concentrate on industry volume (scale economies) to sustain their strategic positions.

The Rule of Three asserts that the exclusive avenue open to smaller firms hoping to sustain a competitive position would involve strategic planning aimed at exploiting a market niche. This theory of industrial evolution, if appropriate to prediction of telecommunications markets, implies that the telephony, broadcasting, and computer sectors would inevitably consolidate after uniform technical standards and infrastructure are fully adopted across national boundaries. Again, if credible, this theory, adjusted for the peculiarities of communications markets in the aftermath of deregulation, suggests that the period extending from 2008 to 2012 will be decisive in restructuring international competition within the industry.

13.7 Hypercompetition and the Knowledge-Based Economy

Scarcely any observer now contests the inevitability of a knowledge-based, customized economy evolved from the current standardized market system. Among other compelling features of the newly shaped economy are these, as recently reported by the International Engineering Consortium:

1. After 2000, in excess of two-thirds of new American jobs will be assumed by highly educated, well-trained workers dependent on the generation, distribution, or interpretation of knowledge.

2. Today, in excess of 60% of Americans can be classified as knowledge-workers.

3. Presently, in excess of 50% of Organization for Economic Cooperation and Development (OECD) nations may be regarded as knowledge-based, post-industrial societies.

If organizations are to become knowledge-based structures, what does this development portend for the management of strategy in telecommunications firms? We have already reviewed the implications of knowledge-based institutions for prospective consumer demand: communications firms are very likely to pursue strategies focused on bundled packages of high value-added services and products. If this contingency holds, then those firms that speed services to market while simultaneously facilitating the facile manipulation of information are likely to achieve greatest success. Knowledge-based institutions managed with high-quality human capital (workers who distill information to effective decision making) will generate a voracious appetite for communications products.

If information cannot be depleted, then the same can be said for knowledge. Strategists must now prepare for a spectacular and exponential growth for their firm's services: We are at the earliest and most primitive stage of a globalized demand for the industry's most enlightened products.

The successful enterprise in this environment is characterized as innovative, consumer-centric, open to rapid political, technological, economic, and social change, nimble in response to competitive thrusts by rivals, and effectively benchmarks its performance relative to its preeminent rivals. These assertions, though fully supported by the empirical evidence, have become something of a bromide for industry analysts. As we peer into the future of a knowledge-based economy, we begin to discern two critical developments impinging upon the strategy of communications companies. These developments extend beyond simple deregulation and will come to play a vital role in strategic priorities: (1) the emergence of the profitable communications customer, who, as yet, is ill-defined and something of a mystery to most firms; and (2) the prospective launch of revenue-aided services (RAS), personalized service menus that enhance the material well-being of customers.

We have learned in recent years, and particularly in the wake of deregulation, that some customers are clearly more valuable to a firm than others. Competition precipitated by deregulation, however unevenly distributed in each communications sector, has combined with advances in one-to-one marketing to reaffirm an enduring feature of Pareto Optimality: 20% of a firm's customer base generates 80% of its revenue. In the uncertain climate of deregulation, most firms have focused their main attention on prospecting, that is, dedicating themselves to acquiring as many consumers as practicable. An important element of Pareto Optimality asserts that this strategy is not necessarily desirable in the present competitive landscape. In a period of extraordinary competitive change, a firm ought not to indiscriminately expand its customer base, as the costs of such strategy may well exceed prospective benefits. In this circumstance, Pareto Optimality argues that the retention of the firm's most profitable (top 20%) customers should be the highest priority; conversely, application of this principle implies that the unprofitable customer base (the bottom 20%) should be diverted to a firm's competitors.

Concurrent with a reaffirmation and application of Pareto Optimality for the modern telecommunications enterprise is the growing validity of RAS. If customer loyalty has become as significant as customer reach for today's enterprise, it has become salient that new incentives must be found to retain consumers. The capacity of a firm, regardless of market share or size, to compete on the basis of price is obviously limited. If today's enterprise is to maximize success, it must do so on the basis of value-added, personalized services. There now emerges a recognition of the most important value-added service that a communications firm can provide to its customers—the capacity to enhance material wealth or streamline time management. If a telecommunications provider can provide services, the effect of which is to stimulate new revenue streams for its customers or alternatively reduce the time invested in the management of information, then it is cultivating RAS.

We are only at the beginning of strategic response marked by the passage of the Telecommunications Act of 1996, groping for new tools to manage the communications enterprise. The procedures elaborated in this summary are intended to provoke an alternative path to conventional strategic management. We can infer from the empirical evidence that, although change always disturbs, a company must nevertheless seek alternatives to convention. The telecommunications sector now exemplifies the broader theme of the unfolding global economy: Legacy firms must learn the strategic rules of a New Economy.

Appendix A

This appendix includes the full text of the Telecommunications Act of 1996 (Appendix A1), and a condensed version of the Communications Act of 1934 (Appendix A2). (Note: The full text of the 1996 Act is available on the Internet at www.fcc.gov, as is the 1934 law; updated regulatory information is also provided via this URL.) The two laws fully elaborate the power of the Federal Communications Commission to regulate interstate commerce as it pertains to the buildout of the Information Superhighway.

As noted in Chapter 2, the significance for the 1996 Act, when connected to the original 1934 law, is its economic ramification not only for the telecommunications industry, but for virtually all other sectors of both the domestic and international economies. When distilled to its core business application, what emerges is the following spectrum of immediate economic and regulatory consequences, as indicated in the right-hand margin:

1. Relaxes, gradually phases out, or eliminates FCC regulatory powers;
2. Identifies areas of present and/or future regulatory contentiousness;
3. Leads to industry consolidation and merger/acquisition activity;
4. Creates and sustains intense competition, with probable price reduction in services and/or equipment;
5. Spurs probable technological innovation or creation of new business formations;
6. Promotes economic convergence through prospective integration of telephony, cable, or computer industries;

7. Diminishes state regulatory power and influence through federal intervention.

The interpretative bar provided at the right-hand margin allows one to quickly identify the long-term economic impact the 1996 law is likely to instigate. In those sections in which the vertical bar is omitted, you may presume that there is no economic or business consequence attached to that section of the law.

The condensed business applications section permits one to link those passages of the 1934 law that continue to influence planning and development of telecommunications systems throughout the nation. Note that the original 1934 Act was amended repeatedly in the years following its enactment. Passages of the 1934 law were revised or eliminated as late as 1994. The 1996 Act did not obviate pertinent elements on the 1934 Act in force today. Thus, one must be acquainted with both the 1934 and 1996 Acts as jointly applied by the FCC in the period following February 1996. For additional interpretation as to the application of regulatory enforcement of these provisions, refer to Chapter 2. Note, too, that the interpretative bar gauges only the *primary* economic impact of deregulatory legislation; in many cases, the full economic effect of various passages reflects the combined influence of two or more variables indicated above.

Appendix A1:
Telecommunications Act of 1996

Telecommunications Act of 1996 (Enrolled Bill (Sent to President))
S.652

One Hundred Fourth Congress
of the
United States of America

AT THE SECOND SESSION

Begun and held at the City of Washington on Wednesday, the third day of January, one thousand nine hundred and ninety-six

An Act

To promote competition and reduce regulation in order to secure lower prices and higher quality services for American telecommunications consumers and encourage the rapid deployment of new telecommunications technologies.

Be it enacted by the Senate and House of Representatives of the United States of America in Congress assembled,

SECTION 1. SHORT TITLE; REFERENCES.

(a) SHORT TITLE- This Act may be cited as the "Telecommunications Act of 1996".

(b) REFERENCES- Except as otherwise expressly provided, whenever in this Act an amendment or repeal is expressed in terms of an amendment to, or repeal of, a section or other provision, the reference shall be considered to be made to a section or other provision of the Communications Act of 1934 (47 U.S.C. 151 et seq.).

SEC. 2. TABLE OF CONTENTS.

The table of contents for this Act is as follows:

Sec. 705. Mobile services direct access to long distance carriers.
Sec. 706. Advanced telecommunications incentives.
Sec. 707. Telecommunications Development Fund.
Sec. 708. National Education Technology Funding Corporation.
Sec. 709. Report on the use of advanced telecommunications services for medical purposes.
Sec. 710. Authorization of appropriations.

SEC. 3. DEFINITIONS.

(a) ADDITIONAL DEFINITIONS- Section 3 (47 U.S.C. 153) is amended—

(1) in subsection (r)—

(A) by inserting "(A)" after "means"; and

(B) by inserting before the period at the end the following: ", or (B) comparable service provided through a system of switches, transmission equipment, or other facilities (or combination thereof) by which a subscriber can originate and terminate a telecommunications service"; and

(2) by adding at the end thereof the following:

"(33) AFFILIATE- The term "affiliate" means a person that (directly or indirectly) owns or controls, is owned or controlled by, or is under common ownership or control with, another person. For purposes of this paragraph, the term "own" means to own an equity interest (or the equivalent thereof) of more than 10 percent.

"(34) AT&T CONSENT DECREE- The term "AT&T Consent Decree" means the order entered August 24, 1982, in the antitrust action styled *United States v. Western Electric*, Civil Action No. 82-0192, in the United States District Court for the District of Columbia, and includes any judgment or order with respect to such action entered on or after August 24, 1982.

"(35) BELL OPERATING COMPANY- The term "Bell operating company"-

"(A) means any of the following companies: Bell Telephone Company of Nevada, Illinois Bell Telephone Company, Indiana Bell Telephone Company, Incorporated, Michigan Bell Telephone Company, New England Telephone and Telegraph Company, New Jersey Bell Telephone Company, New York Telephone Company, US West Communications Company, South Central Bell Telephone Company, Southern Bell Telephone and Telegraph Company, Southwestern Bell Telephone Company, The Bell Telephone Company of Pennsylvania, The Chesapeake and Potomac Telephone Company, The Chesapeake and Potomac Telephone Company of Maryland, The Chesapeake and Potomac Telephone Company of Virginia, The Chesapeake and Potomac Telephone Company of West Virginia, The Diamond State Telephone Company, The Ohio Bell Telephone Company, The Pacific Telephone and Telegraph Company, or Wisconsin Telephone Company; and

"(B) includes any successor or assign of any such company that provides wireline telephone exchange service; but

"(C) does not include an affiliate of any such company, other than an affiliate described in subparagraph (A) or (B).

"(36) CABLE SERVICE- The term "cable service" has the meaning given such term in section 602.

"(37) CABLE SYSTEM- The term "cable system" has the meaning given such term in section 602.

"(38) CUSTOMER PREMISES EQUIPMENT- The term "customer premises equipment" means equipment employed on the premises of a person (other than a carrier) to originate, route, or terminate telecommunications.

"(39) DIALING PARITY- The term "dialing parity" means that a person that is not an affiliate of a local exchange carrier is able to provide telecommunications services in such a manner that customers have the ability to route automatically, without the use of any access code, their telecommunications to the telecommunications services provider of the customer's designation from among 2 or more telecommunications services providers (including such local exchange carrier).

"(40) EXCHANGE ACCESS- The term "exchange access" means the offering of access to telephone exchange services or facilities for the purpose of the origination or termination of telephone toll services.

"(41) INFORMATION SERVICE- The term "information service" means the offering of a capability for generating, acquiring, storing, transforming, processing, retrieving, utilizing, or making available information via telecommunications, and includes electronic publishing, but does not include any use of any such capability for the management, control, or operation of a telecommunications system or the management of a telecommunications service.

"(42) INTERLATA SERVICE- The term "interLATA service" means telecommunications between a point located in a local access and transport area and a point located outside such area.

"(43) LOCAL ACCESS AND TRANSPORT AREA- The term "local access and transport area" or "LATA" means a contiguous geographic area—

"(A) established before the date of enactment of the Telecommunications Act of 1996 by a Bell operating company such that no exchange area includes points within more than 1 metropolitan statistical area, consolidated metropolitan statistical area, or State, except as expressly permitted under the AT&T Consent Decree; or

"(B) established or modified by a Bell operating company after such date of enactment and approved by the Commission.

"(44) LOCAL EXCHANGE CARRIER- The term "local exchange carrier" means any person that is engaged in the provision of telephone exchange service or exchange access. Such term does not include a person insofar as such person is engaged in the provision of a commercial mobile service under section 332(c), except to the extent that the Commission finds that such service should be included in the definition of such term.

"(45) NETWORK ELEMENT- The term "network element" means a facility or equipment used in the provision of a telecommunications service. Such term also includes features, functions, and capabilities that are provided by means of such facility or equipment, including subscriber numbers, databases, signaling systems, and information sufficient for billing and collection or used in the transmission, routing, or other provision of a telecommunications service.

"(46) NUMBER PORTABILITY- The term "number portability" means the ability of users of telecommunications services to retain, at the same location, existing telecommunications numbers without impairment of quality, reliability, or convenience when switching from one telecommunications carrier to another.

"(47) RURAL TELEPHONE COMPANY- The term "rural telephone company" means a local exchange carrier operating entity to the extent that such entity—

"(A) provides common carrier service to any local exchange carrier study area that does not include either—

"(i) any incorporated place of 10,000 inhabitants or more, or any part thereof, based on the most recently available population statistics of the Bureau of the Census; or

"(ii) any territory, incorporated or unincorporated, included in an urbanized area, as defined by the Bureau of the Census as of August 10, 1993;

"(B) provides telephone exchange service, including exchange access, to fewer than 50,000 access lines;

"(C) provides telephone exchange service to any local exchange carrier study area with fewer than 100,000 access lines; or

"(D) has less than 15 percent of its access lines in communities of more than 50,000 on the date of enactment of the Telecommunications Act of 1996.

"(48) TELECOMMUNICATIONS- The term "telecommunications"

means the transmission, between or among points specified by the user, of information of the user's choosing, without change in the form or content of the information as sent and received.

"(49) TELECOMMUNICATIONS CARRIER- The term "telecommunications carrier" means any provider of telecommunications services, except that such term does not include aggregators of telecommunications services (as defined in section 226). A telecommunications carrier shall be treated as a common carrier under this Act only to the extent that it is engaged in providing telecommunications services, except that the Commission shall determine whether the provision of fixed and mobile satellite service shall be treated as common carriage.

"(50) TELECOMMUNICATIONS EQUIPMENT- The term "telecommunications equipment" means equipment, other than customer premises equipment, used by a carrier to provide telecommunications services, and includes software integral to such equipment (including upgrades).

"(51) TELECOMMUNICATIONS SERVICE- The term "telecommunications service" means the offering of telecommunications for a fee directly to the public, or to such classes of users as to be effectively available directly to the public, regardless of the facilities used.".

(b) COMMON TERMINOLOGY- Except as otherwise provided in this Act, the terms used in this Act have the meanings provided in section 3 of the Communications Act of 1934 (47 U.S.C. 153), as amended by this section.

(c) STYLISTIC CONSISTENCY- Section 3 (47 U.S.C. 153) is amended—

(1) in subsections (e) and (n), by redesignating clauses (1), (2), and (3), as clauses (A), (B), and (C), respectively;

(2) in subsection (w), by redesignating paragraphs (1) through (5) as subparagraphs (A) through (E), respectively;

(3) in subsections (y) and (z), by redesignating paragraphs (1) and (2) as subparagraphs (A) and (B), respectively;

(4) by redesignating subsections (a) through (ff) as paragraphs (1) through (32);

(5) by indenting such paragraphs 2 em spaces;

(6) by inserting after the designation of each such paragraph—

(A) a heading, in a form consistent with the form of the heading of this subsection, consisting of the term defined by such paragraph, or the first term so defined if such paragraph defines more than one term; and

(B) the words "The term";

(7) by changing the first letter of each defined term in such paragraphs

from a capital to a lower case letter (except for "United States", "State", "State commission", and "Great Lakes Agreement"); and

(8) by reordering such paragraphs and the additional paragraphs added by subsection (a) in alphabetical order based on the headings of such paragraphs and renumbering such paragraphs as so reordered.

(d) CONFORMING AMENDMENTS- The Act is amended—

(1) in section 225(a)(1), by striking "section 3(h)" and inserting "section 3";

(2) in section 332(d), by striking "section 3(n)" each place it appears and inserting "section 3"; and

(3) in sections 621(d)(3), 636(d), and 637(a)(2), by striking "section 3(v)" and inserting "section 3".

TITLE I—TELECOMMUNICATION SERVICES
SUBTITLE A—TELECOMMUNICATIONS SERVICES
SEC. 101. ESTABLISHMENT OF PART II OF TITLE II.

(a) AMENDMENT- Title II is amended by inserting after section 229 (47 U.S.C. 229) the following new part:

"PART II—DEVELOPMENT OF COMPETITIVE MARKETS
"SEC. 251. INTERCONNECTION.

"(a) GENERAL DUTY OF TELECOMMUNICATIONS CARRIERS- Each telecommunications carrier has the duty—

"(1) to interconnect directly or indirectly with the facilities and equipment of other telecommunications carriers; and

"(2) not to install network features, functions, or capabilities that do not comply with the guidelines and standards established pursuant to section 255 or 256.

"(b) OBLIGATIONS OF ALL LOCAL EXCHANGE CARRIERS- Each local exchange carrier has the following duties:

"(1) RESALE- The duty not to prohibit, and not to impose unreasonable or discriminatory conditions or limitations on, the resale of its telecommunications services.

"(2) NUMBER PORTABILITY- The duty to provide, to the extent technically feasible, number portability in accordance with requirements prescribed by the Commission.

"(3) DIALING PARITY- The duty to provide dialing parity to competing providers of telephone exchange service and telephone toll service, and the duty to permit all such providers to have nondiscriminatory access to telephone numbers, operator services, directory assistance, and directory listing, with no unreasonable dialing delays.

"(4) ACCESS TO RIGHTS-OF-WAY- The duty to afford access to the poles, ducts, conduits, and rights-of-way of such carrier to competing providers of telecommunications services on rates, terms, and conditions that are consistent with section 224.

"(5) RECIPROCAL COMPENSATION- The duty to establish reciprocal compensation arrangements for the transport and termination of telecommunications.

"(c) ADDITIONAL OBLIGATIONS OF INCUMBENT LOCAL EXCHANGE CARRIERS- In addition to the duties contained in subsection (b), each incumbent local exchange carrier has the following duties:

"(1) DUTY TO NEGOTIATE- The duty to negotiate in good faith in accordance with section 252 the particular terms and conditions of agreements to fulfill the duties described in paragraphs (1) through (5) of subsection (b) and this subsection. The requesting telecommunications carrier also has the duty to negotiate in good faith the terms and conditions of such agreements.

"(2) INTERCONNECTION- The duty to provide, for the facilities and equipment of any requesting telecommunications carrier, interconnection with the local exchange carrier's network—

"(A) for the transmission and routing of telephone exchange service and exchange access;

"(B) at any technically feasible point within the carrier's network;

"(C) that is at least equal in quality to that provided by the local exchange carrier to itself or to any subsidiary, affiliate, or any other party to which the carrier provides interconnection; and

"(D) on rates, terms, and conditions that are just, reasonable, and nondiscriminatory, in accordance with the terms and conditions of the agreement and the requirements of this section and section 252.

"(3) UNBUNDLED ACCESS- The duty to provide, to any requesting telecommunications carrier for the provision of a telecommunications service, nondiscriminatory access to network elements on an unbundled basis at any technically feasible point on rates, terms, and conditions that are just, reasonable, and nondiscriminatory in accordance with the terms and conditions of the agreement and the requirements of this section and section 252. An incumbent local exchange carrier shall provide such unbundled network elements in a manner that allows requesting carriers to combine such elements in order to provide such telecommunications service.

"(4) RESALE- The duty—

"(A) to offer for resale at wholesale rates any telecommunications service that the carrier provides at retail to subscribers who are not telecommunications carriers; and

"(B) not to prohibit, and not to impose unreasonable or discriminatory conditions or limitations on, the resale of such telecommunications service, except that a State commission may, consistent with regulations prescribed by the Commission under this section, prohibit a reseller that obtains at wholesale rates a telecommunications service that is available at retail only to a category of subscribers from offering such service to a different category of subscribers.

"(5) NOTICE OF CHANGES- The duty to provide reasonable public notice of changes in the information necessary for the transmission and routing of services using that local exchange carrier's facilities or networks, as well as of any other changes that would affect the interoperability of those facilities and networks.

"(6) COLLOCATION- The duty to provide, on rates, terms, and conditions that are just, reasonable, and nondiscriminatory, for physical collocation of equipment necessary for interconnection or access to unbundled network elements at the premises of the local exchange carrier, except that the carrier may provide for virtual collocation if the local exchange carrier demonstrates to the State commission that physical collocation is not practical for technical reasons or because of space limitations.

"(d) IMPLEMENTATION-

"(1) IN GENERAL- Within 6 months after the date of enactment of the Telecommunications Act of 1996, the Commission shall complete all actions necessary to establish regulations to implement the requirements of this section.

"(2) ACCESS STANDARDS- In determining what network elements should be made available for purposes of subsection (c)(3), the Commission shall consider, at a minimum, whether—

"(A) access to such network elements as are proprietary in nature is necessary; and

"(B) the failure to provide access to such network elements would impair the ability of the telecommunications carrier seeking access to provide the services that it seeks to offer.

"(3) PRESERVATION OF STATE ACCESS REGULATIONS- In prescribing and enforcing regulations to implement the requirements of this section, the Commission shall not preclude the enforcement of any regulation, order, or policy of a State commission that—

"(A) establishes access and interconnection obligations of local exchange carriers;

"(B) is consistent with the requirements of this section; and

"(C) does not substantially prevent implementation of the requirements of this section and the purposes of this part.

"(e) NUMBERING ADMINISTRATION—

"(1) COMMISSION AUTHORITY AND JURISDICTION- The Commission shall create or designate one or more impartial entities to administer telecommunications numbering and to make such numbers available on an equitable basis. The Commission shall have exclusive jurisdiction over those portions of the North American Numbering Plan that pertain to the United States. Nothing in this paragraph shall preclude the Commission from delegating to State commissions or other entities all or any portion of such jurisdiction.

"(2) COSTS- The cost of establishing telecommunications numbering administration arrangements and number portability shall be borne by all telecommunications carriers on a competitively neutral basis as determined by the Commission.

"(f) EXEMPTIONS, SUSPENSIONS, AND MODIFICATIONS-

"(1) EXEMPTION FOR CERTAIN RURAL TELEPHONE COMPANIES-

"(A) EXEMPTION- Subsection (c) of this section shall not apply to a rural telephone company until (i) such company has received a bona fide request for interconnection, services, or network elements, and (ii) the State commission determines (under subparagraph (B)) that such request is not unduly economically burdensome, is technically feasible, and is consistent with section 254 (other than subsections (b)(7) and (c)(1)(D) thereof).

"(B) STATE TERMINATION OF EXEMPTION AND IMPLEMENTATION SCHEDULE- The party making a bona fide request of a rural telephone company for interconnection, services, or network elements shall submit a notice of its request to the State commission. The State commission shall conduct an inquiry for the purpose of determining whether to terminate the exemption under subparagraph (A). Within 120 days after the State commission receives notice of the request, the State commission shall terminate the exemption if the request is not unduly economically burdensome, is technically feasible, and is consistent with section 254 (other than subsections (b)(7) and (c)(1)(D) thereof). Upon termination of the exemption, a State commission shall establish an implementation schedule for compliance with the request that is consistent in time and manner with Commission regulations.

"(C) LIMITATION ON EXEMPTION- The exemption provided by this paragraph shall not apply with respect to a request under subsection

(c) from a cable operator providing video programming, and seeking to provide any telecommunications service, in the area in which the rural telephone company provides video programming. The limitation contained in this subparagraph shall not apply to a rural telephone company that is providing video programming on the date of enactment of the Telecommunications Act of 1996.

"(2) SUSPENSIONS AND MODIFICATIONS FOR RURAL CARRI- ERS- A local exchange carrier with fewer than 2 percent of the Nation's subscriber lines installed in the aggregate nationwide may petition a State commission for a suspension or modification of the application of a requirement or requirements of subsection (b) or (c) to telephone exchange service facilities specified in such petition. The State commission shall grant such petition to the extent that, and for such duration as, the State commission determines that such suspension or modification—

"(A) is necessary—

"(i) to avoid a significant adverse economic impact on users of telecommunications services generally;

"(ii) to avoid imposing a requirement that is unduly economically burdensome; or

"(iii) to avoid imposing a requirement that is technically infeasible; and

"(B) is consistent with the public interest, convenience, and necessity. The State commission shall act upon any petition filed under this paragraph within 180 days after receiving such petition. Pending such action, the State commission may suspend enforcement of the requirement or requirements to which the petition applies with respect to the petitioning carrier or carriers.

"(g) CONTINUED ENFORCEMENT OF EXCHANGE ACCESS AND INTERCONNECTION REQUIREMENTS- On and after the date of enactment of the Telecommunications Act of 1996, each local exchange carrier, to the extent that it provides wireline services, shall provide exchange access, information access, and exchange services for such access to interexchange carriers and information service providers in accordance with the same equal access and nondiscriminatory interconnection restrictions and obligations (including receipt of compensation) that apply to such carrier on the date immediately preceding the date of enactment of the Telecommunications Act of 1996 under any court order, consent decree, or regulation, order, or policy of the Commission, until such restrictions and obligations are explicitly superseded by regulations prescribed by the Commission after such date of enactment. During the period beginning on such date of enactment and until such restrictions and obligations are so superseded, such restrictions and obligations shall be enforceable in the same manner as regulations of the Commission.

"(h) DEFINITION OF INCUMBENT LOCAL EXCHANGE CARRIER-

"(1) DEFINITION- For purposes of this section, the term "incumbent local exchange carrier" means, with respect to an area, the local exchange carrier that—

"(A) on the date of enactment of the Telecommunications Act of 1996, provided telephone exchange service in such area; and

"(B)(i) on such date of enactment, was deemed to be a member of the exchange carrier association pursuant to section 69.601(b) of the Commission's regulations (47 C.F.R. 69.601(b)); or

"(ii) is a person or entity that, on or after such date of enactment, became a successor or assign of a member described in clause (i).

"(2) TREATMENT OF COMPARABLE CARRIERS AS INCUMBENTS- The Commission may, by rule, provide for the treatment of a local exchange carrier (or class or category thereof) as an incumbent local exchange carrier for purposes of this section if—

"(A) such carrier occupies a position in the market for telephone exchange service within an area that is comparable to the position occupied by a carrier described in paragraph (1);

"(B) such carrier has substantially replaced an incumbent local exchange carrier described in paragraph (1); and

"(C) such treatment is consistent with the public interest, convenience, and necessity and the purposes of this section.

"(i) SAVINGS PROVISION- Nothing in this section shall be construed to limit or otherwise affect the Commission's authority under section 201.

"SEC. 252. PROCEDURES FOR NEGOTIATION, ARBITRATION, AND APPROVAL OF AGREEMENTS.

"(a) AGREEMENTS ARRIVED AT THROUGH NEGOTIATION-

"(1) VOLUNTARY NEGOTIATIONS- Upon receiving a request for interconnection, services, or network elements pursuant to section 251, an incumbent local exchange carrier may negotiate and enter into a binding agreement with the requesting telecommunications carrier or carriers without regard to the standards set forth in subsections (b) and (c) of section 251. The agreement shall include a detailed schedule of itemized charges for interconnection and each service or network element included in the agreement. The agreement, including any interconnection agreement negotiated before the date of enactment of the Telecommunications

Act of 1996, shall be submitted to the State commission under subsection (e) of this section.

"(2) MEDIATION- Any party negotiating an agreement under this section may, at any point in the negotiation, ask a State commission to participate in the negotiation and to mediate any differences arising in the course of the negotiation.

"(b) AGREEMENTS ARRIVED AT THROUGH COMPULSORY ARBITRATION-

"(1) ARBITRATION- During the period from the 135th to the 160th day (inclusive) after the date on which an incumbent local exchange carrier receives a request for negotiation under this section, the carrier or any other party to the negotiation may petition a State commission to arbitrate any open issues.

"(2) DUTY OF PETITIONER-

"(A) A party that petitions a State commission under paragraph (1) shall, at the same time as it submits the petition, provide the State commission all relevant documentation concerning—

"(i) the unresolved issues;

"(ii) the position of each of the parties with respect to those issues; and

"(iii) any other issue discussed and resolved by the parties.

"(B) A party petitioning a State commission under paragraph (1) shall provide a copy of the petition and any documentation to the other party or parties not later than the day on which the State commission receives the petition.

"(3) OPPORTUNITY TO RESPOND- A non-petitioning party to a negotiation under this section may respond to the other party's petition and provide such additional information as it wishes within 25 days after the State commission receives the petition.

"(4) ACTION BY STATE COMMISSION-

"(A) The State commission shall limit its consideration of any petition under paragraph (1) (and any response thereto) to the issues set forth in the petition and in the response, if any, filed under paragraph (3).

"(B) The State commission may require the petitioning party and the responding party to provide such information as may be necessary for the State commission to reach a decision on the unresolved issues. If any party refuses or fails unreasonably to respond on a timely basis to any reasonable request from the State commission, then the State commission may proceed on the basis of the best information available to it from whatever source derived.

"(C) The State commission shall resolve each issue set forth in the petition and the response, if any, by imposing appropriate conditions as required to implement subsection (c) upon the parties to the agreement, and shall conclude the resolution of any unresolved issues not later than 9 months after the date on which the local exchange carrier received the request under this section.

"(5) REFUSAL TO NEGOTIATE- The refusal of any other party to the negotiation to participate further in the negotiations, to cooperate with the State commission in carrying out its function as an arbitrator, or to continue to negotiate in good faith in the presence, or with the assistance, of the State commission shall be considered a failure to negotiate in good faith.

"(c) STANDARDS FOR ARBITRATION- In resolving by arbitration under subsection (b) any open issues and imposing conditions upon the parties to the agreement, a State commission shall—

"(1) ensure that such resolution and conditions meet the requirements of section 251, including the regulations prescribed by the Commission pursuant to section 251;

"(2) establish any rates for interconnection, services, or network elements according to subsection (d); and

"(3) provide a schedule for implementation of the terms and conditions by the parties to the agreement.

"(d) PRICING STANDARDS-

"(1) INTERCONNECTION AND NETWORK ELEMENT CHARGES- Determinations by a State commission of the just and reasonable rate for the interconnection of facilities and equipment for purposes of subsection (c)(2) of section 251, and the just and reasonable rate for network elements for purposes of subsection (c)(3) of such section—

"(A) shall be—

"(i) based on the cost (determined without reference to a rate-of-return or other rate-based proceeding) of providing the interconnection or network element (whichever is applicable), and

"(ii) nondiscriminatory, and

"(B) may include a reasonable profit.

"(2) CHARGES FOR TRANSPORT AND TERMINATION OF TRAFFIC-

"(A) IN GENERAL- For the purposes of compliance by an incumbent local exchange carrier with section 251(b)(5), a State commission shall not consider the terms and conditions for reciprocal compensation to be just and reasonable unless—

"(i) such terms and conditions provide for the mutual and reciprocal recovery by each carrier of costs associated with the transport and termination on each carrier's network facilities of calls that originate on the network facilities of the other carrier; and "(ii) such terms and conditions determine such costs on the basis of a reasonable approximation of the additional costs of terminating such calls.

"(B) RULES OF CONSTRUCTION- This paragraph shall not be construed—

"(i) to preclude arrangements that afford the mutual recovery of costs through the offsetting of reciprocal obligations, including arrangements that waive mutual recovery (such as bill-and-keep arrangements); or

"(ii) to authorize the Commission or any State commission to engage in any rate regulation proceeding to establish with particularity the additional costs of transporting or terminating calls, or to require carriers to maintain records with respect to the additional costs of such calls.

"(3) WHOLESALE PRICES FOR TELECOMMUNICATIONS SERVICES- For the purposes of section 251(c)(4), a State commission shall determine wholesale rates on the basis of retail rates charged to subscribers for the telecommunications service requested, excluding the portion thereof attributable to any marketing, billing, collection, and other costs that will be avoided by the local exchange carrier.

"(e) APPROVAL BY STATE COMMISSION-

"(1) APPROVAL REQUIRED- Any interconnection agreement adopted by negotiation or arbitration shall be submitted for approval to the State commission. A State commission to which an agreement is submitted shall approve or reject the agreement, with written findings as to any deficiencies.

"(2) GROUNDS FOR REJECTION- The State commission may only reject—

"(A) an agreement (or any portion thereof) adopted by negotiation under subsection (a) if it finds that—

"(i) the agreement (or portion thereof) discriminates against a telecommunications carrier not a party to the agreement; or

"(ii) the implementation of such agreement or portion is not consistent with the public interest, convenience, and necessity; or

"(B) an agreement (or any portion thereof) adopted by arbitration under subsection (b) if it finds that the agreement does not meet the requirements of section 251, including the regulations prescribed by the Commission pursuant to section 251, or the standards set forth in subsection (d) of this section.

"(3) PRESERVATION OF AUTHORITY- Notwithstanding paragraph (2), but subject to section 253, nothing in this section shall prohibit a State commission from establishing or enforcing other requirements of State law in its review of an agreement, including requiring compliance with intrastate telecommunications service quality standards or requirements.

"(4) SCHEDULE FOR DECISION- If the State commission does not act to approve or reject the agreement within 90 days after submission by the parties of an agreement adopted by negotiation under subsection (a), or within 30 days after submission by the parties of an agreement adopted by arbitration under subsection (b), the agreement shall be deemed approved. No State court shall have jurisdiction to review the action of a State commission in approving or rejecting an agreement under this section.

"(5) COMMISSION TO ACT IF STATE WILL NOT ACT- If a State commission fails to act to carry out its responsibility under this section in any proceeding or other matter under this section, then the Commission shall issue an order preempting the State commission's jurisdiction of that proceeding or matter within 90 days after being notified (or taking notice) of such failure, and shall assume the responsibility of the State commission under this section with respect to the proceeding or matter and act for the State commission.

"(6) REVIEW OF STATE COMMISSION ACTIONS- In a case in which a State fails to act as described in paragraph (5), the proceeding by the Commission under such paragraph and any judicial review of the Commission's actions shall be the exclusive remedies for a State commission's failure to act. In any case in which a State commission makes a determination under this section, any party aggrieved by such determination may bring an action in an appropriate Federal district court to determine whether the agreement or statement meets the requirements of section 251 and this section.

"(f) STATEMENTS OF GENERALLY AVAILABLE TERMS-

"(1) IN GENERAL- A Bell operating company may prepare and file with a State commission a statement of the terms and conditions that such company generally offers within that State to comply with the requirements of section 251 and the regulations thereunder and the standards applicable under this section.

"(2) STATE COMMISSION REVIEW- A State commission may not approve such statement unless such statement complies with subsection (d) of this section and section 251 and the regulations thereunder. Except as provided in section 253, nothing in this section shall prohibit a State commission from establishing or enforcing other requirements of State law in its review of such statement, including requiring compliance with intrastate telecommunications service quality standards or requirements.

"(3) SCHEDULE FOR REVIEW- The State commission to which a

statement is submitted shall, not later than 60 days after the date of such submission—

"(A) complete the review of such statement under paragraph (2) (including any reconsideration thereof), unless the submitting carrier agrees to an extension of the period for such review; or

"(B) permit such statement to take effect.

"(4) AUTHORITY TO CONTINUE REVIEW- Paragraph (3) shall not preclude the State commission from continuing to review a statement that has been permitted to take effect under subparagraph (B) of such paragraph or from approving or disapproving such statement under paragraph (2).

"(5) DUTY TO NEGOTIATE NOT AFFECTED- The submission or approval of a statement under this subsection shall not relieve a Bell operating company of its duty to negotiate the terms and conditions of an agreement under section 251.

"(g) CONSOLIDATION OF STATE PROCEEDINGS- Where not inconsistent with the requirements of this Act, a State commission may, to the extent practical, consolidate proceedings under sections 214(e), 251(f), 253, and this section in order to reduce administrative burdens on telecommunications carriers, other parties to the proceedings, and the State commission in carrying out its responsibilities under this Act.

"(h) FILING REQUIRED- A State commission shall make a copy of each agreement approved under subsection (e) and each statement approved under subsection (f) available for public inspection and copying within 10 days after the agreement or statement is approved. The State commission may charge a reasonable and nondiscriminatory fee to the parties to the agreement or to the party filing the statement to cover the costs of approving and filing such agreement or statement.

"(i) AVAILABILITY TO OTHER TELECOMMUNICATIONS CARRIERS- A local exchange carrier shall make available any interconnection, service, or network element provided under an agreement approved under this section to which it is a party to any other requesting telecommunications carrier upon the same terms and conditions as those provided in the agreement.

"(j) DEFINITION OF INCUMBENT LOCAL EXCHANGE CARRIER- For purposes of this section, the term "incumbent local exchange carrier" has the meaning provided in section 251(h).

"SEC. 253. REMOVAL OF BARRIERS TO ENTRY.

"(a) IN GENERAL- No State or local statute or regulation, or other State or local legal requirement, may prohibit or have the effect of prohibiting the ability of any entity to provide any interstate or intrastate telecommunications service.

"(b) STATE REGULATORY AUTHORITY- Nothing in this section shall affect the ability of a State to impose, on a competitively neutral basis and consistent with section 254, requirements necessary to preserve and advance universal service, protect the public safety and welfare, ensure the continued quality of telecommunications services, and safeguard the rights of consumers.

"(c) STATE AND LOCAL GOVERNMENT AUTHORITY- Nothing in this section affects the authority of a State or local government to manage the public rights-of-way or to require fair and reasonable compensation from telecommunications providers, on a competitively neutral and nondiscriminatory basis, for use of public rights-of-way on a nondiscriminatory basis, if the compensation required is publicly disclosed by such government.

"(d) PREEMPTION- If, after notice and an opportunity for public comment, the Commission determines that a State or local government has permitted or imposed any statute, regulation, or legal requirement that violates subsection (a) or (b), the Commission shall preempt the enforcement of such statute, regulation, or legal requirement to the extent necessary to correct such violation or inconsistency.

"(e) COMMERCIAL MOBILE SERVICE PROVIDERS- Nothing in this section shall affect the application of section 332(c)(3) to commercial mobile service providers.

"(f) RURAL MARKETS- It shall not be a violation of this section for a State to require a telecommunications carrier that seeks to provide telephone exchange service or exchange access in a service area served by a rural telephone company to meet the requirements in section 214(e)(1) for designation as an eligible telecommunications carrier for that area before being permitted to provide such service. This subsection shall not apply—

"(1) to a service area served by a rural telephone company that has obtained an exemption, suspension, or modification of section 251(c)(4) that effectively prevents a competitor from meeting the requirements of section 214(e)(1); and

"(2) to a provider of commercial mobile services.

"SEC. 254. UNIVERSAL SERVICE.

"(a) PROCEDURES TO REVIEW UNIVERSAL SERVICE RE-QUIREMENTS- "(1) FEDERAL-STATE JOINT BOARD ON UNI-VERSAL SERVICE- Within one month after the date of enactment of the Telecommunications Act of 1996, the Commission shall institute and refer to a Federal-State Joint Board under section 410(c) a proceeding to recommend changes to any of its regulations in order to implement sections 214(e) and this section, including the definition of the services that are supported by Federal universal service support mechanisms and a specific timetable for completion of such recommendations. In addition to the members of the Joint Board required under section 410(c), one member of such Joint Board shall be a State-appointed utility consumer advocate nominated by a national organization of State utility consumer advocates. The Joint Board shall, after notice and opportunity for public comment, make its recommendations to the Commission 9 months after the date of enactment of the Telecommunications Act of 1996.

"(2) COMMISSION ACTION- The Commission shall initiate a single proceeding to implement the recommendations from the Joint Board required by paragraph (1) and shall complete such proceeding within 15 months after the date of enactment of the Telecommunications Act of 1996. The rules established by such proceeding shall include a definition of the services that are supported by Federal universal service support mechanisms and a specific timetable for implementation. Thereafter, the Commission shall complete any proceeding to implement subsequent recommendations from any Joint Board on universal service within one year after receiving such recommendations.

"(b) UNIVERSAL SERVICE PRINCIPLES- The Joint Board and the Commission shall base policies for the preservation and advancement of universal service on the following principles:

"(1) QUALITY AND RATES- Quality services should be available at just, reasonable, and affordable rates.

"(2) ACCESS TO ADVANCED SERVICES- Access to advanced telecommunications and information services should be provided in all regions of the Nation.

"(3) ACCESS IN RURAL AND HIGH COST AREAS- Consumers in all regions of the Nation, including low-income consumers and those in rural, insular, and high cost areas, should have access to telecommunications and information services, including interexchange services and advanced telecommunications and information services, that are reasonably comparable to those services provided in urban areas and that are available at rates that are reasonably comparable to rates charged for similar services in urban areas.

"(4) EQUITABLE AND NONDISCRIMINATORY CONTRIBU-TIONS- All providers of telecommunications services should make an equitable and nondiscriminatory contribution to the preservation and advancement of universal service.

"(5) SPECIFIC AND PREDICTABLE SUPPORT MECHANISMS- There should be specific, predictable and sufficient Federal and State mechanisms to preserve and advance universal service.

"(6) ACCESS TO ADVANCED TELECOMMUNICATIONS SERVICES FOR SCHOOLS, HEALTH CARE, AND LIBRARIES- Elementary and secondary schools and classrooms, health care providers, and libraries should have access to advanced telecommunications services as described in subsection (h).

"(7) ADDITIONAL PRINCIPLES- Such other principles as the Joint Board and the Commission determine are necessary and appropriate for the protection of the public interest, convenience, and necessity and are consistent with this Act.

"(c) DEFINITION-

"(1) IN GENERAL- Universal service is an evolving level of telecommunications services that the Commission shall establish periodically under this section, taking into account advances in telecommunications and information technologies and services. The Joint Board in recommending, and the Commission in establishing, the definition of the services that are supported by Federal universal service support mechanisms shall consider the extent to which such telecommunications services—

"(A) are essential to education, public health, or public safety;

"(B) have, through the operation of market choices by customers, been subscribed to by a substantial majority of residential customers;

"(C) are being deployed in public telecommunications networks by telecommunications carriers; and

"(D) are consistent with the public interest, convenience, and necessity.

"(2) ALTERATIONS AND MODIFICATIONS- The Joint Board may, from time to time, recommend to the Commission modifications in the definition of the services that are supported by Federal universal service support mechanisms.

"(3) SPECIAL SERVICES- In addition to the services included in the definition of universal service under paragraph (1), the Commission may designate additional services for such support mechanisms for schools, libraries, and health care providers for the purposes of subsection (h).

"(d) TELECOMMUNICATIONS CARRIER CONTRIBUTION- Every telecommunications carrier that provides interstate telecommunications services shall contribute, on an equitable and nondiscriminatory basis, to the specific, predictable, and sufficient mechanisms established by the Commission to preserve and advance universal service. The Commission may exempt a carrier or class of carriers from this requirement if the carrier's telecommunications activities are limited to such an extent that the level of such carrier's contribution to the preservation and advancement of universal service would be de minimis. Any other provider of interstate telecommunications may be required to contribute to the preservation and advancement of universal service if the public interest so requires.

"(e) UNIVERSAL SERVICE SUPPORT- After the date on which Commission regulations implementing this section take effect, only an eligible telecommunications carrier designated under section 214(e) shall be eligible to receive specific Federal universal service support. A carrier that receives such support shall use that support only for the provision, maintenance, and upgrading of facilities and services for which the support is intended. Any such support should be explicit and sufficient to achieve the purposes of this section.

"(f) STATE AUTHORITY- A State may adopt regulations not inconsistent with the Commission's rules to preserve and advance universal service. Every telecommunications carrier that provides intrastate telecommunications services shall contribute, on an equitable and non-discriminatory basis, in a manner determined by the State to the preservation and advancement of universal service in that State. A State may adopt regulations to provide for additional definitions and standards to preserve and advance universal service within that State only to the extent that such regulations adopt additional specific, predictable, and sufficient mechanisms to support such definitions or standards that do not rely on or burden Federal universal service support mechanisms.

"(g) INTEREXCHANGE AND INTERSTATE SERVICES- Within 6 months after the date of enactment of the Telecommunications Act of 1996, the Commission shall adopt rules to require that the rates charged by providers of interexchange telecommunications services to subscribers in rural and high cost areas shall be no higher than the rates charged by each such provider to its subscribers in urban areas. Such rules shall also require that a provider of interstate interexchange telecommunications services shall provide such services to its subscribers in each State at rates no higher than the rates charged to its subscribers in any other State.

"(h) TELECOMMUNICATIONS SERVICES FOR CERTAIN PROVIDERS-

"(1) IN GENERAL-

"(A) HEALTH CARE PROVIDERS FOR RURAL AREAS- A telecommunications carrier shall, upon receiving a bona fide request, provide telecommunications services which are necessary for the provision of health care services in a State, including instruction relating to such services, to any public or nonprofit health care provider that serves persons who reside in rural areas in that State at rates that are reasonably comparable to rates charged for similar services in urban areas in that State. A telecommunications carrier providing service under this paragraph shall be entitled to have an amount equal to the difference, if any, between the rates for services provided to health care providers for rural areas in a State and the rates for similar services provided to other customers in comparable rural areas in that State treated as a service obligation as a part of its obligation to participate in the mechanisms to preserve and advance universal service.

"(B) EDUCATIONAL PROVIDERS AND LIBRARIES- All telecommunications carriers serving a geographic area shall, upon a bona fide request for any of its services that are within the definition of universal service under subsection (c)(3), provide such services to elementary schools, secondary schools, and libraries for educational purposes at rates less than the amounts charged for similar services to other parties. The discount shall be an amount that the Commission, with respect to interstate services, and the States, with respect to intrastate services, determine is appropriate and necessary to ensure affordable access to and use of such services by such entities. A telecommunications carrier providing service under this paragraph shall—

"(i) have an amount equal to the amount of the discount treated as an offset to its obligation to contribute to the mechanisms to preserve and advance universal service, or

"(ii) notwithstanding the provisions of subsection

(e) of this section, receive reimbursement utilizing the support mechanisms to preserve and advance universal service.

"(2) ADVANCED SERVICES- The Commission shall establish competitively neutral rules—

"(A) to enhance, to the extent technically feasible and economically reasonable, access to advanced telecommunications and information services for all public and nonprofit elementary and secondary school classrooms, health care providers, and libraries; and

"(B) to define the circumstances under which a telecommunications carrier may be required to connect its network to such public institutional telecommunications users.

"(3) TERMS AND CONDITIONS- Telecommunications services and network capacity provided to a public institutional telecommunications user under this subsection may not be sold, resold, or otherwise transferred by such user in consideration for money or any other thing of value.

"(4) ELIGIBILITY OF USERS- No entity listed in this subsection shall be entitled to preferential rates or treatment as required by this subsection, if such entity operates as a for-profit business, is a school described in paragraph (5)(A) with an endowment of more than $50,000,000, or is a library not eligible for participation in State-based plans for funds under title III of the Library Services and Construction Act (20 U.S.C. 335c et seq.).

"(5) DEFINITIONS- For purposes of this subsection:

"(A) ELEMENTARY AND SECONDARY SCHOOLS- The term "elementary and secondary schools" means elementary schools and secondary schools, as defined in paragraphs (14) and (25), respectively, of section 14101 of the Elementary and Secondary Education Act of 1965 (20 U.S.C. 8801).

"(B) HEALTH CARE PROVIDER- The term "health care provider" means—

"(i) post-secondary educational institutions offering health care instruction, teaching hospitals, and medical schools;

"(ii) community health centers or health centers providing health care to migrants;

"(iii) local health departments or agencies;

"(iv) community mental health centers;

"(v) not-for-profit hospitals;

"(vi) rural health clinics; and

"(vii) consortia of health care providers consisting of one or more entities described in clauses (i) through (vi).

"(C) PUBLIC INSTITUTIONAL TELECOMMUNICATIONS USER- The term "public institutional telecommunications user" means an elementary or secondary school, a library, or a health care provider as those terms are defined in this paragraph.

"(i) CONSUMER PROTECTION- The Commission and the States should ensure that universal service is available at rates that are just, reasonable, and affordable.

"(j) LIFELINE ASSISTANCE- Nothing in this section shall affect the collection, distribution, or administration of the Lifeline Assistance Program provided for by the Commission under regulations set forth in section 69.117 of title 47, Code of Federal Regulations, and other related sections of such title.

"(k) SUBSIDY OF COMPETITIVE SERVICES PROHIBITED- A telecommunications carrier may not use services that are not competitive to subsidize services that are subject to competition. The Commission, with respect to interstate services, and the States, with respect to intrastate services, shall establish any necessary cost allocation rules, accounting safeguards, and guidelines to ensure that services included in the definition of universal service bear no more than a reasonable share of the joint and common costs of facilities used to provide those services.

"SEC. 255. ACCESS BY PERSONS WITH DISABILITIES.

"(a) DEFINITIONS- As used in this section—

"(1) DISABILITY- The term "disability" has the meaning given to it by section 3(2)(A) of the Americans with Disabilities Act of 1990 (42 U.S.C. 12102(2)(A)).

"(2) READILY ACHIEVABLE- The term "readily achievable" has the meaning given to it by section 301(9) of that Act (42 U.S.C. 12181(9)).

"(b) MANUFACTURING- A manufacturer of telecommunications equipment or customer premises equipment shall ensure that the equipment is designed, developed, and fabricated to be accessible to and usable by individuals with disabilities, if readily achievable.

"(c) TELECOMMUNICATIONS SERVICES- A provider of telecommunications service shall ensure that the service is accessible to and usable by individuals with disabilities, if readily achievable.

"(d) COMPATIBILITY- Whenever the requirements of subsections (b) and (c) are not readily achievable, such a manufacturer or provider shall ensure that the equipment or service is compatible with existing peripheral devices or specialized customer premises equipment commonly used by individuals with disabilities to achieve access, if readily achievable.

"(e) GUIDELINES- Within 18 months after the date of enactment of the Telecommunications Act of 1996, the Architectural and Transportation Barriers Compliance Board shall develop guidelines for accessibility of telecommunications equipment and customer premises equipment in conjunction with the Commission. The Board shall review and update the guidelines periodically.

"(f) NO ADDITIONAL PRIVATE RIGHTS AUTHORIZED- Nothing in this section shall be construed to authorize any private right of action to enforce any requirement of this section or any regulation thereunder. The Commission shall have exclusive jurisdiction with respect to any complaint under this section.

"SEC. 256. COORDINATION FOR INTERCONNECTIVITY.

"(a) PURPOSE- It is the purpose of this section—

"(1) to promote nondiscriminatory accessibility by the broadest number of users and vendors of communications products and services to public telecommunications networks used to provide telecommunications service through—

"(A) coordinated public telecommunications network planning and design by telecommunications carriers and other providers of telecommunications service; and

"(B) public telecommunications network interconnectivity, and interconnectivity of devices with such networks used to provide telecommunications service; and

"(2) to ensure the ability of users and information providers to seamlessly and transparently transmit and receive information between and across telecommunications networks.

"(b) COMMISSION FUNCTIONS- In carrying out the purposes of this section, the Commission—

"(1) shall establish procedures for Commission oversight of coordinated network planning by telecommunications carriers and other providers of telecommunications service for the effective and efficient interconnection of public telecommunications networks used to provide telecommunications service; and

"(2) may participate, in a manner consistent with its authority and practice prior to the date of enactment of this section, in the development by appropriate industry standards-setting organizations of public telecommunications network interconnectivity standards that promote access to—

"(A) public telecommunications networks used to provide telecommunications service;

"(B) network capabilities and services by individuals with disabilities; and

"(C) information services by subscribers of rural telephone companies.

"(c) COMMISSION'S AUTHORITY- Nothing in this section shall be construed as expanding or limiting any authority that the Commission may have under law in effect before the date of enactment of the Telecommunications Act of 1996.

"(d) DEFINITION- As used in this section, the term "public telecommunications network interconnectivity" means the ability of two or more public telecommunications networks used to provide telecommunications service to communicate and exchange information without degeneration, and to interact in concert with one another.

"SEC. 257. MARKET ENTRY BARRIERS PROCEEDING.

"(a) ELIMINATION OF BARRIERS- Within 15 months after the date of enactment of the Telecommunications Act of 1996, the Commission shall complete a proceeding for the purpose of identifying and eliminating, by regulations pursuant to its authority under this Act (other than this section), market entry barriers for entrepreneurs and other small businesses in the provision and ownership of telecommunications services and information services, or in the provision of parts or services to providers of telecommunications services and information services.

"(b) NATIONAL POLICY- In carrying out subsection (a), the Commission shall seek to promote the policies and purposes of this Act favoring diversity of media voices, vigorous economic competition, technological advancement, and promotion of the public interest, convenience, and necessity.

"(c) PERIODIC REVIEW- Every 3 years following the completion of the proceeding required by subsection (a), the Commission shall review and report to Congress on—

"(1) any regulations prescribed to eliminate barriers within its jurisdiction that are identified under subsection (a) and that can be prescribed consistent with the public interest, convenience, and necessity; and

"(2) the statutory barriers identified under subsection (a) that the Commission recommends be eliminated, consistent with the public interest, convenience, and necessity.

"SEC. 258. ILLEGAL CHANGES IN SUBSCRIBER CARRIER SELECTIONS.

"(a) PROHIBITION- No telecommunications carrier shall submit or execute a change in a subscriber's selection of a provider of telephone exchange service or telephone toll service except in accordance with such verification procedures as the Commission shall prescribe. Nothing in this section shall preclude any State commission from enforcing such procedures with respect to intrastate services.

"(b) LIABILITY FOR CHARGES- Any telecommunications carrier that violates the verification procedures described in subsection (a) and that collects charges for telephone exchange service or telephone toll service from a subscriber shall be liable to the carrier previously selected by the subscriber in an amount equal to all charges paid by such subscriber after such violation, in accordance with such procedures as the Commission may prescribe. The remedies provided by this subsection are in addition to any other remedies available by law.

"SEC. 259. INFRASTRUCTURE SHARING.

"(a) REGULATIONS REQUIRED- The Commission shall prescribe, within one year after the date of enactment of the Telecommunications

Act of 1996, regulations that require incumbent local exchange carriers (as defined in section 251(h)) to make available to any qualifying carrier such public switched network infrastructure, technology, information, and telecommunications facilities and functions as may be requested by such qualifying carrier for the purpose of enabling such qualifying carrier to provide telecommunications services, or to provide access to information services, in the service area in which such qualifying carrier has requested and obtained designation as an eligible telecommunications carrier under section 214(e).

"(b) TERMS AND CONDITIONS OF REGULATIONS- The regulations prescribed by the Commission pursuant to this section shall—

"(1) not require a local exchange carrier to which this section applies to take any action that is economically unreasonable or that is contrary to the public interest;

"(2) permit, but shall not require, the joint ownership or operation of public switched network infrastructure and services by or among such local exchange carrier and a qualifying carrier;

"(3) ensure that such local exchange carrier will not be treated by the Commission or any State as a common carrier for hire or as offering common carrier services with respect to any infrastructure, technology, information, facilities, or functions made available to a qualifying carrier in accordance with regulations issued pursuant to this section;

"(4) ensure that such local exchange carrier makes such infrastructure, technology, information, facilities, or functions available to a qualifying carrier on just and reasonable terms and conditions that permit such qualifying carrier to fully benefit from the economies of scale and scope of such local exchange carrier, as determined in accordance with guidelines prescribed by the Commission in regulations issued pursuant to this section;

"(5) establish conditions that promote cooperation between local exchange carriers to which this section applies and qualifying carriers;

"(6) not require a local exchange carrier to which this section applies to engage in any infrastructure sharing agreement for any services or access which are to be provided or offered to consumers by the qualifying carrier in such local exchange carrier's telephone exchange area; and

"(7) require that such local exchange carrier file with the Commission or State for public inspection, any tariffs, contracts, or other arrangements showing the rates, terms, and conditions under which such carrier is making available public switched network infrastructure and functions under this section.

"(c) INFORMATION CONCERNING DEPLOYMENT OF NEW

SERVICES AND EQUIPMENT- A local exchange carrier to which this section applies that has entered into an infrastructure sharing agreement under this section shall provide to each party to such agreement timely information on the planned deployment of telecommunications services and equipment, including any software or upgrades of software integral to the use or operation of such telecommunications equipment.

"(d) DEFINITION- For purposes of this section, the term "qualifying carrier" means a telecommunications carrier that—

"(1) lacks economies of scale or scope, as determined in accordance with regulations prescribed by the Commission pursuant to this section; and

"(2) offers telephone exchange service, exchange access, and any other service that is included in universal service, to all consumers without preference throughout the service area for which such carrier has been designated as an eligible telecommunications carrier under section 214(e).

"SEC. 260. PROVISION OF TELEMESSAGING SERVICE.

"(a) NONDISCRIMINATION SAFEGUARDS- Any local exchange carrier subject to the requirements of section 251(c) that provides tele-messaging service—

"(1) shall not subsidize its telemessaging service directly or indirectly from its telephone exchange service or its exchange access; and

"(2) shall not prefer or discriminate in favor of its telemessaging service operations in its provision of telecommunications services.

"(b) EXPEDITED CONSIDERATION OF COMPLAINTS- The Commission shall establish procedures for the receipt and review of complaints concerning violations of subsection (a) or the regulations thereunder that result in material financial harm to a provider of telemessaging service. Such procedures shall ensure that the Commission will make a final determination with respect to any such complaint within 120 days after receipt of the complaint. If the complaint contains an appropriate showing that the alleged violation occurred, the Commission shall, within 60 days after receipt of the complaint, order the local exchange carrier and any affiliates to cease engaging in such violation pending such final determination.

"(c) DEFINITION- As used in this section, the term "telemessaging service" means voice mail and voice storage and retrieval services, any live operator services used to record, transcribe, or relay messages (other than telecommunications relay services), and any ancillary services offered in combination with these services.

"SEC. 261. EFFECT ON OTHER REQUIREMENTS.

"(a) COMMISSION REGULATIONS- Nothing in this part shall be construed to prohibit the Commission from enforcing regulations prescribed prior to the date of enactment of the Telecommunications Act of 1996 in fulfilling the requirements of this part, to the extent that such regulations are not inconsistent with the provisions of this part.

"(b) EXISTING STATE REGULATIONS- Nothing in this part shall be construed to prohibit any State commission from enforcing regulations prescribed prior to the date of enactment of the Telecommunications Act of 1996, or from prescribing regulations after such date of enactment, in fulfilling the requirements of this part, if such regulations are not inconsistent with the provisions of this part.

"(c) ADDITIONAL STATE REQUIREMENTS- Nothing in this part precludes a State from imposing requirements on a telecommunications carrier for intrastate services that are necessary to further competition in the provision of telephone exchange service or exchange access, as long as the State's requirements are not inconsistent with this part or the Commission's regulations to implement this part.".

(b) DESIGNATION OF PART I- Title II of the Act is further amended by inserting before the heading of section 201 the following new heading: "PART I—COMMON CARRIER REGULATION".

(c) STYLISTIC CONSISTENCY- The Act is amended so that—

(1) the designation and heading of each title of the Act shall be in the form and typeface of the designation and heading of this title of this Act; and

(2) the designation and heading of each part of each title of the Act shall be in the form and typeface of the designation and heading of part I of title II of the Act, as amended by subsection (a).

SEC. 102. ELIGIBLE TELECOMMUNICATIONS CARRIERS.

(a) IN GENERAL- Section 214 (47 U.S.C. 214) is amended by adding at the end thereof the following new subsection:

"(e) PROVISION OF UNIVERSAL SERVICE-

"(1) ELIGIBLE TELECOMMUNICATIONS CARRIERS- A common carrier designated as an eligible telecommunications carrier under paragraph (2) or (3) shall be eligible to receive universal service support in accordance with section 254 and shall, throughout the service area for which the designation is received—

"(A) offer the services that are supported by Federal universal service support mechanisms under section 254(c), either using its own facilities or a combination of its own facilities and resale of another carrier's services (including the services offered by another eligible telecommunications carrier); and

"(B) advertise the availability of such services and the charges therefor using media of general distribution.

"(2) DESIGNATION OF ELIGIBLE TELECOMMUNICATIONS CARRIERS- A State commission shall upon its own motion or upon request designate a common carrier that meets the requirements of paragraph (1) as an eligible telecommunications carrier for a service area designated by the State commission. Upon request and consistent with the public interest, convenience, and necessity, the State commission may, in the case of an area served by a rural telephone company, and shall, in the case of all other areas, designate more than one common carrier as an eligible telecommunications carrier for a service area designated by the State commission, so long as each additional requesting carrier meets the requirements of paragraph (1). Before designating an additional eligible telecommunications carrier for an area served by a rural telephone company, the State commission shall find that the designation is in the public interest.

"(3) DESIGNATION OF ELIGIBLE TELECOMMUNICATIONS CARRIERS FOR UNSERVED AREAS- If no common carrier will provide the services that are supported by Federal universal service support mechanisms under section 254(c) to an unserved community or any portion thereof that requests such service, the Commission, with respect to interstate services, or a State commission, with respect to intrastate services, shall determine which common carrier or carriers are best able to provide such service to the requesting unserved community or portion thereof and shall order such carrier or carriers to provide such service for that unserved community or portion thereof. Any carrier or carriers ordered to provide such service under this paragraph shall meet the requirements of paragraph (1) and shall be designated as an eligible telecommunications carrier for that community or portion thereof.

"(4) RELINQUISHMENT OF UNIVERSAL SERVICE- A State commission shall permit an eligible telecommunications carrier to relinquish its designation as such a carrier in any area served by more than one eligible telecommunications carrier. An eligible telecommunications carrier that seeks to relinquish its eligible telecommunications carrier designation for an area served by more than one eligible telecommunications carrier shall give advance notice to the State commission of such relinquishment. Prior to permitting a telecommunications carrier designated as an eligible telecommunications carrier to cease providing universal service in an area served by more than one eligible telecommunications carrier, the State commission shall require the remaining eligible telecommunications carrier or carriers to ensure that all customers served by the relinquishing carrier will continue to be served, and shall require sufficient notice to permit the purchase or construction of adequate facilities by any remaining eligible telecommunications carrier. The State commission shall establish a time, not to exceed one year after the State commission approves such relinquishment under this paragraph, within which such purchase or construction shall be completed.

"(5) SERVICE AREA DEFINED- The term "service area" means a geographic area established by a State commission for the purpose of determining universal service obligations and support mechanisms. In the case of an area served by a rural telephone company, "service area" means such company's "study area" unless and until the Commission and the States, after taking into account recommendations of a Federal-State Joint Board instituted under section 410(c), establish a different definition of service area for such company.".

SEC. 103. EXEMPT TELECOMMUNICATIONS COMPANIES.

The Public Utility Holding Company Act of 1935 (15 U.S.C. 79 and following) is amended by redesignating sections 34 and 35 as sections 35 and 36, respectively, and by inserting the following new section after section 33:

"SEC. 34. EXEMPT TELECOMMUNICATIONS COMPANIES.

"(a) DEFINITIONS- For purposes of this section—

"(1) EXEMPT TELECOMMUNICATIONS COMPANY- The term "exempt telecommunications company" means any person determined by the Federal Communications Commission to be engaged directly or indirectly, wherever located, through one or more affiliates (as defined in section 2(a)(11)(B)), and exclusively in the business of providing—

"(A) telecommunications services;

"(B) information services;

"(C) other services or products subject to the jurisdiction of the Federal Communications Commission; or

"(D) products or services that are related or incidental to the provision of a product or service described in subparagraph (A), (B), or (C). No person shall be deemed to be an exempt telecommunications company under this section unless such person has applied to the Federal Communications Commission for a determination under this paragraph. A person applying in good faith for such a determination shall be deemed an exempt telecommunications company under this section, with all of the exemptions provided by this section, until the Federal Communications Commission makes such determination. The Federal Communications Commission shall make such determination within 60 days of its receipt of any such application filed after the enactment of this section and shall notify the Commission whenever a determination is made under this paragraph that any person is an exempt telecommunications company. Not later than 12 months after the date of enactment of this section, the Federal Communications Commission shall promulgate rules implementing the provisions of this paragraph which shall be applicable to applications filed under this paragraph after the effective date of such rules.

"(2) OTHER TERMS- For purposes of this section, the terms "telecommunications services" and "information services" shall have the same meanings as provided in the Communications Act of 1934.

"(b) STATE CONSENT FOR SALE OF EXISTING RATE-BASED FACILITIES- If a rate or charge for the sale of electric energy or natural gas (other than any portion of a rate or charge which represents recovery of the cost of a wholesale rate or charge) for, or in connection with, assets of a public utility company that is an associate company or affiliate of a registered holding company was in effect under the laws of any State as of December 19, 1995, the public utility company owning such assets may not sell such assets to an exempt telecommunications company that is an associate company or affiliate unless State commissions having jurisdiction over such public utility company approve such sale. Nothing in this subsection shall preempt the otherwise applicable authority of any State to approve or disapprove the sale of such assets. The approval of the Commission under this Act shall not be required for the sale of assets as provided in this subsection.

"(c) OWNERSHIP OF ETCS BY EXEMPT HOLDING COMPA-NIES- Notwithstanding any provision of this Act, a holding company that is exempt under section 3 of this Act shall be permitted, without condition or limitation under this Act, to acquire and maintain an interest in the business of one or more exempt telecommunications companies.

"(d) OWNERSHIP OF ETCS BY REGISTERED HOLDING COM-PANIES- Notwithstanding any provision of this Act, a registered holding company shall be permitted (without the need to apply for, or receive, approval from the Commission, and otherwise without condition under this Act) to acquire and hold the securities, or an interest in the business, of one or more exempt telecommunications companies.

"(e) FINANCING AND OTHER RELATIONSHIPS BETWEEN

ETCS AND REGISTERED HOLDING COMPANIES- The relationship between an exempt telecommunications company and a registered holding company, its affiliates and associate companies, shall remain subject to the jurisdiction of the Commission under this Act: *Provided, That*—

"(1) section 11 of this Act shall not prohibit the ownership of an interest in the business of one or more exempt telecommunications companies by a registered holding company (regardless of activities engaged in or where facilities owned or operated by such exempt telecommunications companies are located), and such ownership by a registered holding company shall be deemed consistent with the operation of an integrated public utility system;

"(2) the ownership of an interest in the business of one or more exempt telecommunications companies by a registered holding company (regardless of activities engaged in or where facilities owned or operated by such exempt telecommunications companies are located) shall be considered as reasonably incidental, or economically necessary or appropriate, to the operations of an integrated public utility system;

"(3) the Commission shall have no jurisdiction under this Act over, and there shall be no restriction or approval required under this Act with respect to (A) the issue or sale of a security by a registered holding company for purposes of financing the acquisition of an exempt telecommunications company, or (B) the guarantee of a security of an exempt telecommunications company by a registered holding company; and

"(4) except for costs that should be fairly and equitably allocated among companies that are associate companies of a registered holding company, the Commission shall have no jurisdiction under this Act over the sales, service, and construction contracts between an exempt telecommunications company and a registered holding company, its affiliates and associate companies.

"(f) REPORTING OBLIGATIONS CONCERNING INVESTMENTS AND ACTIVITIES OF REGISTERED PUBLIC-UTILITY HOLDING COMPANY SYSTEMS-

"(1) OBLIGATIONS TO REPORT INFORMATION- Any registered holding company or subsidiary thereof that acquires or holds the securities, or an interest in the business, of an exempt telecommunications company shall file with the Commission such information as the Commission, by rule, may prescribe concerning—

"(A) investments and activities by the registered holding company, or any subsidiary thereof, with respect to exempt telecommunications companies, and

"(B) any activities of an exempt telecommunications company within the holding company system, that are reasonably likely to have a material impact on the financial or operational condition of the holding company system.

"(2) AUTHORITY TO REQUIRE ADDITIONAL INFORMATION- If, based on reports provided to the Commission pursuant to paragraph (1) of this subsection or other available information, the Commission reasonably concludes that it has concerns regarding the financial or operational condition of any registered holding company or any subsidiary thereof (including an exempt telecommunications company), the Commission may require such registered holding company to make additional reports and provide additional information.

"(3) AUTHORITY TO LIMIT DISCLOSURE OF INFORMATION- Notwithstanding any other provision of law, the Commission shall not be compelled to disclose any information required to be reported under this subsection. Nothing in this subsection shall authorize the Commission to withhold the information from Congress, or prevent the Commission from complying with a request for information from any other Federal or State department or agency requesting the information for purposes within the scope of its jurisdiction. For purposes of section 552 of title 5, United States Code, this subsection shall be considered a statute described in subsection (b)(3)(B) of such section 552.

"(g) ASSUMPTION OF LIABILITIES- Any public utility company that is an associate company, or an affiliate, of a registered holding company and that is subject to the jurisdiction of a State commission with respect to its retail electric or gas rates shall not issue any security for the purpose of financing the acquisition, ownership, or operation of an exempt telecommunications company. Any public utility company that is an associate company, or an affiliate, of a registered holding company and that is subject to the jurisdiction of a State commission with respect to its retail electric or gas rates shall not assume any obligation or liability as guarantor, endorser, surety, or otherwise by the public utility company in respect of any security of an exempt telecommunications company.

"(h) PLEDGING OR MORTGAGING OF ASSETS- Any public utility company that is an associate company, or affiliate, of a registered holding company and that is subject to the jurisdiction of a State commission with respect to its retail electric or gas rates shall not pledge, mortgage, or otherwise use as collateral any assets of the public utility company or assets of any subsidiary company thereof for the benefit of an exempt telecommunications company.

"(i) PROTECTION AGAINST ABUSIVE AFFILIATE TRANSAC-

TIONS- A public utility company may enter into a contract to purchase services or products described in subsection (a)(1) from an exempt telecommunications company that is an affiliate or associate company of the public utility company only if—

"(1) every State commission having jurisdiction over the retail rates of such public utility company approves such contract; or

"(2) such public utility company is not subject to State commission retail rate regulation and the purchased services or products—

"(A) would not be resold to any affiliate or associate company; or

"(B) would be resold to an affiliate or associate company and every State commission having jurisdiction over the retail rates of such affiliate or associate company makes the determination required by subparagraph (A). The requirements of this subsection shall not apply in any case in which the State or the State commission concerned publishes a notice that the State or State commission waives its authority under this subsection.

"(j) NONPREEMPTION OF RATE AUTHORITY- Nothing in this Act shall preclude the Federal Energy Regulatory Commission or a State commission from exercising its jurisdiction under otherwise applicable law to determine whether a public utility company may recover in rates the costs of products or services purchased from or sold to an associate company or affiliate that is an exempt telecommunications company, regardless of whether such costs are incurred through the direct or indirect purchase or sale of products or services from such associate company or affiliate.

"(k) RECIPROCAL ARRANGEMENTS PROHIBITED- Reciprocal arrangements among companies that are not affiliates or associate companies of each other that are entered into in order to avoid the provisions of this section are prohibited.

"(l) BOOKS AND RECORDS- (1) Upon written order of a State commission, a State commission may examine the books, accounts, memoranda, contracts, and records of—

"(A) a public utility company subject to its regulatory authority under State law;

"(B) any exempt telecommunications company selling products or services to such public utility company or to an associate company of such public utility company; and

"(C) any associate company or affiliate of an exempt telecommunications company which sells products or services to a public utility company referred to in subparagraph (A), wherever located, if such examination is required for the effective discharge of the State commission's regulatory responsibilities affecting the provision of electric or gas service in connection with the activities of such exempt telecommunications company.

"(2) Where a State commission issues an order pursuant to paragraph (1), the State commission shall not publicly disclose trade secrets or sensitive commercial information.

"(3) Any United States district court located in the State in which the State commission referred to in paragraph (1) is located shall have jurisdiction to enforce compliance with this subsection.

"(4) Nothing in this section shall—

"(A) preempt applicable State law concerning the provision of records and other information; or

"(B) in any way limit rights to obtain records and other information under Federal law, contracts, or otherwise.

"(m) INDEPENDENT AUDIT AUTHORITY FOR STATE COMMISSIONS-

"(1) STATE MAY ORDER AUDIT- Any State commission with jurisdiction over a public utility company that—

"(A) is an associate company of a registered holding company; and

"(B) transacts business, directly or indirectly, with a subsidiary company, an affiliate or an associate company that is an exempt telecommunications company, may order an independent audit to be performed, no more frequently than on an annual basis, of all matters deemed relevant by the selected auditor that reasonably relate to retail rates: *Provided,* That such matters relate, directly or indirectly, to transactions or transfers between the public utility company subject to its jurisdiction and such exempt telecommunications company.

"(2) SELECTION OF FIRM TO CONDUCT AUDIT- (A) If a State commission orders an audit in accordance with paragraph (1), the public utility company and the State commission shall jointly select, within 60 days, a firm to perform the audit. The firm selected to perform the audit shall possess demonstrated qualifications relating to—

"(i) competency, including adequate technical training and professional proficiency in each discipline necessary to carry out the audit; and

"(ii) independence and objectivity, including that the firm be free from personal or external impairments to independence, and should assume an independent position with the State commission and auditee, making certain that the audit is based upon an impartial consideration of all pertinent facts and responsible opinions.

"(B) The public utility company and the exempt telecommunications company shall cooperate fully with all reasonable requests necessary to perform the audit and the public utility company shall bear all costs of having the audit performed.

"(3) AVAILABILITY OF AUDITOR'S REPORT- The auditor's report shall be provided to the State commission not later than 6 months after the selection of the auditor, and provided to the public utility company not later than 60 days thereafter.

"(n) APPLICABILITY OF TELECOMMUNICATIONS REGULA-TION- Nothing in this section shall affect the authority of the Federal Communications Commission under the Communications Act of 1934, or the authority of State commissions under State laws concerning the provision of telecommunications services, to regulate the activities of an exempt telecommunications company.".

SEC. 104. NONDISCRIMINATION PRINCIPLE.
Section 1 (47 U.S.C. 151) is amended by inserting after "to all the people of the United States" the following: ", without discrimination on the basis of race, color, religion, national origin, or sex,".

SUBTITLE B—SPECIAL PROVISIONS CONCERNING BELL OPERATING COMPANIES
SEC. 151. BELL OPERATING COMPANY PROVISIONS.

(a) ESTABLISHMENT OF PART III OF TITLE II- Title II is amended by adding at the end of part II (as added by section 101) the following new part:

"PART III—SPECIAL PROVISIONS CONCERNING BELL OPERATING COMPANIES
"SEC. 271. BELL OPERATING COMPANY ENTRY INTO INTER-LATA SERVICES."(a) GENERAL LIMITATION- Neither a Bell operating company, nor any affiliate of a Bell operating company, may provide interLATA services except as provided in this section.

"(b) INTERLATA SERVICES TO WHICH THIS SECTION APPLIES-

"(1) IN-REGION SERVICES- A Bell operating company, or any affiliate of that Bell operating company, may provide interLATA services originating in any of its in-region States (as defined in subsection (i)) if the Commission approves the application of such company for such State under subsection (d)(3).

"(2) OUT-OF-REGION SERVICES- A Bell operating company, or any affiliate of that Bell operating company, may provide interLATA services originating outside its in-region States after the date of enactment of the Telecommunications Act of 1996, subject to subsection (j).

"(3) INCIDENTAL INTERLATA SERVICES- A Bell operating company, or any affiliate of a Bell operating company, may provide incidental interLATA services (as defined in subsection (g)) originating in any State after the date of enactment of the Telecommunications Act of 1996.

"(4) TERMINATION- Nothing in this section prohibits a Bell operating company or any of its affiliates from providing termination for inter-LATA services, subject to subsection (j).

"(c) REQUIREMENTS FOR PROVIDING CERTAIN IN-REGION INTERLATA SERVICES-

"(1) AGREEMENT OR STATEMENT- A Bell operating company meets the requirements of this paragraph if it meets the requirements of subparagraph (A) or subparagraph (B) of this paragraph for each State for which the authorization is sought.

"(A) PRESENCE OF A FACILITIES-BASED COMPETITOR- A Bell operating company meets the requirements of this subparagraph if it has entered into one or more binding agreements that have been approved under section 252 specifying the terms and conditions under which the Bell operating company is providing access and interconnection to its network facilities for the network facilities of one or more unaffiliated competing providers of telephone exchange service (as defined in section 3(47)(A), but excluding exchange access) to residential and business subscribers. For the purpose of this subparagraph, such telephone exchange service may be offered by such competing providers either exclusively over their own telephone exchange service facilities or predominantly over their own telephone exchange service facilities in combination with the resale of the telecommunications services of another carrier. For the purpose of this subparagraph, services provided pursuant to subpart K of part 22 of the Commission's regulations (47 C.F.R. 22.901 et seq.) shall not be considered to be telephone exchange services.

"(B) FAILURE TO REQUEST ACCESS- A Bell operating company meets the requirements of this subparagraph if, after 10 months after the date of enactment of the Telecommunications Act of 1996, no such provider has requested the access and interconnection described in subparagraph (A) before the date which is 3 months before the date the company makes its application under subsection (d)(1), and a statement of the terms and conditions that the company generally offers to provide such access and interconnection has been approved or permitted to take effect by the State commission under section 252(f). For purposes of this subparagraph, a Bell operating company shall be considered not to have received any request for access and interconnection if the State commission of such State certifies that the only provider or providers making such a request have (i) failed to negotiate in good faith as required by section 252, or (ii) violated the terms of an agreement approved under section 252 by the provider's failure to comply, within a reasonable period of time, with the implementation schedule contained in such agreement.

"(2) SPECIFIC INTERCONNECTION REQUIREMENTS-

"(A) AGREEMENT REQUIRED- A Bell operating company meets the requirements of this paragraph if, within the State for which the authorization is sought—

"(i)(I) such company is providing access and interconnection pursuant to one or more agreements described in paragraph (1)(A), or

"(II) such company is generally offering access and interconnection pursuant to a statement described in paragraph (1)(B), and

"(ii) such access and interconnection meets the requirements of subparagraph (B) of this paragraph.

"(B) COMPETITIVE CHECKLIST- Access or interconnection provided or generally offered by a Bell operating company to other telecommunications carriers meets the requirements of this subparagraph if such access and interconnection includes each of the following:

"(i) Interconnection in accordance with the requirements of sections 251(c)(2) and 252(d)(1).

"(ii) Nondiscriminatory access to network elements in accordance with the requirements of sections 251(c)(3) and 252(d)(1).

"(iii) Nondiscriminatory access to the poles, ducts, conduits, and rights-of-way owned or controlled by the Bell operating company at just and reasonable rates in accordance with the requirements of section 224.

"(iv) Local loop transmission from the central office to the customer's premises, unbundled from local switching or other services.

"(v) Local transport from the trunk side of a wireline local exchange carrier switch unbundled from switching or other services.

"(vi) Local switching unbundled from transport, local loop transmission, or other services.

"(vii) Nondiscriminatory access to—

"(I) 911 and E911 services;

"(II) directory assistance services to allow the other carrier's customers to obtain telephone numbers; and

"(III) operator call completion services.

"(viii) White pages directory listings for customers of the other carrier's telephone exchange service.

"(ix) Until the date by which telecommunications numbering administration guidelines, plan, or rules are established, nondiscriminatory access to telephone numbers for assignment to the other carrier's telephone exchange service customers. After that date, compliance with such guidelines, plan, or rules.

"(x) Nondiscriminatory access to databases and associated signaling necessary for call routing and completion.

"(xi) Until the date by which the Commission issues regulations pursuant to section 251 to require number portability, interim telecommunications number portability through remote call forwarding, direct inward dialing trunks, or other comparable arrangements, with as little impairment of functioning, quality, reliability, and convenience as possible. After that date, full compliance with such regulations.

"(xii) Nondiscriminatory access to such services or information as are necessary to allow the requesting carrier to implement local dialing parity in accordance with the requirements of section 251(b)(3).

"(xiii) Reciprocal compensation arrangements in accordance with the requirements of section 252(d)(2).

"(xiv) Telecommunications services are available for resale in accordance with the requirements of sections 251(c)(4) and 252(d)(3).

"(d) ADMINISTRATIVE PROVISIONS-

"(1) APPLICATION TO COMMISSION- On and after the date of enactment of the Telecommunications Act of 1996, a Bell operating company or its affiliate may apply to the Commission for authorization to provide interLATA services originating in any in-region State. The application shall identify each State for which the authorization is sought.

"(2) CONSULTATION-

"(A) CONSULTATION WITH THE ATTORNEY GENERAL- The Commission shall notify the Attorney General promptly of any application under paragraph (1). Before making any determination under this subsection, the Commission shall consult with the Attorney General, and if the Attorney General submits any comments in writing, such comments shall be included in the record of the Commission's decision. In consulting with and submitting comments to the Commission under this paragraph, the Attorney General shall provide to the Commission an evaluation of the application using any standard the Attorney General considers appropriate. The Commission shall give substantial weight to the Attorney General's evaluation, but such evaluation shall not have any preclusive effect on any Commission decision under paragraph (3).

"(B) CONSULTATION WITH STATE COMMISSIONS- Before making any determination under this subsection, the Commission shall consult with the State commission of any State that is the subject of the application in order to verify the compliance of the Bell operating company with the requirements of subsection (c).

"(3) DETERMINATION- Not later than 90 days after receiving an application under paragraph (1), the Commission shall issue a written determination approving or denying the authorization requested in the application for each State. The Commission shall not approve the authorization requested in an application submitted under paragraph (1) unless it finds that—

"(A) the petitioning Bell operating company has met the requirements of subsection (c)(1) and—

"(i) with respect to access and interconnection provided pursuant to subsection (c)(1)(A), has fully implemented the competitive checklist in subsection (c)(2)(B); or

"(ii) with respect to access and interconnection generally offered pursuant to a statement under subsection (c)(1)(B), such statement offers all of the items included in the competitive checklist in subsection (c)(2)(B);

"(B) the requested authorization will be carried out in accordance with the requirements of section 272; and

"(C) the requested authorization is consistent with the public interest, convenience, and necessity. The Commission shall state the basis for its approval or denial of the application.

"(4) LIMITATION ON COMMISSION- The Commission may not, by rule or otherwise, limit or extend the terms used in the competitive checklist set forth in subsection (c)(2)(B).

"(5) PUBLICATION- Not later than 10 days after issuing a determination under paragraph (3), the Commission shall publish in the Federal Register a brief description of the determination.

"(6) ENFORCEMENT OF CONDITIONS-

"(A) COMMISSION AUTHORITY- If at any time after the approval of an application under paragraph (3), the Commission determines that a Bell operating company has ceased to meet any of the conditions required for such approval, the Commission may, after notice and opportunity for a hearing—

"(i) issue an order to such company to correct the deficiency;

"(ii) impose a penalty on such company pursuant to title V; or

"(iii) suspend or revoke such approval.

"(B) RECEIPT AND REVIEW OF COMPLAINTS- The Commission shall establish procedures for the review of complaints concerning failures by Bell operating companies to meet conditions required for approval under paragraph (3). Unless the parties otherwise agree, the Commission shall act on such complaint within 90 days.

"(e) LIMITATIONS-

"(1) JOINT MARKETING OF LOCAL AND LONG DISTANCE SERVICES- a Bell operating company is authorized pursuant to subsection (d) to provide interLATA services in an in-region State, or until 36 months have passed since the date of enactment of the Telecommunications Act of 1996, whichever is earlier, a telecommunications carrier that serves greater than 5 percent of the Nation's presubscribed access lines may not jointly market in such State telephone exchange service obtained from such company pursuant to section 251(c)(4) with interLATA services offered by that telecommunications carrier.

"(2) INTRALATA TOLL DIALING PARITY-

"(A) PROVISION REQUIRED- A Bell operating company granted authority to provide interLATA services under subsection (d) shall provide intraLATA toll dialing parity throughout that State coincident with its exercise of that authority.

"(B) LIMITATION- Except for single-LATA States and States that have issued an order by December 19, 1995, requiring a Bell operating company to implement intraLATA toll dialing parity, a State may not require a Bell operating company to implement intraLATA toll dialing parity in that State before a Bell operating company has been granted authority under this section to provide interLATA services originating in that State or before 3 years after the date of enactment of the Telecommunications Act of 1996, whichever is earlier. Nothing in this subparagraph precludes a State from issuing an order requiring intraLATA toll dialing parity in that State prior to either such date so long as such order does not take effect until after the earlier of either such dates.

"(f) EXCEPTION FOR PREVIOUSLY AUTHORIZED ACTIVITIES- Neither subsection (a) nor section 273 shall prohibit a Bell operating company or affiliate from engaging, at any time after the date of enactment of the Telecommunications Act of 1996, in any activity to the extent authorized by, and subject to the terms and conditions contained in, an order entered by the United States District Court for the District of Columbia pursuant to section VII or VIII(C) of the AT&T Consent Decree if such order was entered on or before such date of enactment, to the extent such order is not reversed or vacated on appeal. Nothing in this subsection shall be construed to limit, or to impose terms or conditions on, an activity in which a Bell operating company is otherwise authorized to engage under any other provision of this section.

"(g) DEFINITION OF INCIDENTAL INTERLATA SERVICES- For purposes of this section, the term "incidental interLATA services" means the interLATA provision by a Bell operating company or its affiliate—

"(1)(A) of audio programming, video programming, or other programming services to subscribers to such services of such company or affiliate;

"(B) of the capability for interaction by such subscribers to select or respond to such audio programming, video programming, or other programming services;

"(C) to distributors of audio programming or video programming that such company or affiliate owns or controls, or is licensed by the copyright owner of such programming (or by an assignee of such owner) to distribute; or

"(D) of alarm monitoring services;

"(2) of two-way interactive video services or Internet services over dedicated facilities to or for elementary and secondary schools as defined in section 254(h)(5);

"(3) of commercial mobile services in accordance with section 332(c) of this Act and with the regulations prescribed by the Commission pursuant to paragraph (8) of such section;

"(4) of a service that permits a customer that is located in one LATA to retrieve stored information from, or file information for storage in, information storage facilities of such company that are located in another LATA;

"(5) of signaling information used in connection with the provision of telephone exchange services or exchange access by a local exchange carrier; or

"(6) of network control signaling information to, and receipt of such signaling information from, common carriers offering interLATA services at any location within the area in which such Bell operating company provides telephone exchange services or exchange access.

"(h) LIMITATIONS- The provisions of subsection (g) are intended to be narrowly construed. The interLATA services provided under subparagraph (A), (B), or (C) of subsection (g)(1) are limited to those interLATA transmissions incidental to the provision by a Bell operating company or its affiliate of video, audio, and other programming services that the company or its affiliate is engaged in providing to the public. The Commission shall ensure that the provision of services authorized under subsection (g) by a Bell operating company or its affiliate will not adversely affect telephone exchange service ratepayers or competition in any telecommunications market.

"(i) ADDITIONAL DEFINITIONS- As used in this section—

"(1) IN-REGION STATE- The term "in-region State" means a State in which a Bell operating company or any of its affiliates was authorized to provide wireline telephone exchange service pursuant to the reorganization plan approved under the AT&T Consent Decree, as in effect on the day before the date of enactment of the Telecommunications Act of 1996.

"(2) AUDIO PROGRAMMING SERVICES- The term "audio programming services" means programming provided by, or generally considered to be comparable to programming provided by, a radio broadcast station.

"(3) VIDEO PROGRAMMING SERVICES; OTHER PROGRAMMING SERVICES- The terms "video programming service" and "other programming services" have the same meanings as such terms have under section 602 of this Act.

"(j) CERTAIN SERVICE APPLICATIONS TREATED AS IN-REGION SERVICE APPLICATIONS- For purposes of this section, a Bell operating company application to provide 800 service, private line service, or their equivalents that—

"(1) terminate in an in-region State of that Bell operating company, and

"(2) allow the called party to determine the interLATA carrier, shall be considered an in-region service subject to the requirements of subsection (b)(1).

"SEC. 272. SEPARATE AFFILIATE; SAFEGUARDS.

"(a) SEPARATE AFFILIATE REQUIRED FOR COMPETITIVE ACTIVITIES-

"(1) IN GENERAL- A Bell operating company (including any affiliate) which is a local exchange carrier that is subject to the requirements of section 251(c) may not provide any service described in paragraph (2) unless it provides that service through one or more affiliates that—

"(A) are separate from any operating company entity that is subject to the requirements of section 251(c); and

"(B) meet the requirements of subsection (b).

"(2) SERVICES FOR WHICH A SEPARATE AFFILIATE IS REQUIRED- The services for which a separate affiliate is required by paragraph (1) are:

"(A) Manufacturing activities (as defined in section 273(h)).

"(B) Origination of interLATA telecommunications services, other than—

"(i) incidental interLATA services described in paragraphs (1), (2), (3), (5), and (6) of section 271(g);

"(ii) out-of-region services described in section 271(b)(2); or

"(iii) previously authorized activities described in section 271(f).

"(C) InterLATA information services, other than electronic publishing (as defined in section 274(h)) and alarm monitoring services (as defined in section 275(e)).

"(b) STRUCTURAL AND TRANSACTIONAL REQUIREMENTS- The separate affiliate required by this section—

"(1) shall operate independently from the Bell operating company;

"(2) shall maintain books, records, and accounts in the manner prescribed by the Commission which shall be separate from the books, records, and accounts maintained by the Bell operating company of which it is an affiliate;

"(3) shall have separate officers, directors, and employees from the Bell operating company of which it is an affiliate;

"(4) may not obtain credit under any arrangement that would permit a creditor, upon default, to have recourse to the assets of the Bell operating company; and

"(5) shall conduct all transactions with the Bell operating company of which it is an affiliate on an arm's length basis with any such transactions reduced to writing and available for public inspection.

"(c) NONDISCRIMINATION SAFEGUARDS- In its dealings with its affiliate described in subsection (a), a Bell operating company—

"(1) may not discriminate between that company or affiliate and any other entity in the provision or procurement of goods, services, facilities, and information, or in the establishment of standards; and

"(2) shall account for all transactions with an affiliate described in subsection (a) in accordance with accounting principles designated or approved by the Commission.

"(d) BIENNIAL AUDIT-

"(1) GENERAL REQUIREMENT- A company required to operate a separate affiliate under this section shall obtain and pay for a joint Federal/State audit every 2 years conducted by an independent auditor to determine whether such company has complied with this section and the regulations promulgated under this section, and particularly whether such company has complied with the separate accounting requirements under subsection (b).

"(2) RESULTS SUBMITTED TO COMMISSION; STATE COMMISSIONS- The auditor described in paragraph (1) shall submit the results of the audit to the Commission and to the State commission of each State in which the company audited provides service, which shall make such results available for public inspection. Any party may submit comments on the final audit report.

"(3) ACCESS TO DOCUMENTS- For purposes of conducting audits and reviews under this subsection—

"(A) the independent auditor, the Commission, and the State commission shall have access to the financial accounts and records of each company and of its affiliates necessary to verify transactions conducted with that company that are relevant to the specific activities permitted under this section and that are necessary for the regulation of rates;

"(B) the Commission and the State commission shall have access to the working papers and supporting materials of any auditor who performs an audit under this section; and

"(C) the State commission shall implement appropriate procedures to ensure the protection of any proprietary information submitted to it under this section.

"(e) FULFILLMENT OF CERTAIN REQUESTS- A Bell operating company and an affiliate that is subject to the requirements of section 251(c)—

"(1) shall fulfill any requests from an unaffiliated entity for telephone exchange service and exchange access within a period no longer than the period in which it provides such telephone exchange service and exchange access to itself or to its affiliates;

"(2) shall not provide any facilities, services, or information concerning its provision of exchange access to the affiliate described in subsection (a) unless such facilities, services, or information are made available to other providers of interLATA services in that market on the same terms and conditions;

"(3) shall charge the affiliate described in subsection (a), or impute to itself (if using the access for its provision of its own services), an amount for access to its telephone exchange service and exchange access that is no less than the amount charged to any unaffiliated interexchange carriers for such service; and

"(4) may provide any interLATA or intraLATA facilities or services to its interLATA affiliate if such services or facilities are made available to all carriers at the same rates and on the same terms and conditions, and so long as the costs are appropriately allocated.

"(f) SUNSET-

"(1) MANUFACTURING AND LONG DISTANCE- The provisions of this section (other than subsection (e)) shall cease to apply with respect to the manufacturing activities or the interLATA telecommunications services of a Bell operating company 3 years after the date such Bell operating company or any Bell operating company affiliate is authorized to provide interLATA telecommunications services under section 271(d), unless the Commission extends such 3-year period by rule or order.

"(2) INTERLATA INFORMATION SERVICES- The provisions of this section (other than subsection (e)) shall cease to apply with respect to the interLATA information services of a Bell operating company 4 years after the date of enactment of the Telecommunications Act of 1996, unless the Commission extends such 4-year period by rule or order.

"(3) PRESERVATION OF EXISTING AUTHORITY- Nothing in this subsection shall be construed to limit the authority of the Commission under any other section of this Act to prescribe safeguards consistent with the public interest, convenience, and necessity.

"(g) JOINT MARKETING-

"(1) AFFILIATE SALES OF TELEPHONE EXCHANGE SERVICES- A Bell operating company affiliate required by this section may not market or sell telephone exchange services provided by the Bell operating company unless that company permits other entities offering the same or similar service to market and sell its telephone exchange services.

"(2) BELL OPERATING COMPANY SALES OF AFFILIATE SERVICES- A Bell operating company may not market or sell interLATA service provided by an affiliate required by this section within any of its in-region States until such company is authorized to provide interLATA services in such State under section 271(d).

"(3) RULE OF CONSTRUCTION- The joint marketing and sale of services permitted under this subsection shall not be considered to violate the nondiscrimination provisions of subsection (c).

"(h) TRANSITION- With respect to any activity in which a Bell operating company is engaged on the date of enactment of the Telecommunications Act of 1996, such company shall have one year from such date of enactment to comply with the requirements of this section.

"SEC. 273. MANUFACTURING BY BELL OPERATING COMPANIES.

"(a) AUTHORIZATION- A Bell operating company may manufacture and provide telecommunications equipment, and manufacture customer premises equipment, if the Commission authorizes that Bell operating company or any Bell operating company affiliate to provide interLATA services under section 271(d), subject to the requirements of this section and the regulations prescribed thereunder, except that neither a Bell operating company nor any of its affiliates may engage in such manufactur-

ing in conjunction with a Bell operating company not so affiliated or any of its affiliates.

"(b) COLLABORATION; RESEARCH AND ROYALTY AGREE-MENTS-

"(1) COLLABORATION- Subsection (a) shall not prohibit a Bell operating company from engaging in close collaboration with any manufacturer of customer premises equipment or telecommunications equipment during the design and development of hardware, software, or combinations thereof related to such equipment.

"(2) CERTAIN RESEARCH ARRANGEMENTS; ROYALTY AGREE-MENTS-

Subsection (a) shall not prohibit a Bell operating company from—

"(A) engaging in research activities related to manufacturing, and

"(B) entering into royalty agreements with manufacturers of telecommunications equipment.

"(c) INFORMATION REQUIREMENTS-

"(1) INFORMATION ON PROTOCOLS AND TECHNICAL RE-QUIREMENTS-

Each Bell operating company shall, in accordance with regulations prescribed by the Commission, maintain and file with the Commission full and complete information with respect to the protocols and technical requirements for connection with and use of its telephone exchange service facilities. Each such company shall report promptly to the Commission any material changes or planned changes to such protocols and requirements, and the schedule for implementation of such changes or planned changes.

"(2) DISCLOSURE OF INFORMATION- A Bell operating company shall not disclose any information required to be filed under paragraph (1) unless that information has been filed promptly, as required by regulation by the Commission.

"(3) ACCESS BY COMPETITORS TO INFORMATION- The Commission may prescribe such additional regulations under this subsection as may be necessary to ensure that manufacturers have access to the information with respect to the protocols and technical requirements for connection with and use of telephone exchange service facilities that a Bell operating company makes available to any manufacturing affiliate or any unaffiliated manufacturer.

"(4) PLANNING INFORMATION- Each Bell operating company shall provide, to interconnecting carriers providing telephone exchange service, timely information on the planned deployment of telecommunications equipment.

"(d) MANUFACTURING LIMITATIONS FOR STANDARD-SETTING ORGANIZATIONS-

"(1) APPLICATION TO BELL COMMUNICATIONS RESEARCH OR MANUFACTURERS- Bell Communications Research, Inc., or any successor entity or affiliate—

"(A) shall not be considered a Bell operating company or a successor or assign of a Bell operating company at such time as it is no longer an affiliate of any Bell operating company; and

"(B) notwithstanding paragraph (3), shall not engage in manufacturing telecommunications equipment or customer premises equipment as long as it is an affiliate of more than 1 otherwise unaffiliated Bell operating company or successor or assign of any such company. Nothing in this subsection prohibits Bell Communications Research, Inc., or any successor entity, from engaging in any activity in which it is lawfully engaged on the date of enactment of the Telecommunications Act of 1996. Nothing provided in this subsection shall render Bell Communications Research, Inc., or any successor entity, a common carrier under title II of this Act. Nothing in this subsection restricts any manufacturer from engaging in any activity in which it is lawfully engaged on the date of enactment of the Telecommunications Act of 1996.

"(2) PROPRIETARY INFORMATION- Any entity which establishes standards for telecommunications equipment or customer premises equipment, or generic network requirements for such equipment, or certifies telecommunications equipment or customer premises equipment, shall be prohibited from releasing or otherwise using any proprietary information, designated as such by its owner, in its possession as a result of such activity, for any purpose other than purposes authorized in writing by the owner of such information, even after such entity ceases to be so engaged.

"(3) MANUFACTURING SAFEGUARDS- (A) Except as prohibited in paragraph (1), and subject to paragraph (6), any entity which certifies telecommunications equipment or customer premises equipment manufactured by an unaffiliated entity shall only manufacture a particular class of telecommunications equipment or customer premises equipment for which it is undertaking or has undertaken, during the previous 18 months, certification activity for such class of equipment through a separate affiliate.

"(B) Such separate affiliate shall—

"(i) maintain books, records, and accounts separate from those of the entity that certifies such equipment, consistent with generally acceptable accounting principles;

"(ii) not engage in any joint manufacturing activities with such entity; and

"(iii) have segregated facilities and separate employees with such entity.

"(C) Such entity that certifies such equipment shall—

"(i) not discriminate in favor of its manufacturing affiliate in the establishment of standards, generic requirements, or product certification;

"(ii) not disclose to the manufacturing affiliate any proprietary information that has been received at any time from an unaffiliated manufacturer, unless authorized in writing by the owner of the information; and

"(iii) not permit any employee engaged in product certification for telecommunications equipment or customer premises equipment to engage jointly in sales or marketing of any such equipment with the affiliated manufacturer.

"(4) STANDARD-SETTING ENTITIES- Any entity that is not an accredited standards development organization and that establishes industry-wide standards for telecommunications equipment or customer premises equipment, or industry-wide generic network requirements for such equipment, or that certifies telecommunications equipment or customer premises equipment manufactured by an unaffiliated entity, shall—

"(A) establish and publish any industry-wide standard for, industry-wide generic requirement for, or any substantial modification of an existing industry-wide standard or industry-wide generic requirement for, telecommunications equipment or customer premises equipment only in compliance with the following procedure—

"(i) such entity shall issue a public notice of its consideration of a proposed industry-wide standard or industry-wide generic requirement;

"(ii) such entity shall issue a public invitation to interested industry parties to fund and participate in such efforts on a reasonable and nondiscriminatory basis, administered in such a manner as not to unreasonably exclude any interested industry party;

"(iii) such entity shall publish a text for comment by such parties as have agreed to participate in the process pursuant to clause (ii), provide such parties a full opportunity to submit comments, and respond to comments from such parties;

"(iv) such entity shall publish a final text of the industry-wide standard or industry-wide generic requirement, including the comments in their entirety, of any funding party which requests to have its comments so published; and

"(v) such entity shall attempt, prior to publishing a text for comment, to agree with the funding parties as a group on a mutually satisfactory dispute resolution process which such parties shall utilize as their sole recourse in the event of a dispute on technical issues as to which there is disagreement between any funding party and the entity conducting such activities, except that if no dispute resolution process is agreed to by all the parties, a funding party may utilize the dispute resolution procedures established pursuant to paragraph (5) of this subsection;

"(B) engage in product certification for telecommunications equipment or customer premises equipment manufactured by unaffiliated entities only if—

"(i) such activity is performed pursuant to published criteria;

"(ii) such activity is performed pursuant to auditable criteria; and

"(iii) such activity is performed pursuant to available industry-accepted testing methods and standards, where applicable, unless otherwise agreed upon by the parties funding and performing such activity;

"(C) not undertake any actions to monopolize or attempt to monopolize the market for such services; and

"(D) not preferentially treat its own telecommunications equipment or customer premises equipment, or that of its affiliate, over that of any other entity in establishing and publishing industry-wide standards or industry-wide generic requirements for, and in certification of, telecommunications equipment and customer premises equipment.

"(5) ALTERNATE DISPUTE RESOLUTION- Within 90 days after the date of enactment of the Telecommunications Act of 1996, the Commission shall prescribe a dispute resolution process to be utilized in the event that a dispute resolution process is not agreed upon by all the parties when establishing and publishing any industry-wide standard or industry-wide generic requirement for telecommunications equipment or customer premises equipment, pursuant to paragraph (4)(A)(v). The Commission shall not establish itself as a party to the dispute resolution process. Such dispute resolution process shall permit any funding party to resolve a dispute with the entity conducting the activity that significantly affects such funding party's interests, in an open, nondiscriminatory, and unbiased fashion, within 30 days after the filing of such dispute. Such disputes may be filed within 15 days after the date the funding party receives a response to its comments from the entity conducting the activity. The Commission shall establish penalties to be assessed for delays caused by referral of frivolous disputes to the dispute resolution process.

"(6) SUNSET- The requirements of paragraphs (3) and (4) shall terminate for the particular relevant activity when the Commission determines that there are alternative sources of industry-wide standards, industry-wide generic requirements, or product certification for a particular class of telecommunications equipment or customer premises equipment available in the United States. Alternative sources shall be deemed to exist when such sources provide commercially viable alternatives that are providing such services to customers. The Commission shall act on any application for such a determination within 90 days after receipt of such application, and shall receive public comment on such application.

"(7) ADMINISTRATION AND ENFORCEMENT AUTHORITY- For the purposes of administering this subsection and the regulations prescribed thereunder, the Commission shall have the same remedial authority as the Commission has in administering and enforcing the provisions of this title with respect to any common carrier subject to this Act.

"(8) DEFINITIONS- For purposes of this subsection:

"(A) The term "affiliate" shall have the same meaning as in section 3 of this Act, except that, for purposes of paragraph (1)(B)—

"(i) an aggregate voting equity interest in Bell Communications Research, Inc., of at least 5 percent of its total voting equity, owned directly or indirectly by more than 1 otherwise unaffiliated Bell operating company, shall constitute an affiliate relationship; and

"(ii) a voting equity interest in Bell Communications Research, Inc., by any otherwise unaffiliated Bell operating company of less than 1 percent of Bell Communications Research's total voting equity shall not be considered to be an equity interest under this paragraph.

"(B) The term "generic requirement" means a description of acceptable product attributes for use by local exchange carriers in establishing product specifications for the purchase of telecommunications equipment, customer premises equipment, and software integral thereto.

"(C) The term "industry-wide" means activities funded by or performed on behalf of local exchange carriers for use in providing wireline telephone exchange service whose combined total of deployed access lines in the United States constitutes at least 30 percent of all access lines deployed by telecommunications carriers in the United States as of the date of enactment of the Telecommunications Act of 1996.

"(D) The term "certification" means any technical process whereby a party determines whether a product, for use by more than one local exchange carrier, conforms with the specified requirements pertaining to such product.

"(E) The term "accredited standards development organization" means an entity composed of industry members which has been accredited by an institution vested with the responsibility for standards accreditation by the industry.

"(e) BELL OPERATING COMPANY EQUIPMENT PROCURE-MENT AND SALES-

"(1) NONDISCRIMINATION STANDARDS FOR MANUFAC-TURING- In the procurement or awarding of supply contracts for telecommunications equipment, a Bell operating company, or any entity acting on its behalf, for the duration of the requirement for a separate subsidiary including manufacturing under this Act—

"(A) shall consider such equipment, produced or supplied by unrelated persons; and

"(B) may not discriminate in favor of equipment produced or supplied by an affiliate or related person.

"(2) PROCUREMENT STANDARDS- Each Bell operating company or any entity acting on its behalf shall make procurement decisions and award all supply contracts for equipment, services, and software on the basis of an objective assessment of price, quality, delivery, and other commercial factors.

"(3) NETWORK PLANNING AND DESIGN- A Bell operating company shall, to the extent consistent with the antitrust laws, engage in joint network planning and design with local exchange carriers operating in the same area of interest. No participant in such planning shall be allowed to delay the introduction of new technology or the deployment of facilities to provide telecommunications services, and agreement with such other carriers shall not be required as a prerequisite for such introduction or deployment.

"(4) SALES RESTRICTIONS- Neither a Bell operating company engaged in manufacturing nor a manufacturing affiliate of such a company shall restrict sales to any local exchange carrier of telecommunications equipment, including software integral to the operation of such equipment and related upgrades.

"(5) PROTECTION OF PROPRIETARY INFORMATION- A Bell operating company and any entity it owns or otherwise controls shall protect the proprietary information submitted for procurement decisions from release not specifically authorized by the owner of such information.

"(f) ADMINISTRATION AND ENFORCEMENT AUTHORITY- For the purposes of administering and enforcing the provisions of this section and the regulations prescribed thereunder, the Commission shall have the same authority, power, and functions with respect to any Bell operating company or any affiliate thereof as the Commission has in administering and enforcing the provisions of this title with respect to any common carrier subject to this Act.

"(g) ADDITIONAL RULES AND REGULATIONS- The Commission

may prescribe such additional rules and regulations as the Commission determines are necessary to carry out the provisions of this section, and otherwise to prevent discrimination and cross-subsidization in a Bell operating company's dealings with its affiliate and with third parties.

"(h) DEFINITION- As used in this section, the term "manufacturing" has the same meaning as such term has under the AT&T Consent Decree.

"SEC. 274. ELECTRONIC PUBLISHING BY BELL OPERATING COMPANIES.

"(a) LIMITATIONS- No Bell operating company or any affiliate may engage in the provision of electronic publishing that is disseminated by means of such Bell operating company's or any of its affiliates' basic telephone service, except that nothing in this section shall prohibit a separated affiliate or electronic publishing joint venture operated in accordance with this section from engaging in the provision of electronic publishing.

"(b) SEPARATED AFFILIATE OR ELECTRONIC PUBLISHING JOINT VENTURE REQUIREMENTS- A separated affiliate or electronic publishing joint venture shall be operated independently from the Bell operating company. Such separated affiliate or joint venture and the Bell operating company with which it is affiliated shall—

"(1) maintain separate books, records, and accounts and prepare separate financial statements;

"(2) not incur debt in a manner that would permit a creditor of the separated affiliate or joint venture upon default to have recourse to the assets of the Bell operating company;

"(3) carry out transactions (A) in a manner consistent with such independence, (B) pursuant to written contracts or tariffs that are filed with the Commission and made publicly available, and (C) in a manner that is auditable in accordance with generally accepted auditing standards;

"(4) value any assets that are transferred directly or indirectly from the Bell operating company to a separated affiliate or joint venture, and record any transactions by which such assets are transferred, in accordance with such regulations as may be prescribed by the Commission or a State commission to prevent improper cross subsidies;

"(5) between a separated affiliate and a Bell operating company—

"(A) have no officers, directors, and employees in common after the effective date of this section; and

"(B) own no property in common;

"(6) not use for the marketing of any product or service of the separated affiliate or joint venture, the name, trademarks, or service marks of an existing Bell operating company except for names, trademarks, or service marks that are owned by the entity that owns or controls the Bell operating company;

"(7) not permit the Bell operating company—

"(A) to perform hiring or training of personnel on behalf of a separated affiliate;

"(B) to perform the purchasing, installation, or maintenance of equipment on behalf of a separated affiliate, except for telephone service that it provides under tariff or contract subject to the provisions of this section; or

"(C) to perform research and development on behalf of a separated affiliate;

"(8) each have performed annually a compliance review—

"(A) that is conducted by an independent entity for the purpose of determining compliance during the preceding calendar year with any provision of this section; and

"(B) the results of which are maintained by the separated affiliate or joint venture and the Bell operating company for a period of 5 years subject to review by any lawful authority; and

"(9) within 90 days of receiving a review described in paragraph (8), file a report of any exceptions and corrective action with the Commission and allow any person to inspect and copy such report subject to reasonable safeguards to protect any proprietary information contained in such report from being used for purposes other than to enforce or pursue remedies under this section.

"(c) JOINT MARKETING-

"(1) IN GENERAL- Except as provided in paragraph (2)—

"(A) a Bell operating company shall not carry out any promotion, marketing, sales, or advertising for or in conjunction with a separated affiliate; and

"(B) a Bell operating company shall not carry out any promotion, marketing, sales, or advertising for or in conjunction with an affiliate that is related to the provision of electronic publishing.

"(2) PERMISSIBLE JOINT ACTIVITIES-

"(A) JOINT TELEMARKETING- A Bell operating company may provide inbound telemarketing or referral services related to the provision of electronic publishing for a separated affiliate, electronic publishing joint venture, affiliate, or unaffiliated electronic publisher: *Provided*, That if such services are provided to a separated affiliate, electronic publishing joint venture, or affiliate, such services shall be made available to all electronic publishers on request, on nondiscriminatory terms.

"(B) TEAMING ARRANGEMENTS- A Bell operating company may engage in nondiscriminatory teaming or business arrangements to engage in electronic publishing with any separated affiliate or with any other electronic publisher if (i) the Bell operating company only provides facilities, services, and basic telephone service information as authorized by this section, and (ii) the Bell operating company does not own such teaming or business arrangement.

"(C) ELECTRONIC PUBLISHING JOINT VENTURES- A Bell operating company or affiliate may participate on a nonexclusive basis in electronic publishing joint ventures with entities that are not a Bell operating company, affiliate, or separated affiliate to provide electronic publishing services, if the Bell operating company or affiliate has not more than a 50 percent direct or indirect equity interest (or the equivalent thereof) or the right to more than 50 percent of the gross revenues under a revenue sharing or royalty agreement in any electronic publishing joint venture. Officers and employees of a Bell operating company or affiliate participating in an electronic publishing joint venture may not have more than 50 percent of the voting control over the electronic publishing joint venture. In the case of joint ventures with small, local electronic publishers, the Commission for good cause shown may authorize the Bell operating company or affiliate to have a larger equity interest, revenue share, or voting control but not to exceed 80 percent. A Bell operating company participating in an electronic publishing joint venture may provide promotion, marketing, sales, or advertising personnel and services to such joint venture.

"(d) BELL OPERATING COMPANY REQUIREMENT- A Bell operating company under common ownership or control with a separated affiliate or electronic publishing joint venture shall provide network access and interconnections for basic telephone service to electronic publishers at just and reasonable rates that are tariffed (so long as rates for such services are subject to regulation) and that are not higher on a per-unit basis than those charged for such services to any other electronic publisher or any separated affiliate engaged in electronic publishing.

"(e) PRIVATE RIGHT OF ACTION-

"(1) DAMAGES- Any person claiming that any act or practice of any Bell operating company, affiliate, or separated affiliate constitutes a violation of this section may file a complaint with the Commission or bring suit as provided in section 207 of this Act, and such Bell operating company, affiliate, or separated affiliate shall be liable as provided in section

206 of this Act; except that damages may not be awarded for a violation that is discovered by a compliance review as required by subsection (b)(7) of this section and corrected within 90 days.

"(2) CEASE AND DESIST ORDERS- In addition to the provisions of paragraph (1), any person claiming that any act or practice of any Bell operating company, affiliate, or separated affiliate constitutes a violation of this section may make application to the Commission for an order to cease and desist such violation or may make application in any district court of the United States of competent jurisdiction for an order enjoining such acts or practices or for an order compelling compliance with such requirement.

"(f) SEPARATED AFFILIATE REPORTING REQUIREMENT- Any separated affiliate under this section shall file with the Commission annual reports in a form substantially equivalent to the Form 10-K required by regulations of the Securities and Exchange Commission.

"(g) EFFECTIVE DATES-

"(1) TRANSITION- Any electronic publishing service being offered to the public by a Bell operating company or affiliate on the date of enactment of the Telecommunications Act of 1996 shall have one year from such date of enactment to comply with the requirements of this section.

"(2) SUNSET- The provisions of this section shall not apply to conduct occurring after 4 years after the date of enactment of the Telecommunications Act of 1996.

"(h) DEFINITION OF ELECTRONIC PUBLISHING-

"(1) IN GENERAL- The term "electronic publishing" means the dissemination, provision, publication, or sale to an unaffiliated entity or person, of any one or more of the following: news (including sports); entertainment (other than interactive games); business, financial, legal, consumer, or credit materials; editorials, columns, or features; advertising; photos or images; archival or research material; legal notices or public records; scientific, educational, instructional, technical, professional, trade, or other literary materials; or other like or similar information.

"(2) EXCEPTIONS- The term "electronic publishing" shall not include the following services:

"(A) Information access, as that term is defined by the AT&T Consent Decree.

"(B) The transmission of information as a common carrier.

"(C) The transmission of information as part of a gateway to an information service that does not involve the generation or alteration of the content of information, including data transmission, address translation, protocol conversion, billing management, introductory information content, and navigational systems that enable users to access electronic publishing services, which do not affect the presentation of such electronic publishing services to users.

"(D) Voice storage and retrieval services, including voice messaging and electronic mail services.

"(E) Data processing or transaction processing services that do not involve the generation or alteration of the content of information.

"(F) Electronic billing or advertising of a Bell operating company's regulated telecommunications services.

"(G) Language translation or data format conversion.

"(H) The provision of information necessary for the management, control, or operation of a telephone company telecommunications system.

"(I) The provision of directory assistance that provides names, addresses, and telephone numbers and does not include advertising.

"(J) Caller identification services.

"(K) Repair and provisioning databases and credit card and billing validation for telephone company operations.

"(L) 911-E and other emergency assistance databases.

"(M) Any other network service of a type that is like or similar to these network services and that does not involve the generation or alteration of the content of information.

"(N) Any upgrades to these network services that do not involve the generation or alteration of the content of information.

"(O) Video programming or full motion video entertainment on demand.

"(i) ADDITIONAL DEFINITIONS- As used in this section—

"(1) The term "affiliate" means any entity that, directly or indirectly, owns or controls, is owned or controlled by, or is under common ownership or control with, a Bell operating company. Such term shall not include a separated affiliate.

"(2) The term "basic telephone service" means any wireline telephone exchange service, or wireline telephone exchange service facility, provided by a Bell operating company in a telephone exchange area, except that such term does not include—

"(A) a competitive wireline telephone exchange service provided in a telephone exchange area where another entity provides a wireline telephone exchange service that was provided on January 1, 1984, or

"(B) a commercial mobile service.

"(3) The term "basic telephone service information" means network and customer information of a Bell operating company and other information acquired by a Bell operating company as a result of its engaging in the provision of basic telephone service.

"(4) The term "control" has the meaning that it has in 17 C.F.R. 240.12b-2, the regulations promulgated by the Securities and Exchange Commission pursuant to the Securities Exchange Act of 1934 (15 U.S.C. 78a et seq.) or any successor provision to such section.

"(5) The term "electronic publishing joint venture" means a joint venture owned by a Bell operating company or affiliate that engages in the provision of electronic publishing which is disseminated by means of such Bell operating company's or any of its affiliates' basic telephone service.

"(6) The term "entity" means any organization, and includes corporations, partnerships, sole proprietorships, associations, and joint ventures.

"(7) The term "inbound telemarketing" means the marketing of property, goods, or services by telephone to a customer or potential customer who initiated the call.

"(8) The term "own" with respect to an entity means to have a direct or indirect equity interest (or the equivalent thereof) of more than 10 percent of an entity, or the right to more than 10 percent of the gross revenues of an entity under a revenue sharing or royalty agreement.

"(9) The term "separated affiliate" means a corporation under common ownership or control with a Bell operating company that does not own or control a Bell operating company and is not owned or controlled by a Bell operating company and that engages in the provision of electronic publishing which is disseminated by means of such Bell operating company's or any of its affiliates' basic telephone service.

"(10) The term "Bell operating company" has the meaning provided in section 3, except that such term includes any entity or corporation that is owned or controlled by such a company (as so defined) but does not include an electronic publishing joint venture owned by such an entity or corporation.

"SEC. 275. ALARM MONITORING SERVICES.

"(a) DELAYED ENTRY INTO ALARM MONITORING-

"(1) PROHIBITION- No Bell operating company or affiliate thereof shall engage in the provision of alarm monitoring services before the date which is 5 years after the date of enactment of the Telecommunications Act of 1996.

"(2) EXISTING ACTIVITIES- Paragraph (1) does not prohibit or limit the provision, directly or through an affiliate, of alarm monitoring services by a Bell operating company that was engaged in providing alarm monitoring services as of November 30, 1995, directly or through an affiliate. Such Bell operating company or affiliate may not acquire any equity interest in, or obtain financial control of, any unaffiliated alarm monitoring service entity after November 30, 1995, and until 5 years after the date of enactment of the Telecommunications Act of 1996, except that this sentence shall not prohibit an exchange of customers for the customers of an unaffiliated alarm monitoring service entity.

"(b) NONDISCRIMINATION- An incumbent local exchange carrier (as defined in section 251(h)) engaged in the provision of alarm monitoring services shall—

"(1) provide nonaffiliated entities, upon reasonable request, with the network services it provides to its own alarm monitoring operations, on nondiscriminatory terms and conditions; and

"(2) not subsidize its alarm monitoring services either directly or indirectly from telephone exchange service operations.

"(c) EXPEDITED CONSIDERATION OF COMPLAINTS- The Commission shall establish procedures for the receipt and review of complaints concerning violations of subsection (b) or the regulations thereunder that result in material financial harm to a provider of alarm monitoring service. Such procedures shall ensure that the Commission will make a final determination with respect to any such complaint within 120 days after receipt of the complaint. If the complaint contains an appropriate showing that the alleged violation occurred, as determined by the Commission in accordance with such regulations, the Commission shall, within 60 days after receipt of the complaint, order the incumbent local exchange carrier (as defined in section 251(h)) and its affiliates to cease engaging in such violation pending such final determination.

"(d) USE OF DATA- A local exchange carrier may not record or use in any fashion the occurrence or contents of calls received by providers of alarm monitoring services for the purposes of marketing such services on behalf of such local exchange carrier, or any other entity. Any regulations necessary to enforce this subsection shall be issued initially within 6 months after the date of enactment of the Telecommunications Act of 1996.

"(e) DEFINITION OF ALARM MONITORING SERVICE- The term "alarm monitoring service" means a service that uses a device located at a residence, place of business, or other fixed premises—

"(1) to receive signals from other devices located at or about such premises regarding a possible threat at such premises to life, safety, or property, from burglary, fire, vandalism, bodily injury, or other emergency, and

"(2) to transmit a signal regarding such threat by means of transmission facilities of a local exchange carrier or one of its affiliates to a remote monitoring center to alert a person at such center of the need to inform the customer or another person or police, fire, rescue, security, or public safety personnel of such threat, but does not include a service that uses a medical monitoring device attached to an individual for the automatic surveillance of an ongoing medical condition.

"SEC. 276. PROVISION OF PAYPHONE SERVICE.

"(a) NONDISCRIMINATION SAFEGUARDS- After the effective date of the rules prescribed pursuant to subsection (b), any Bell operating company that provides payphone service—

"(1) shall not subsidize its payphone service directly or indirectly from its telephone exchange service operations or its exchange access operations; and

"(2) shall not prefer or discriminate in favor of its payphone service.

"(b) REGULATIONS-

"(1) CONTENTS OF REGULATIONS- In order to promote competition among payphone service providers and promote the widespread deployment of payphone services to the benefit of the general public, within 9 months after the date of enactment of the Telecommunications Act of 1996, the Commission shall take all actions necessary (including any reconsideration) to prescribe regulations that—

"(A) establish a per call compensation plan to ensure that all payphone service providers are fairly compensated for each and every completed intrastate and interstate call using their payphone, except that emergency calls and telecommunications relay service calls for hearing disabled individuals shall not be subject to such compensation;

"(B) discontinue the intrastate and interstate carrier access charge payphone service elements and payments in effect on such date of enactment, and all intrastate and interstate payphone subsidies from basic exchange and exchange access revenues, in favor of a compensation plan as specified in subparagraph (A);

"(C) prescribe a set of nonstructural safeguards for Bell operating company payphone service to implement the provisions of paragraphs (1) and (2) of subsection (a), which safeguards shall, at a minimum, include the nonstructural safeguards equal to those adopted in the Computer Inquiry-III (CC Docket No. 90-623) proceeding;

"(D) provide for Bell operating company payphone service providers to have the same right that independent payphone providers have to negotiate with the location provider on the location provider's selecting and contracting with, and, subject to the terms of any agreement with the location provider, to select and contract with, the carriers that carry inter-LATA calls from their payphones, unless the Commission determines in the rulemaking pursuant to this section that it is not in the public interest; and

"(E) provide for all payphone service providers to have the right to negotiate with the location provider on the location provider's selecting and contracting with, and, subject to the terms of any agreement with the location provider, to select and contract with, the carriers that carry intra-LATA calls from their payphones.

"(2) PUBLIC INTEREST TELEPHONES- In the rulemaking conducted pursuant to paragraph (1), the Commission shall determine whether public interest payphones, which are provided in the interest of public health, safety, and welfare, in locations where there would otherwise not be a payphone, should be maintained, and if so, ensure that such public interest payphones are supported fairly and equitably.

"(3) EXISTING CONTRACTS- Nothing in this section shall affect any existing contracts between location providers and payphone service providers or interLATA or intraLATA carriers that are in force and effect as of the date of enactment of the Telecommunications Act of 1996.

"(c) STATE PREEMPTION- To the extent that any State requirements are inconsistent with the Commission's regulations, the Commission's regulations on such matters shall preempt such State requirements.

"(d) DEFINITION- As used in this section, the term "payphone service" means the provision of public or semi-public pay telephones, the provision of inmate telephone service in correctional institutions, and any ancillary services.".

(b) REVIEW OF ENTRY DECISIONS- Section 402(b) (47 U.S.C. 402(b)) is amended—

(1) in paragraph (6), by striking "(3), and (4)" and inserting "(3), (4), and (9)"; and

(2) by adding at the end the following new paragraph:

"(9) By any applicant for authority to provide interLATA services under section 271 of this Act whose application is denied by the Commission.".

TITLE II—BROADCAST SERVICES

SEC. 201. BROADCAST SPECTRUM FLEXIBILITY.

Title III is amended by inserting after section 335 (47 U.S.C. 335) the following new section:

"SEC. 336. BROADCAST SPECTRUM FLEXIBILITY.

"(a) COMMISSION ACTION- If the Commission determines to issue additional licenses for advanced television services, the Commission—

"(1) should limit the initial eligibility for such licenses to persons that, as of the date of such issuance, are licensed to operate a television broadcast station or hold a permit to construct such a station (or both); and

"(2) shall adopt regulations that allow the holders of such licenses to offer such ancillary or supplementary services on designated frequencies as may be consistent with the public interest, convenience, and necessity.

"(b) CONTENTS OF REGULATIONS- In prescribing the regulations required by subsection (a), the Commission shall—

"(1) only permit such licensee or permittee to offer ancillary or supplementary services if the use of a designated frequency for such services is consistent with the technology or method designated by the Commission for the provision of advanced television services;

"(2) limit the broadcasting of ancillary or supplementary services on designated frequencies so as to avoid derogation of any advanced television services, including high definition television broadcasts, that the Commission may require using such frequencies;

"(3) apply to any other ancillary or supplementary service such of the Commission's regulations as are applicable to the offering of analogous services by any other person, except that no ancillary or supplementary service shall have any rights to carriage under section 614 or 615 or be deemed a multichannel video programming distributor for purposes of section 628;

"(4) adopt such technical and other requirements as may be necessary or appropriate to assure the quality of the signal used to provide advanced television services, and may adopt regulations that stipulate the minimum number of hours per day that such signal must be transmitted; and

"(5) prescribe such other regulations as may be necessary for the protection of the public interest, convenience, and necessity.

"(c) RECOVERY OF LICENSE- If the Commission grants a license for advanced television services to a person that, as of the date of such issuance, is licensed to operate a television broadcast station or holds a permit to construct such a station (or both), the Commission shall, as a condition of such license, require that either the additional license or the original license held by the licensee be surrendered to the Commission for reallocation or reassignment (or both) pursuant to Commission regulation.

"(d) PUBLIC INTEREST REQUIREMENT- Nothing in this section shall be construed as relieving a television broadcasting station from its obligation to serve the public interest, convenience, and necessity. In the Commission's review of any application for renewal of a broadcast license for a television station that provides ancillary or supplementary services, the television licensee shall establish that all of its program services on the existing or advanced television spectrum are in the public interest. Any violation of the Commission rules applicable to ancillary or supplementary services shall reflect upon the licensee's qualifications for renewal of its license.

"(e) FEES-

"(1) SERVICES TO WHICH FEES APPLY- If the regulations prescribed pursuant to subsection (a) permit a licensee to offer ancillary or supplementary services on a designated frequency—

"(A) for which the payment of a subscription fee is required in order to receive such services, or

"(B) for which the licensee directly or indirectly receives compensation from a third party in return for transmitting material furnished by such third party (other than commercial advertisements used to support broadcasting for which a subscription fee is not required), the Commission shall establish a program to assess and collect from the licensee for such designated frequency an annual fee or other schedule or method of payment that promotes the objectives described in subparagraphs (A) and (B) of paragraph (2).

"(2) COLLECTION OF FEES- The program required by paragraph (1) shall—

"(A) be designed (i) to recover for the public a portion of the value of the public spectrum resource made available for such commercial use, and (ii) to avoid unjust enrichment through the method employed to permit such uses of that resource;

"(B) recover for the public an amount that, to the extent feasible, equals but does not exceed (over the term of the license) the amount that would have been recovered had such services been licensed pursuant to the provisions of section 309(j) of this Act and the Commission's regulations thereunder; and

"(C) be adjusted by the Commission from time to time in order to continue to comply with the requirements of this paragraph.

"(3) TREATMENT OF REVENUES-

"(A) GENERAL RULE- Except as provided in subparagraph (B), all proceeds obtained pursuant to the regulations required by this subsection shall be deposited in the Treasury in accordance with chapter 33 of title 31, United States Code.

"(B) RETENTION OF REVENUES- Notwithstanding subparagraph

(A), the salaries and expenses account of the Commission shall retain as an offsetting collection such sums as may be necessary from such proceeds for the costs of developing and implementing the program required by this section and regulating and supervising advanced television services. Such offsetting collections shall be available for obligation subject to the terms and conditions of the receiving appropriations account, and shall be deposited in such accounts on a quarterly basis.

"(4) REPORT- Within 5 years after the date of enactment of the Telecommunications Act of 1996, the Commission shall report to the Congress on the implementation of the program required by this subsection, and shall annually thereafter advise the Congress on the amounts collected pursuant to such program.

"(f) EVALUATION- Within 10 years after the date the Commission first issues additional licenses for advanced television services, the Commission shall conduct an evaluation of the advanced television services program. Such evaluation shall include—

"(1) an assessment of the willingness of consumers to purchase the television receivers necessary to receive broadcasts of advanced television services;

"(2) an assessment of alternative uses, including public safety use, of the frequencies used for such broadcasts; and

"(3) the extent to which the Commission has been or will be able to reduce the amount of spectrum assigned to licensees.

"(g) DEFINITIONS- As used in this section:

"(1) ADVANCED TELEVISION SERVICES- The term "advanced television services" means television services provided using digital or other advanced technology as further defined in the opinion, report, and order of the Commission entitled "Advanced Television Systems and Their Impact Upon the Existing Television Broadcast Service", MM Docket 87-268, adopted September 17, 1992, and successor proceedings.

"(2) DESIGNATED FREQUENCIES- The term "designated frequency" means each of the frequencies designated by the Commission for licenses for advanced television services.

"(3) HIGH DEFINITION TELEVISION- The term "high definition television" refers to systems that offer approximately twice the vertical and horizontal resolution of receivers generally available on the date of enactment of the Telecommunications Act of 1996, as further defined in the proceedings described in paragraph (1) of this subsection.".

SEC. 202. BROADCAST OWNERSHIP.
(a) NATIONAL RADIO STATION OWNERSHIP RULE CHANGES

REQUIRED- The Commission shall modify section 73.3555 of its regulations (47 C.F.R. 73.3555) by eliminating any provisions limiting the number of AM or FM broadcast stations which may be owned or controlled by one entity nationally.

(b) LOCAL RADIO DIVERSITY-

(1) APPLICABLE CAPS- The Commission shall revise section 73.3555(a) of its regulations (47 C.F.R. 73.3555) to provide that—

(A) in a radio market with 45 or more commercial radio stations, a party may own, operate, or control up to 8 commercial radio stations, not more than 5 of which are in the same service (AM or FM);

(B) in a radio market with between 30 and 44 (inclusive) commercial radio stations, a party may own, operate, or control up to 7 commercial radio stations, not more than 4 of which are in the same service (AM or FM);

(C) in a radio market with between 15 and 29 (inclusive) commercial radio stations, a party may own, operate, or control up to 6 commercial radio stations, not more than 4 of which are in the same service (AM or FM); and

(D) in a radio market with 14 or fewer commercial radio stations, a party may own, operate, or control up to 5 commercial radio stations, not more than 3 of which are in the same service (AM or FM), except that a party may not own, operate, or control more than 50 percent of the stations in such market.

(2) EXCEPTION- Notwithstanding any limitation authorized by this subsection, the Commission may permit a person or entity to own, operate, or control, or have a cognizable interest in, radio broadcast stations if the Commission determines that such ownership, operation, control, or interest will result in an increase in the number of radio broadcast stations in operation.

(c) TELEVISION OWNERSHIP LIMITATIONS-

(1) NATIONAL OWNERSHIP LIMITATIONS- The Commission shall modify its rules for multiple ownership set forth in section 73.3555 of its regulations (47 C.F.R. 73.3555)—

(A) by eliminating the restrictions on the number of television stations that a person or entity may directly or indirectly own, operate, or control, or have a cognizable interest in, nationwide; and

(B) by increasing the national audience reach limitation for television stations to 35 percent.

(2) LOCAL OWNERSHIP LIMITATIONS- The Commission shall conduct a rulemaking proceeding to determine whether to retain, modify, or eliminate its limitations on the number of television stations that

a person or entity may own, operate, or control, or have a cognizable interest in, within the same television market.

(d) RELAXATION OF ONE-TO-A-MARKET- With respect to its enforcement of its one-to-a-market ownership rules under section 73.3555 of its regulations, the Commission shall extend its waiver policy to any of the top 50 markets, consistent with the public interest, convenience, and necessity.

(e) DUAL NETWORK CHANGES- The Commission shall revise section 73.658(g) of its regulations (47 C.F.R. 658(g)) to permit a television broadcast station to affiliate with a person or entity that maintains 2 or more networks of television broadcast stations unless such dual or multiple networks are composed of—

(1) two or more persons or entities that, on the date of enactment of the Telecommunications Act of 1996, are "networks" as defined in section 73.3613(a)(1) of the Commission's regulations (47 C.F.R. 73.3613(a)(1)); or

(2) any network described in paragraph (1) and an English-language program distribution service that, on such date, provides 4 or more hours of programming per week on a national basis pursuant to network affiliation arrangements with local television broadcast stations in markets reaching more than 75 percent of television homes (as measured by a national ratings service).

(f) CABLE CROSS OWNERSHIP-

(1) ELIMINATION OF RESTRICTIONS- The Commission shall revise section 76.501 of its regulations (47 C.F.R. 76.501) to permit a person or entity to own or control a network of broadcast stations and a cable system.

(2) SAFEGUARDS AGAINST DISCRIMINATION- The Commission shall revise such regulations if necessary to ensure carriage, channel positioning, and nondiscriminatory treatment of nonaffiliated broadcast stations by a cable system described in paragraph (1).

(g) LOCAL MARKETING AGREEMENTS- Nothing in this section shall be construed to prohibit the origination, continuation, or renewal of any television local marketing agreement that is in compliance with the regulations of the Commission.

(h) FURTHER COMMISSION REVIEW- The Commission shall review its rules adopted pursuant to this section and all of its ownership rules biennially as part of its regulatory reform review under section 11 of the Communications Act of 1934 and shall determine whether any of such rules are necessary in the public interest as the result of competition.

The Commission shall repeal or modify any regulation it determines to be no longer in the public interest.

(i) ELIMINATION OF STATUTORY RESTRICTION- Section 613(a) (47 U.S.C. 533(a)) is amended—

(1) by striking paragraph (1);

(2) by redesignating paragraph (2) as subsection (a);

(3) by redesignating subparagraphs (A) and (B) as paragraphs (1) and (2), respectively;

(4) by striking "and" at the end of paragraph (1) (as so redesignated);

(5) by striking the period at the end of paragraph (2) (as so redesignated) and inserting "; and"; and

(6) by adding at the end the following new paragraph:

"(3) shall not apply the requirements of this subsection to any cable operator in any franchise area in which a cable operator is subject to effective competition as determined under section 623(l).".

SEC. 203. TERM OF LICENSES.

Section 307(c) (47 U.S.C. 307(c)) is amended to read as follows:

"(c) TERMS OF LICENSES-

"(1) INITIAL AND RENEWAL LICENSES- Each license granted for the operation of a broadcasting station shall be for a term of not to exceed 8 years. Upon application therefor, a renewal of such license may be granted from time to time for a term of not to exceed 8 years from the date of expiration of the preceding license, if the Commission finds that public interest, convenience, and necessity would be served thereby. Consistent with the foregoing provisions of this subsection, the Commission may by rule prescribe the period or periods for which licenses shall be granted and renewed for particular classes of stations, but the Commission may not adopt or follow any rule which would preclude it, in any case involving a station of a particular class, from granting or renewing a license for a shorter period than that prescribed for stations of such class if, in its judgment, the public interest, convenience, or necessity would be served by such action.

"(2) MATERIALS IN APPLICATION- In order to expedite action on applications for renewal of broadcasting station licenses and in order to avoid needless expense to applicants for such renewals, the Commission shall not require any such applicant to file any information which previously has been furnished to the Commission or which is not directly material to the considerations that affect the granting or denial of such application, but the Commission may require any new or additional facts it deems necessary to make its findings.

"(3) CONTINUATION PENDING DECISION- Pending any hearing and final decision on such an application and the disposition of any petition for rehearing pursuant to section 405, the Commission shall continue such license in effect.".

SEC. 204. BROADCAST LICENSE RENEWAL PROCEDURES.
(a) RENEWAL PROCEDURES-
(1) AMENDMENT- Section 309 (47 U.S.C. 309) is amended by adding at the end thereof the following new subsection:
"(k) BROADCAST STATION RENEWAL PROCEDURES-
"(1) STANDARDS FOR RENEWAL- If the licensee of a broadcast station submits an application to the Commission for renewal of such license, the Commission shall grant the application if it finds, with respect to that station, during the preceding term of its license—
"(A) the station has served the public interest, convenience, and necessity;
"(B) there have been no serious violations by the licensee of this Act or the rules and regulations of the Commission; and
"(C) there have been no other violations by the licensee of this Act or the rules and regulations of the Commission which, taken together, would constitute a pattern of abuse.
"(2) CONSEQUENCE OF FAILURE TO MEET STANDARD- If any licensee of a broadcast station fails to meet the requirements of this subsection, the Commission may deny the application for renewal in accordance with paragraph (3), or grant such application on terms and conditions as are appropriate, including renewal for a term less than the maximum otherwise permitted.
"(3) STANDARDS FOR DENIAL- If the Commission determines, after notice and opportunity for a hearing as provided in subsection (e), that a licensee has failed to meet the requirements specified in paragraph (1) and that no mitigating factors justify the imposition of lesser sanctions, the Commission shall—
"(A) issue an order denying the renewal application filed by such licensee under section 308; and
"(B) only thereafter accept and consider such applications for a construction permit as may be filed under section 308 specifying the channel or broadcasting facilities of the former licensee.
"(4) COMPETITOR CONSIDERATION PROHIBITED- In making the determinations specified in paragraph (1) or (2), the Commission shall not consider whether the public interest, convenience, and necessity might be served by the grant of a license to a person other than the renewal applicant.".

(2) CONFORMING AMENDMENT- Section 309(d) (47 U.S.C. 309(d)) is amended by inserting after "with subsection (a)" each place it appears the following: "(or subsection (k) in the case of renewal of any broadcast station license)".

(b) SUMMARY OF COMPLAINTS ON VIOLENT PROGRAM-MING- Section 308 (47 U.S.C. 308) is amended by adding at the end the following new subsection:

"(d) SUMMARY OF COMPLAINTS- Each applicant for the renewal of a commercial or noncommercial television license shall attach as an exhibit to the application a summary of written comments and suggestions received from the public and maintained by the licensee (in accordance with Commission regulations) that comment on the applicant's programming, if any, and that are characterized by the commentor as constituting violent programming.".

(c) EFFECTIVE DATE- The amendments made by this section apply to applications filed after May 1, 1995.

SEC. 205. DIRECT BROADCAST SATELLITE SERVICE.

(a) DBS SIGNAL SECURITY- Section 705(e)(4) (47 U.S.C. 605(e)(4)) is amended by inserting "or direct-to-home satellite services," after "programming,".

(b) FCC JURISDICTION OVER DIRECT-TO-HOME SATELLITE SERVICES- Section 303 (47 U.S.C. 303) is amended by adding at the end thereof the following new subsection:

"(v) Have exclusive jurisdiction to regulate the provision of direct-to-home satellite services. As used in this subsection, the term "direct-to-home satellite services" means the distribution or broadcasting of programming or services by satellite directly to the subscriber's premises without the use of ground receiving or distribution equipment, except at the subscriber's premises or in the uplink process to the satellite.".

SEC. 206. AUTOMATED SHIP DISTRESS AND SAFETY SYSTEMS.

Part II of title III is amended by inserting after section 364 (47 U.S.C. 362) the following new section:

"SEC. 365. AUTOMATED SHIP DISTRESS AND SAFETY SYS-TEMS. "Notwithstanding any provision of this Act or any other provision of law or regulation, a ship documented under the laws of the United States operating in accordance with the Global Maritime Distress and Safety System provisions of the Safety of Life at Sea Convention shall not be required to be equipped with a radio telegraphy station operated by one or more radio officers or operators. This section shall take effect for each vessel upon a determination by the United States Coast Guard that such vessel has the equipment required to implement the Global Maritime Distress and Safety System installed and operating in good working condition.".

SEC. 207. RESTRICTIONS ON OVER-THE-AIR RECEPTION DEVICES.

Within 180 days after the date of enactment of this Act, the Commission shall, pursuant to section 303 of the Communications Act of 1934, promulgate regulations to prohibit restrictions that impair a viewer's ability to receive video programming services through devices designed for over-the-air reception of television broadcast signals, multichannel multipoint distribution service, or direct broadcast satellite services.

TITLE III—CABLE SERVICES
SEC. 301. CABLE ACT REFORM.

(a) DEFINITIONS-

(1) DEFINITION OF CABLE SERVICE- Section 602(6)(B) (47 U.S.C. 522(6)(B)) is amended by inserting "or use" after "the selection".

(2) CHANGE IN DEFINITION OF CABLE SYSTEM- Section 602(7) (47 U.S.C. 522(7)) is amended by striking "(B) a facility that serves only subscribers in 1 or more multiple unit dwellings under common ownership, control, or management, unless such facility or facilities uses any public right-of-way;" and inserting "(B) a facility that serves subscribers without using any public right-of-way;".

(b) RATE DEREGULATION-

(1) UPPER TIER REGULATION- Section 623(c) (47 U.S.C. 543(c)) is amended—

(A) in paragraph (1)(B), by striking "subscriber, franchising authority, or other relevant State or local government entity" and inserting "franchising authority (in accordance with paragraph (3))";

(B) in paragraph (1)(C), by striking "such complaint" and inserting "the first complaint filed with the franchising authority under paragraph (3)"; and

(C) by striking paragraph (3) and inserting the following:

"(3) REVIEW OF RATE CHANGES- The Commission shall review any complaint submitted by a franchising authority after the date of enactment of the Telecommunications Act of 1996 concerning an increase in rates for cable programming services and issue a final order within 90 days after it receives such a complaint, unless the parties agree to extend the period for such review. A franchising authority may not file a complaint under this paragraph unless, within 90 days after such increase becomes effective it receives subscriber complaints.

"(4) SUNSET OF UPPER TIER RATE REGULATION- This subsection shall not apply to cable programming services provided after March 31, 1999.".

(2) SUNSET OF UNIFORM RATE STRUCTURE IN MARKETS WITH EFFECTIVE COMPETITION- Section 623(d) (47 U.S.C. 543(d)) is amended by adding at the end thereof the following: "This subsection does not apply to (1) a cable operator with respect to the provision of cable service over its cable system in any geographic area in which the video programming services offered by the operator in that area are subject to effective competition, or (2) any video programming offered on a per channel or per program basis. Bulk discounts to multiple dwelling units shall not be subject to this subsection, except that a cable operator of a cable system that is not subject to effective competition may not charge predatory prices to a multiple dwelling unit. Upon a prima facie showing by a complainant that there are reasonable grounds to believe that the discounted price is predatory, the cable system shall have the burden of showing that its discounted price is not predatory.".

(3) EFFECTIVE COMPETITION- Section 623(l)(1) (47 U.S.C. 543(l)(1)) is amended—

(A) by striking "or" at the end of subparagraph (B);

(B) by striking the period at the end of subparagraph (C) and inserting "; or"; and

(C) by adding at the end the following:

"(D) a local exchange carrier or its affiliate (or any multichannel video programming distributor using the facilities of such carrier or its affiliate) offers video programming services directly to subscribers by any means (other than direct-to-home satellite services) in the franchise area of an unaffiliated cable operator which is providing cable service in that franchise area, but only if the video programming services so offered in that area are comparable to the video programming services provided by the unaffiliated cable operator in that area.".

(c) GREATER DEREGULATION FOR SMALLER CABLE COMPANIES- Section 623

(47 U.S.C 543) is amended by adding at the end thereof the following:

"(m) SPECIAL RULES FOR SMALL COMPANIES-

"(1) IN GENERAL- Subsections (a), (b), and (c) do not apply to a small cable operator with respect to—

"(A) cable programming services, or

"(B) a basic service tier that was the only service tier subject to regulation as of December 31, 1994, in any franchise area in which that operator services 50,000 or fewer subscribers.

"(2) DEFINITION OF SMALL CABLE OPERATOR- For purposes of this subsection, the term "small cable operator" means a cable operator that, directly or through an affiliate, serves in the aggregate fewer than 1 percent of all subscribers in the United States and is not affiliated with any entity or entities whose gross annual revenues in the aggregate exceed $250,000,000.".

(d) MARKET DETERMINATIONS-

(1) MARKET DETERMINATIONS; EXPEDITED DECISIONMAK-ING- Section 614(h)(1)(C) (47 U.S.C. 534(h)(1)(C)) is amended—

(A) by striking "in the manner provided in section 73.3555(d)(3)(i) of title 47, Code of Federal Regulations, as in effect on May 1, 1991," in clause (i) and inserting "by the Commission by regulation or order using, where available, commercial publications which delineate television markets based on viewing patterns,"; and

(B) by striking clause (iv) and inserting the following:

"(iv) Within 120 days after the date on which a request is filed under this subparagraph (or 120 days after the date of enactment of the Telecommunications Act of 1996, if later), the Commission shall grant or deny the request.".

(2) APPLICATION TO PENDING REQUESTS- The amendment made by paragraph (1) shall apply to—

(A) any request pending under section 614(h)(1)(C) of the Communications Act of 1934 (47 U.S.C. 534(h)(1)(C)) on the date of enactment of this Act; and

(B) any request filed under that section after that date.

(e) TECHNICAL STANDARDS- Section 624(e) (47 U.S.C. 544(e)) is amended by striking the last two sentences and inserting the following: "No State or franchising authority may prohibit, condition, or restrict a cable system's use of any type of subscriber equipment or any transmission technology.".

(f) CABLE EQUIPMENT COMPATIBILITY- Section 624A (47 U.S.C. 544A) is amended—

(1) in subsection (a) by striking "and" at the end of paragraph (2), by striking the period at the end of paragraph (3) and inserting "; and"; and by adding at the end the following new paragraph:

"(4) compatibility among televisions, video cassette recorders, and cable systems can be assured with narrow technical standards that mandate a minimum degree of common design and operation, leaving all features, functions, protocols, and other product and service options for selection through open competition in the market.";

(2) in subsection (c)(1)—

(A) by redesignating subparagraphs (A) and (B) as subparagraphs (B) and (C), respectively; and

(B) by inserting before such redesignated subparagraph (B) the following new subparagraph:

"(A) the need to maximize open competition in the market for all features, functions, protocols, and other product and service options of converter boxes and other cable converters unrelated to the descrambling or decryption of cable television signals;"; and

(3) in subsection (c)(2)—

(A) by redesignating subparagraphs (D) and (E) as subparagraphs (E) and (F), respectively; and

(B) by inserting after subparagraph (C) the following new subparagraph:

"(D) to ensure that any standards or regulations developed under the authority of this section to ensure compatibility between televisions, video cassette recorders, and cable systems do not affect features, functions, protocols, and other product and service options other than those specified in paragraph (1)(B), including telecommunications interface equipment, home automation communications, and computer network services;".

(g) SUBSCRIBER NOTICE- Section 632 (47 U.S.C. 552) is amended—

(1) by redesignating subsection (c) as subsection (d); and

(2) by inserting after subsection (b) the following new subsection:

"(c) SUBSCRIBER NOTICE- A cable operator may provide notice of service and rate changes to subscribers using any reasonable written means at its sole discretion. Notwithstanding section 623(b)(6) or any other provision of this Act, a cable operator shall not be required to provide prior notice of any rate change that is the result of a regulatory fee, franchise fee, or any other fee, tax, assessment, or charge of any kind imposed by any Federal agency, State, or franchising authority on the transaction between the operator and the subscriber.".

(h) PROGRAM ACCESS- Section 628 (47 U.S.C. 548) is amended by adding at the end the following:

"(j) COMMON CARRIERS- Any provision that applies to a cable operator under this section shall apply to a common carrier or its affiliate that provides video programming by any means directly to subscribers. Any such provision that applies to a satellite cable programming vendor in which a cable operator has an attributable interest shall apply to any satellite cable programming vendor in which such common carrier has an attributable interest. For the purposes of this subsection, two or fewer common officers or directors shall not by itself establish an attributable interest by a common carrier in a satellite cable programming vendor (or its parent company).".

(i) ANTITRAFFICKING- Section 617 (47 U.S.C. 537) is amended—

(1) by striking subsections (a) through (d); and

(2) in subsection (e), by striking "(e)" and all that follows through "a franchising authority" and inserting "A franchising authority".

(j) AGGREGATION OF EQUIPMENT COSTS- Section 623(a) (47 U.S.C. 543(a)) is amended by adding at the end the following new paragraph:

"(7) AGGREGATION OF EQUIPMENT COSTS-

"(A) IN GENERAL- The Commission shall allow cable operators, pursuant to any rules promulgated under subsection (b)(3), to aggregate, on a franchise, system, regional, or company level, their equipment costs into broad categories, such as converter boxes, regardless of the varying levels of functionality of the equipment within each such broad category. Such aggregation shall not be permitted with respect to equipment used by subscribers who receive only a rate regulated basic service tier.

"(B) REVISION TO COMMISSION RULES; FORMS- Within 120 days of the date of enactment of the Telecommunications Act of 1996, the Commission shall issue revisions to the appropriate rules and forms necessary to implement subparagraph (A).".

(k) TREATMENT OF PRIOR YEAR LOSSES-

(1) AMENDMENT- Section 623 (48 U.S.C. 543) is amended by adding at the end thereof the following:

"(n) TREATMENT OF PRIOR YEAR LOSSES- Notwithstanding any other provision of this section or of section 612, losses associated with a cable system (including losses associated with the grant or award of a franchise) that were incurred prior to September 4, 1992, with respect to a cable system that is owned and operated by the original franchisee of such system shall not be disallowed, in whole or in part, in the determination of whether the rates for any tier of service or any type of equipment that is subject to regulation under this section are lawful.".

(2) EFFECTIVE DATE- The amendment made by paragraph (1) shall take effect on the date of enactment of this Act and shall be applicable to any rate proposal filed on or after September 4, 1993, upon which no final action has been taken by December 1, 1995.

SEC. 302. CABLE SERVICE PROVIDED BY TELEPHONE COMPANIES.

(a) PROVISIONS FOR REGULATION OF CABLE SERVICE PROVIDED BY TELEPHONE COMPANIES- Title VI (47 U.S.C. 521 et seq.) is amended by adding at the end the following new part:

"PART V—VIDEO PROGRAMMING SERVICES PROVIDED BY TELEPHONE COMPANIES

"SEC. 651. REGULATORY TREATMENT OF VIDEO PROGRAMMING SERVICES.

"(a) LIMITATIONS ON CABLE REGULATION-

"(1) RADIO-BASED SYSTEMS- To the extent that a common carrier (or any other person) is providing video programming to subscribers using radio communication, such carrier (or other person) shall be subject to the requirements of title III and section 652, but shall not otherwise be subject to the requirements of this title.

"(2) COMMON CARRIAGE OF VIDEO TRAFFIC- To the extent that a common carrier is providing transmission of video programming on a common carrier basis, such carrier shall be subject to the requirements of title II and section 652, but shall not otherwise be subject to the requirements of this title. This paragraph shall not affect the treatment under section 602(7)(C) of a facility of a common carrier as a cable system.

"(3) CABLE SYSTEMS AND OPEN VIDEO SYSTEMS- To the extent that a common carrier is providing video programming to its subscribers in any manner other than that described in paragraphs (1) and (2)—

"(A) such carrier shall be subject to the requirements of this title, unless such programming is provided by means of an open video system for which the Commission has approved a certification under section 653; or

"(B) if such programming is provided by means of an open video system for which the Commission has approved a certification under section 653, such carrier shall be subject to the requirements of this part, but shall be subject to parts I through IV of this title only as provided in 653(c).

"(4) ELECTION TO OPERATE AS OPEN VIDEO SYSTEM- A common carrier that is providing video programming in a manner described in paragraph (1) or (2), or a combination thereof, may elect to provide such programming by means of an open video system that complies with section 653. If the Commission approves such carrier's certification under section 653, such carrier shall be subject to the requirements of this part, but shall be subject to parts I through IV of this title only as provided in 653(c).

"(b) LIMITATIONS ON INTERCONNECTION OBLIGATIONS- A local exchange carrier that provides cable service through an open video system or a cable system shall not be required, pursuant to title II of this Act, to make capacity available on a nondiscriminatory basis to any other person for the provision of cable service directly to subscribers.

"(c) ADDITIONAL REGULATORY RELIEF- A common carrier shall not be required to obtain a certificate under section 214 with respect to the establishment or operation of a system for the delivery of video programming.

"SEC. 652. PROHIBITION ON BUY OUTS.

"(a) ACQUISITIONS BY CARRIERS- No local exchange carrier or any affiliate of such carrier owned by, operated by, controlled by, or under common control with such carrier may purchase or otherwise acquire directly or indirectly more than a 10 percent financial interest, or any management interest, in any cable operator providing cable service within the local exchange carrier's telephone service area.

"(b) ACQUISITIONS BY CABLE OPERATORS- No cable operator or affiliate of a cable operator that is owned by, operated by, controlled by, or under common ownership with such cable operator may purchase or otherwise acquire, directly or indirectly, more than a 10 percent financial interest, or any management interest, in any local exchange carrier providing telephone exchange service within such cable operator's franchise area.

"(c) JOINT VENTURES- A local exchange carrier and a cable operator whose telephone service area and cable franchise area, respectively, are in the same market may not enter into any joint venture or partnership to provide video programming directly to subscribers or to provide telecommunications services within such market.

"(d) EXCEPTIONS-

"(1) RURAL SYSTEMS- Notwithstanding subsections (a), (b), and (c) of this section, a local exchange carrier (with respect to a cable system located in its telephone service area) and a cable operator (with respect to the facilities of a local exchange carrier used to provide telephone exchange service in its cable franchise area) may obtain a controlling interest in, management interest in, or enter into a joint venture or partnership with the operator of such system or facilities for the use of such system or facilities to the extent that—

"(A) such system or facilities only serve incorporated or unincorporated—

"(i) places or territories that have fewer than 35,000 inhabitants; and

"(ii) are outside an urbanized area, as defined by the Bureau of the Census; and

"(B) in the case of a local exchange carrier, such system, in the aggregate with any other system in which such carrier has an interest, serves less than 10 percent of the households in the telephone service area of such carrier.

"(2) JOINT USE- Notwithstanding subsection (c), a local exchange carrier may obtain, with the concurrence of the cable operator on the rates, terms, and conditions, the use of that part of the transmission facilities of a cable system extending from the last multi-user terminal to the premises of the end user, if such use is reasonably limited in scope and duration, as determined by the Commission.

"(3) ACQUISITIONS IN COMPETITIVE MARKETS- Notwithstanding subsections (a) and (c), a local exchange carrier may obtain a controlling interest in, or form a joint venture or other partnership with, or provide financing to, a cable system (hereinafter in this paragraph referred to as "the subject cable system"), if—

"(A) the subject cable system operates in a television market that is not in the top 25 markets, and such market has more than 1 cable system operator, and the subject cable system is not the cable system with the most subscribers in such television market;

"(B) the subject cable system and the cable system with the most subscribers in such television market held on May 1, 1995, cable television franchises from the largest municipality in the television market and the boundaries of such franchises were identical on such date;

"(C) the subject cable system is not owned by or under common ownership or control of any one of the 50 cable system operators with the most subscribers as such operators existed on May 1, 1995; and

"(D) the system with the most subscribers in the television market is owned by or under common ownership or control of any one of the 10 largest cable system operators as such operators existed on May 1, 1995.

"(4) EXEMPT CABLE SYSTEMS- Subsection (a) does not apply to any cable system if—

"(A) the cable system serves no more than 17,000 cable subscribers, of which no less than 8,000 live within an urban area, and no less than 6,000 live within a nonurbanized area as of June 1, 1995;

"(B) the cable system is not owned by, or under common ownership or control with, any of the 50 largest cable system operators in existence on June 1, 1995; and

"(C) the cable system operates in a television market that was not in the top 100 television markets as of June 1, 1995.

"(5) SMALL CABLE SYSTEMS IN NONURBAN AREAS- Notwithstanding subsections (a) and (c), a local exchange carrier with less than $100,000,000 in annual operating revenues (or any affiliate of such carrier owned by, operated by, controlled by, or under common control with such carrier) may purchase or otherwise acquire more than a 10 percent financial interest in, or any management interest in, or enter into a joint venture or partnership with, any cable system within the local exchange carrier's telephone service area that serves no more than 20,000 cable subscribers, if no more than 12,000 of those subscribers live within an urbanized area, as defined by the Bureau of the Census.

"(6) WAIVERS- The Commission may waive the restrictions of subsections (a), (b), or (c) only if—

"(A) the Commission determines that, because of the nature of the market served by the affected cable system or facilities used to provide telephone exchange service—

"(i) the affected cable operator or local exchange carrier would be subjected to undue economic distress by the enforcement of such provisions;

"(ii) the system or facilities would not be economically viable if such provisions were enforced; or

"(iii) the anticompetitive effects of the proposed transaction are clearly outweighed in the public interest by the probable effect of the transaction in meeting the convenience and needs of the community to be served; and

"(B) the local franchising authority approves of such waiver.

"(e) DEFINITION OF TELEPHONE SERVICE AREA- For purposes of this section, the term "telephone service area" when used in connection with a common carrier subject in whole or in part to title II of this Act means the area within which such carrier provided telephone exchange service as of January 1, 1993, but if any common carrier after such date transfers its telephone exchange service facilities to another common carrier, the area to which such facilities provide telephone exchange service shall be treated as part of the telephone service area of the acquiring common carrier and not of the selling common carrier.

"SEC. 653. ESTABLISHMENT OF OPEN VIDEO SYSTEMS.

"(a) OPEN VIDEO SYSTEMS-

"(1) CERTIFICATES OF COMPLIANCE- A local exchange carrier

may provide cable service to its cable service subscribers in its telephone service area through an open video system that complies with this section. To the extent permitted by such regulations as the Commission may prescribe consistent with the public interest, convenience, and necessity, an operator of a cable system or any other person may provide video programming through an open video system that complies with this section. An operator of an open video system shall qualify for reduced regulatory burdens under subsection (c) of this section if the operator of such system certifies to the Commission that such carrier complies with the Commission's regulations under subsection (b) and the Commission approves such certification. The Commission shall publish notice of the receipt of any such certification and shall act to approve or disapprove any such certification within 10 days after receipt of such certification.

"(2) DISPUTE RESOLUTION- The Commission shall have the authority to resolve disputes under this section and the regulations prescribed thereunder. Any such dispute shall be resolved within 180 days after notice of such dispute is submitted to the Commission. At that time or subsequently in a separate damages proceeding, the Commission may, in the case of any violation of this section, require carriage, award damages to any person denied carriage, or any combination of such sanctions. Any aggrieved party may seek any other remedy available under this Act.

"(b) COMMISSION ACTIONS-

"(1) REGULATIONS REQUIRED- Within 6 months after the date of enactment of the Telecommunications Act of 1996, the Commission shall complete all actions necessary (including any reconsideration) to prescribe regulations that—

"(A) except as required pursuant to section 611, 614, or 615, prohibit an operator of an open video system from discriminating among video programming providers with regard to carriage on its open video system, and ensure that the rates, terms, and conditions for such carriage are just and reasonable, and are not unjustly or unreasonably discriminatory;

"(B) if demand exceeds the channel capacity of the open video system, prohibit an operator of an open video system and its affiliates from selecting the video programming services for carriage on more than one-third of the activated channel capacity on such system, but nothing in this subparagraph shall be construed to limit the number of channels that the carrier and its affiliates may offer to provide directly to subscribers;

"(C) permit an operator of an open video system to carry on only one channel any video programming service that is offered by more than one video programming provider (including the local exchange carrier's video programming affiliate): *Provided*, That subscribers have ready and immediate access to any such video programming service;

"(D) extend to the distribution of video programming over open video systems the Commission's regulations concerning sports exclusivity (47 C.F.R. 76.67), network nonduplication (47 C.F.R. 76.92 et seq.), and syndicated exclusivity (47 C.F.R. 76.151 et seq.); and

"(E)(i) prohibit an operator of an open video system from unreasonably discriminating in favor of the operator or its affiliates with regard to material or information (including advertising) provided by the operator to subscribers for the purposes of selecting programming on the open video system, or in the way such material or information is presented to subscribers;

"(ii) require an operator of an open video system to ensure that video programming providers or copyright holders (or both) are able suitably and uniquely to identify their programming services to subscribers;

"(iii) if such identification is transmitted as part of the programming signal, require the carrier to transmit such identification without change or alteration; and

"(iv) prohibit an operator of an open video system from omitting television broadcast stations or other unaffiliated video programming services carried on such system from any navigational device, guide, or menu.

"(2) CONSUMER ACCESS- Subject to the requirements of paragraph (1) and the regulations thereunder, nothing in this section prohibits a common carrier or its affiliate from negotiating mutually agreeable terms and conditions with over-the-air broadcast stations and other unaffiliated video programming providers to allow consumer access to their signals on any level or screen of any gateway, menu, or other program guide, whether provided by the carrier or its affiliate.

"(c) REDUCED REGULATORY BURDENS FOR OPEN VIDEO SYSTEMS-

"(1) IN GENERAL- Any provision that applies to a cable operator under—

"(A) sections 613 (other than subsection (a) thereof), 616, 623(f), 628, 631, and 634 of this title, shall apply,

"(B) sections 611, 614, and 615 of this title, and section 325 of title III, shall apply in accordance with the regulations prescribed under paragraph (2), and

"(C) sections 612 and 617, and parts III and IV (other than sections 623(f), 628, 631, and 634), of this title shall not apply, to any operator of an open video system for which the Commission has approved a certification under this section.

"(2) IMPLEMENTATION-

"(A) COMMISSION ACTION- In the rulemaking proceeding to prescribe the regulations required by subsection (b)(1), the Commission shall, to the extent possible, impose obligations that are no greater or lesser than the obligations contained in the provisions described in paragraph (1)(B) of this subsection. The Commission shall complete all action (including any reconsideration) to prescribe such regulations no later than 6 months after the date of enactment of the Telecommunications Act of 1996.

"(B) FEES- An operator of an open video system under this part may be subject to the payment of fees on the gross revenues of the operator for the provision of cable service imposed by a local franchising authority or other governmental entity, in lieu of the franchise fees permitted under section 622. The rate at which such fees are imposed shall not exceed the rate at which franchise fees are imposed on any cable operator transmitting video programming in the franchise area, as determined in accordance with regulations prescribed by the Commission. An operator of an open video system may designate that portion of a subscriber's bill attributable to the fee under this subparagraph as a separate item on the bill.

"(3) REGULATORY STREAMLINING- With respect to the establishment and operation of an open video system, the requirements of this section shall apply in lieu of, and not in addition to, the requirements of title II.

"(4) TREATMENT AS CABLE OPERATOR- Nothing in this Act precludes a video programming provider making use of an open video system from being treated as an operator of a cable system for purposes of section 111 of title 17, United States Code.

"(d) DEFINITION OF TELEPHONE SERVICE AREA- For purposes of this section, the term "telephone service area" when used in connection with a common carrier subject in whole or in part to title II of this Act means the area within which such carrier is offering telephone exchange service.".

(b) CONFORMING AND TECHNICAL AMENDMENTS-

(1) REPEAL- Subsection (b) of section 613 (47 U.S.C. 533(b)) is repealed.

(2) DEFINITIONS- Section 602 (47 U.S.C. 531) is amended—

(A) in paragraph (7), by striking ", or (D)" and inserting the following: ", unless the extent of such use is solely to provide interactive on-demand services; (D) an open video system that complies with section 653 of this title; or (E)";

(B) by redesignating paragraphs (12) through (19) as paragraphs (13) through (20), respectively; and

(C) by inserting after paragraph (11) the following new paragraph: "(12) the term "interactive on-demand services" means a service providing video programming to subscribers over switched networks on an on-demand, point-to-point basis, but does not include services providing video programming prescheduled by the programming provider;".

(3) TERMINATION OF VIDEO-DIALTONE REGULATIONS- The Commission's regulations and policies with respect to video dialtone requirements issued in CC Docket No. 87-266 shall cease to be effective on the date of enactment of this Act. This paragraph shall not be construed to require the termination of any video-dialtone system that the Commission has approved before the date of enactment of this Act.

SEC. 303. PREEMPTION OF FRANCHISING AUTHORITY REGULATION OF TELECOMMUNICATIONS SERVICES.

(a) PROVISION OF TELECOMMUNICATIONS SERVICES BY A CABLE OPERATOR- Section 621(b) (47 U.S.C. 541(b)) is amended by adding at the end thereof the following new paragraph:

"(3)(A) If a cable operator or affiliate thereof is engaged in the provision of telecommunications services—

"(i) such cable operator or affiliate shall not be required to obtain a franchise under this title for the provision of telecommunications services; and

"(ii) the provisions of this title shall not apply to such cable operator or affiliate for the provision of telecommunications services.

"(B) A franchising authority may not impose any requirement under this title that has the purpose or effect of prohibiting, limiting, restricting, or conditioning the provision of a telecommunications service by a cable operator or an affiliate thereof.

"(C) A franchising authority may not order a cable operator or affiliate thereof—

"(i) to discontinue the provision of a telecommunications service, or

"(ii) to discontinue the operation of a cable system, to the extent such cable system is used for the provision of a telecommunications service, by reason of the failure of such cable operator or affiliate thereof to obtain a franchise or franchise renewal under this title with respect to the provision of such telecommunications service.

"(D) Except as otherwise permitted by sections 611 and 612, a franchising authority may not require a cable operator to provide any telecommunications service or facilities, other than institutional networks, as a condition of the initial grant of a franchise, a franchise renewal, or a transfer of a franchise.".

(b) FRANCHISE FEES- Section 622(b) (47 U.S.C. 542(b)) is amended by inserting "to provide cable services" immediately before the period at the end of the first sentence thereof.

SEC. 304. COMPETITIVE AVAILABILITY OF NAVIGATION DEVICES.

Part III of title VI is amended by inserting after section 628 (47 U.S.C. 548) the following new section:

"SEC. 629. COMPETITIVE AVAILABILITY OF NAVIGATION DEVICES.

"(a) COMMERCIAL CONSUMER AVAILABILITY OF EQUIPMENT USED TO ACCESS SERVICES PROVIDED BY MULTICHANNEL VIDEO PROGRAMMING DISTRIBUTORS- The Commission shall, in consultation with appropriate industry standard-setting organizations, adopt regulations to assure the commercial availability, to consumers of multichannel video programming and other services offered over multichannel video programming systems, of converter boxes, interactive communications equipment, and other equipment used by consumers to access multichannel video programming and other services offered over multichannel video programming systems, from manufacturers, retailers, and other vendors not affiliated with any multichannel video programming distributor. Such regulations shall not prohibit any multichannel video programming distributor from also offering converter boxes, interactive communications equipment, and other equipment used by consumers to access multichannel video programming and other services offered over multichannel video programming systems, to consumers, if the system operator's charges to consumers for such devices and equipment are separately stated and not subsidized by charges for any such service.

"(b) PROTECTION OF SYSTEM SECURITY- The Commission shall not prescribe regulations under subsection (a) which would jeopardize security of multichannel video programming and other services offered over multichannel video programming systems, or impede the legal rights of a provider of such services to prevent theft of service.

"(c) WAIVER- The Commission shall waive a regulation adopted under subsection (a) for a limited time upon an appropriate showing by a provider of multichannel video programming and other services offered over multichannel video programming systems, or an equipment provider, that such waiver is necessary to assist the development or introduction of a new or improved multichannel video programming or other service offered over multichannel video programming systems, technology, or products. Upon an appropriate showing, the Commission shall grant any such waiver request within 90 days of any application filed under this subsection, and such waiver shall be effective for all service providers and products in that category and for all providers of services and products.

"(d) AVOIDANCE OF REDUNDANT REGULATIONS-

"(1) COMMERCIAL AVAILABILITY DETERMINATIONS- Determinations made or regulations prescribed by the Commission with respect to commercial availability to consumers of converter boxes, interactive communications equipment, and other equipment used by consumers to access multichannel video programming and other services offered over multichannel video programming systems, before the date of enactment of the Telecommunications Act of 1996 shall fulfill the requirements of this section.

"(2) REGULATIONS- Nothing in this section affects section 64.702(e) of the Commission's regulations (47 C.F.R. 64.702(e)) or other Commission regulations governing interconnection and competitive provision of customer premises equipment used in connection with basic common carrier communications services.

"(e) SUNSET- The regulations adopted under this section shall cease to apply when the Commission determines that—

"(1) the market for the multichannel video programming distributors is fully competitive;

"(2) the market for converter boxes, and interactive communications equipment, used in conjunction with that service is fully competitive; and

"(3) elimination of the regulations would promote competition and the public interest.

"(f) COMMISSION'S AUTHORITY- Nothing in this section shall be construed as expanding or limiting any authority that the Commission may have under law in effect before the date of enactment of the Telecommunications Act of 1996.".

SEC. 305. VIDEO PROGRAMMING ACCESSIBILITY.

Title VII is amended by inserting after section 712 (47 U.S.C. 612) the following new section:

"SEC. 713. VIDEO PROGRAMMING ACCESSIBILITY.

"(a) COMMISSION INQUIRY- Within 180 days after the date of enactment of the Telecommunications Act of 1996, the Federal Communications Commission shall complete an inquiry to ascertain the level at which video programming is closed captioned. Such inquiry shall examine the extent to which existing or previously published programming is closed captioned, the size of the video programming provider or programming owner providing closed captioning, the size of the market served, the relative audience shares achieved, or any other related factors. The Commission shall submit to the Congress a report on the results of such inquiry.

"(b) ACCOUNTABILITY CRITERIA- Within 18 months after such date of enactment, the Commission shall prescribe such regulations as are necessary to implement this section. Such regulations shall ensure that—

"(1) video programming first published or exhibited after the effective date of such regulations is fully accessible through the provision of closed captions, except as provided in subsection (d); and

"(2) video programming providers or owners maximize the accessibility of video programming first published or exhibited prior to the effective date of such regulations through the provision of closed captions, except as provided in subsection (d).

"(c) DEADLINES FOR CAPTIONING- Such regulations shall include an appropriate schedule of deadlines for the provision of closed captioning of video programming.

"(d) EXEMPTIONS- Notwithstanding subsection (b)—

"(1) the Commission may exempt by regulation programs, classes of programs, or services for which the Commission has determined that the provision of closed captioning would be economically burdensome to the provider or owner of such programming;

"(2) a provider of video programming or the owner of any program carried by the provider shall not be obligated to supply closed captions if such action would be inconsistent with contracts in effect on the date of enactment of the Telecommunications Act of 1996, except that nothing in this section shall be construed to relieve a video programming provider of its obligations to provide services required by Federal law; and

"(3) a provider of video programming or program owner may petition the Commission for an exemption from the requirements of this section, and the Commission may grant such petition upon a showing that the requirements contained in this section would result in an undue burden.

"(e) UNDUE BURDEN- The term "undue burden" means significant difficulty or expense. In determining whether the closed captions necessary to comply with the requirements of this paragraph would result in an undue economic burden, the factors to be considered include—

"(1) the nature and cost of the closed captions for the programming;

"(2) the impact on the operation of the provider or program owner;

"(3) the financial resources of the provider or program owner; and

"(4) the type of operations of the provider or program owner.

"(f) VIDEO DESCRIPTIONS INQUIRY- Within 6 months after the date of enactment of the Telecommunications Act of 1996, the Commission shall commence an inquiry to examine the use of video descriptions on video programming in order to ensure the accessibility of video programming to persons with visual impairments, and report to Congress on its findings. The Commission's report shall assess appropriate methods and schedules for phasing video descriptions into the marketplace, technical and quality standards for video descriptions, a definition of programming for which video descriptions would apply, and other technical and legal issues that the Commission deems appropriate.

"(g) VIDEO DESCRIPTION- For purposes of this section, "video description" means the insertion of audio narrated descriptions of a television program's key visual elements into natural pauses between the program's dialogue.

"(h) PRIVATE RIGHTS OF ACTIONS PROHIBITED- Nothing in this section shall be construed to authorize any private right of action to enforce any requirement of this section or any regulation thereunder. The Commission shall have exclusive jurisdiction with respect to any complaint under this section.".

TITLE IV—REGULATORY REFORM
SEC. 401. REGULATORY FORBEARANCE.
Title I is amended by inserting after section 9 (47 U.S.C. 159) the following new section:

"SEC. 10. COMPETITION IN PROVISION OF
TELECOMMUNICATIONS SERVICE.
"(a)   REGULATORY   FLEXIBILITY-   Notwithstanding   section 332(c)(1)(A) of this Act, the Commission shall forbear from applying any regulation or any provision of this Act to a telecommunications carrier or telecommunications service, or class of telecommunications carriers or telecommunications services, in any or some of its or their geographic markets, if the Commission determines that—

"(1) enforcement of such regulation or provision is not necessary to ensure that the charges, practices, classifications, or regulations by, for, or in connection with that telecommunications carrier or telecommunications service are just and reasonable and are not unjustly or unreasonably discriminatory;

"(2) enforcement of such regulation or provision is not necessary for the protection of consumers; and

"(3) forbearance from applying such provision or regulation is consistent with the public interest.

"(b) COMPETITIVE EFFECT TO BE WEIGHED- In making the determination under subsection (a)(3), the Commission shall consider whether forbearance from enforcing the provision or regulation will promote competitive market conditions, including the extent to which such forbearance will enhance competition among providers of telecommunications services. If the Commission determines that such forbearance will promote competition among providers of telecommunications services, that determination may be the basis for a Commission finding that forbearance is in the public interest.

"(c) PETITION FOR FORBEARANCE- Any telecommunications carrier, or class of telecommunications carriers, may submit a petition to the Commission requesting that the Commission exercise the authority granted under this section with respect to that carrier or those carriers, or any service offered by that carrier or carriers. Any such petition shall be deemed granted if the Commission does not deny the petition for failure to meet the requirements for forbearance under subsection (a) within one year after the Commission receives it, unless the one-year period is extended by the Commission. The Commission may extend the initial one-year period by an additional 90 days if the Commission finds that an extension is necessary to meet the requirements of subsection (a). The Commission may grant or deny a petition in whole or in part and shall explain its decision in writing.

"(d) LIMITATION- Except as provided in section 251(f), the Commission may not forbear from applying the requirements of section 251(c) or 271 under subsection (a) of this section until it determines that those requirements have been fully implemented.

"(e) STATE ENFORCEMENT AFTER COMMISSION FORBEARANCE- A State commission may not continue to apply or enforce any provision of this Act that the Commission has determined to forbear from applying under subsection (a).".

SEC. 402. BIENNIAL REVIEW OF REGULATIONS;
REGULATORY RELIEF.

(a) BIENNIAL REVIEW- Title I is amended by inserting after section 10 (as added by section 401) the following new section:

"SEC. 11. REGULATORY REFORM.

"(a) BIENNIAL REVIEW OF REGULATIONS- In every even-numbered year (beginning with 1998), the Commission—

"(1) shall review all regulations issued under this Act in effect at the time of the review that apply to the operations or activities of any provider of telecommunications service; and

"(2) shall determine whether any such regulation is no longer necessary in the public interest as the result of meaningful economic competition between providers of such service.

"(b) EFFECT OF DETERMINATION- The Commission shall repeal or modify any regulation it determines to be no longer necessary in the public interest.".

(b) REGULATORY RELIEF-

(1) Streamlined procedures for changes in charges, classifications, regulations, or practices-

(A) Section 204(a) (47 U.S.C. 204(a)) is amended—

(i) by striking "12 months" the first place it appears in paragraph (2)(A) and inserting "5 months";

(ii) by striking "effective," and all that follows in paragraph (2)(A) and inserting "effective."; and

(iii) by adding at the end thereof the following:

"(3) A local exchange carrier may file with the Commission a new or revised charge, classification, regulation, or practice on a streamlined basis. Any such charge, classification, regulation, or practice shall be deemed lawful and shall be effective 7 days (in the case of a reduction in rates) or 15 days (in the case of an increase in rates) after the date on which it is filed with the Commission unless the Commission takes action under paragraph (1) before the end of that 7-day or 15-day period, as is appropriate.".

(B) Section 208(b) (47 U.S.C. 208(b)) is amended—

(i) by striking "12 months" the first place it appears in paragraph (1) and inserting "5 months"; and

(ii) by striking "filed,' and all that follows in paragraph (1) and inserting "filed.'.

(2) EXTENSIONS OF LINES UNDER SECTION 214; ARMIS RE-PORTS- The Commission shall permit any common carrier—

(A) to be exempt from the requirements of section 214 of the Communications Act of 1934 for the extension of any line; and

(B) to file cost allocation manuals and ARMIS reports annually, to the extent such carrier is required to file such manuals or reports.

(3) FORBEARANCE AUTHORITY NOT LIMITED- Nothing in this

subsection shall be construed to limit the authority of the Commission to waive, modify, or forbear from applying any of the requirements to which reference is made in paragraph (1) under any other provision of this Act or other law.

(4) EFFECTIVE DATE OF AMENDMENTS- The amendments made by paragraph (1) of this subsection shall apply with respect to any charge, classification, regulation, or practice filed on or after one year after the date of enactment of this Act.

(c) CLASSIFICATION OF CARRIERS- In classifying carriers according to section 32.11 of its regulations (47 C.F.R. 32.11) and in establishing reporting requirements pursuant to part 43 of its regulations (47 C.F.R. part 43) and section 64.903 of its regulations (47 C.F.R. 64.903), the Commission shall adjust the revenue requirements to account for inflation as of the release date of the Commission's Report and Order in CC Docket No. 91-141, and annually thereafter. This subsection shall take effect on the date of enactment of this Act.

SEC. 403. ELIMINATION OF UNNECESSARY COMMISSION REGULATIONS AND FUNCTIONS.

(a) MODIFICATION OF AMATEUR RADIO EXAMINATION PROCEDURES- Section 4(f)(4) (47 U.S.C. 154(f)(4)) is amended—
(1) in subparagraph (A)—
(A) by inserting "or administering" after "for purposes of preparing";
(B) by inserting "of" after "than the class"; and
(C) by inserting "or administered" after "for which the examination is being prepared";
(2) by striking subparagraph (B);
(3) in subparagraph (H), by striking "(A), (B), and (C)" and inserting "(A) and (B)";
(4) in subparagraph (J)—
(A) by striking "or (B)"; and
(B) by striking the last sentence; and
(5) by redesignating subparagraphs (C) through (J) as subparagraphs (B) through (I), respectively.

(b) AUTHORITY TO DESIGNATE ENTITIES TO INSPECT- Section 4(f)(3) (47 U.S.C. 154(f)(3)) is amended by inserting before the period at the end the following: ": and *Provided further*, That, in the alternative, an entity designated by the Commission may make the inspections referred to in this paragraph".

(c) EXPEDITING INSTRUCTIONAL TELEVISION FIXED SERVICE PROCESSING- Section 5(c)(1) (47 U.S.C. 155(c)(1)) is amended by striking the last sentence and inserting the following: "Except for cases involving the authorization of service in the instructional television fixed service, or as otherwise provided in this Act, nothing in this paragraph shall authorize the Commission to provide for the conduct, by any person or persons other than persons referred to in paragraph (2) or (3) of section 556(b) of title 5, United States Code, of any hearing to which such section applies.".

(d) REPEAL SETTING OF DEPRECIATION RATES- The first sentence of section 220(b) (47 U.S.C. 220(b)) is amended by striking "shall prescribe for such carriers" and inserting "may prescribe, for such carriers as it determines to be appropriate,".

(e) USE OF INDEPENDENT AUDITORS- Section 220(c) (47 U.S.C. 220(c)) is amended by adding at the end thereof the following: "The Commission may obtain the services of any person licensed to provide public accounting services under the law of any State to assist with, or conduct, audits under this section. While so employed or engaged in conducting an audit for the Commission under this section, any such person shall have the powers granted the Commission under this subsection and shall be subject to subsection (f) in the same manner as if that person were an employee of the Commission.".

(f) DELEGATION OF EQUIPMENT TESTING AND CERTIFICATION TO PRIVATE LABORATORIES- Section 302 (47 U.S.C. 302) is amended by adding at the end the following:

"(e) The Commission may—

"(1) authorize the use of private organizations for testing and certifying the compliance of devices or home electronic equipment and systems with regulations promulgated under this section;

"(2) accept as prima facie evidence of such compliance the certification by any such organization; and

"(3) establish such qualifications and standards as it deems appropriate for such private organizations, testing, and certification.".

(g) MAKING LICENSE MODIFICATION UNIFORM- Section 303(f) (47 U.S.C. 303(f)) is amended by striking "unless, after a public hearing," and inserting "unless".

(h) ELIMINATE FCC JURISDICTION OVER GOVERNMENT-OWNED SHIP RADIO STATIONS-

(1) Section 305 (47 U.S.C. 305) is amended by striking subsection (b) and redesignating subsections (c) and (d) as (b) and (c), respectively.

(2) Section 382(2) (47 U.S.C. 382(2)) is amended by striking "except a vessel of the United States Maritime Administration, the Inland and Coastwise Waterways Service, or the Panama Canal Company,".

(i) PERMIT OPERATION OF DOMESTIC SHIP AND AIRCRAFT RADIOS WITHOUT LICENSE- Section 307(e) (47 U.S.C. 307(e)) is amended to read as follows:

"(e)(1) Notwithstanding any license requirement established in this Act, if the Commission determines that such authorization serves the public interest, convenience, and necessity, the Commission may by rule authorize the operation of radio stations without individual licenses in the following radio services: (A) the citizens band radio service; (B) the radio control service; (C) the aviation radio service for aircraft stations operated on domestic flights when such aircraft are not otherwise required to carry a radio station; and (D) the maritime radio service for ship stations navigated on domestic voyages when such ships are not otherwise required to carry a radio station.

"(2) Any radio station operator who is authorized by the Commission to operate without an individual license shall comply with all other provisions of this Act and with rules prescribed by the Commission under this Act.

"(3) For purposes of this subsection, the terms "citizens band radio service", "radio control service", "aircraft station" and "ship station" shall have the meanings given them by the Commission by rule.".

(j) EXPEDITED LICENSING FOR FIXED MICROWAVE SERVICE- Section 309(b)(2) (47 U.S.C. 309(b)(2)) is amended by striking subparagraph (A) and redesignating subparagraphs (B) through (G) as subparagraphs (A) through (F), respectively.

(k) FOREIGN DIRECTORS- Section 310(b) (47 U.S.C. 310(b)) is amended—

(1) in paragraph (3), by striking "of which any officer or director is an alien or"; and

(2) in paragraph (4), by striking "of which any officer or more than one-fourth of the directors are aliens, or".

(l) LIMITATION ON SILENT STATION AUTHORIZATIONS- Section 312 (47 U.S.C. 312) is amended by adding at the end the following:

"(g) If a broadcasting station fails to transmit broadcast signals for any consecutive 12-month period, then the station license granted for the operation of that broadcast station expires at the end of that period, notwithstanding any provision, term, or condition of the license to the contrary.".

(m) MODIFICATION OF CONSTRUCTION PERMIT REQUIREMENT- Section 319(d) is amended by striking the last two sentences and inserting the following: "With respect to any broadcasting station, the Commission shall not have any authority to waive the requirement of a permit for construction, except that the Commission may by regulation determine that a permit shall not be required for minor changes in the facilities of authorized broadcast stations. With respect to any other station or class of stations, the Commission shall not waive the requirement for a construction permit unless the Commission determines that the public interest, convenience, and necessity would be served by such a waiver.".

(n) CONDUCT OF INSPECTIONS- Section 362(b) (47 U.S.C. 362(b)) is amended to read as follows:

"(b) Every ship of the United States that is subject to this part shall have the equipment and apparatus prescribed therein inspected at least once each year by the Commission or an entity designated by the Commission. If, after such inspection, the Commission is satisfied that all relevant provisions of this Act and the station license have been complied with, the fact shall be so certified on the station license by the Commission. The Commission shall make such additional inspections at frequent intervals as the Commission determines may be necessary to ensure compliance with the requirements of this Act. The Commission may, upon a finding that the public interest could be served thereby—

"(1) waive the annual inspection required under this section for a period of up to 90 days for the sole purpose of enabling a vessel to complete its voyage and proceed to a port in the United States where an inspection can be held; or

"(2) waive the annual inspection required under this section for a vessel that is in compliance with the radio provisions of the Safety Convention and that is operating solely in waters beyond the jurisdiction of the United States: *Provided*, That such inspection shall be performed within 30 days of such vessel's return to the United States.".

(o) INSPECTION BY OTHER ENTITIES- Section 385 (47 U.S.C. 385) is amended—

(1) by inserting "or an entity designated by the Commission" after "The Commission"; and

(2) by adding at the end thereof the following: "In accordance with such other provisions of law as apply to Government contracts, the Commission may enter into contracts with any person for the purpose of carrying out such inspections and certifying compliance with those requirements, and may, as part of any such contract, allow any such person to accept reimbursement from the license holder for travel and expense costs of any employee conducting an inspection or certification.".

TITLE V—OBSCENITY AND VIOLENCE
SUBTITLE A—OBSCENE, HARASSING, AND WRONGFUL UTILIZATION OF TELECOMMUNICATIONS FACILITIES
SEC. 501. SHORT TITLE.

This title may be cited as the "Communications Decency Act of 1996".

SEC. 502. OBSCENE OR HARASSING USE OF TELECOMMUNICATIONS FACILITIES UNDER THE COMMUNICATIONS ACT OF 1934.

Section 223 (47 U.S.C. 223) is amended—

(1) by striking subsection (a) and inserting in lieu thereof:

"(a) Whoever—

"(1) in interstate or foreign communications—

"(A) by means of a telecommunications device knowingly—

"(i) makes, creates, or solicits, and

"(ii) initiates the transmission of, any comment, request, suggestion, proposal, image, or other communication which is obscene, lewd, lascivious, filthy, or indecent, with intent to annoy, abuse, threaten, or harass another person;

"(B) by means of a telecommunications device knowingly—

"(i) makes, creates, or solicits, and

"(ii) initiates the transmission of, any comment, request, suggestion, proposal, image, or other communication which is obscene or indecent, knowing that the recipient of the communication is under 18 years of age, regardless of whether the maker of such communication placed the call or initiated the communication;

"(C) makes a telephone call or utilizes a telecommunications device, whether or not conversation or communication ensues, without disclosing his identity and with intent to annoy, abuse, threaten, or harass any person at the called number or who receives the communications;

"(D) makes or causes the telephone of another repeatedly or continuously to ring, with intent to harass any person at the called number; or

"(E) makes repeated telephone calls or repeatedly initiates communication with a telecommunications device, during which conversation or communication ensues, solely to harass any person at the called number or who receives the communication; or

"(2) knowingly permits any telecommunications facility under his control to be used for any activity prohibited by paragraph (1) with the intent that it be used for such activity, shall be fined under title 18,

United States Code, or imprisoned not more than two years, or both.";
and

(2) by adding at the end the following new subsections:

"(d) Whoever—

"(1) in interstate or foreign communications knowingly—

"(A) uses an interactive computer service to send to a specific person or
persons under 18 years of age, or

"(B) uses any interactive computer service to display in a manner available to a person under 18 years of age, any comment, request, suggestion,
proposal, image, or other communication that, in context, depicts or
describes, in terms patently offensive as measured by contemporary community standards, sexual or excretory activities or organs, regardless of
whether the user of such service placed the call or initiated the communication; or

"(2) knowingly permits any telecommunications facility under such person's control to be used for an activity prohibited by paragraph (1) with
the intent that it be used for such activity, shall be fined under title 18,
United States Code, or imprisoned not more than two years, or both.

"(e) In addition to any other defenses available by law:

"(1) No person shall be held to have violated subsection (a) or (d) solely
for providing access or connection to or from a facility, system, or network not under that person's control, including transmission, download-
ing, intermediate storage, access software, or other related capabilities
that are incidental to providing such access or connection that does not
include the creation of the content of the communication.

"(2) The defenses provided by paragraph (1) of this subsection shall not
be applicable to a person who is a conspirator with an entity actively
involved in the creation or knowing distribution of communications that
violate this section, or who knowingly advertises the availability of such
communications.

"(3) The defenses provided in paragraph (1) of this subsection shall not
be applicable to a person who provides access or connection to a facility,
system, or network engaged in the violation of this section that is owned
or controlled by such person.

"(4) No employer shall be held liable under this section for the actions
of an employee or agent unless the employee's or agent's conduct is
within the scope of his or her employment or agency and the employer
(A) having knowledge of such conduct, authorizes or ratifies such conduct, or (B) recklessly disregards such conduct.

"(5) It is a defense to a prosecution under subsection (a)(1)(B) or (d), or under subsection (a)(2) with respect to the use of a facility for an activity under subsection (a)(1)(B) that a person—

"(A) has taken, in good faith, reasonable, effective, and appropriate actions under the circumstances to restrict or prevent access by minors to a communication specified in such subsections, which may involve any appropriate measures to restrict minors from such communications, including any method which is feasible under available technology; or

"(B) has restricted access to such communication by requiring use of a verified credit card, debit account, adult access code, or adult personal identification number.

"(6) The Commission may describe measures which are reasonable, effective, and appropriate to restrict access to prohibited communications under subsection (d). Nothing in this section authorizes the Commission to enforce, or is intended to provide the Commission with the authority to approve, sanction, or permit, the use of such measures. The Commission shall have no enforcement authority over the failure to utilize such measures. The Commission shall not endorse specific products relating to such measures. The use of such measures shall be admitted as evidence of good faith efforts for purposes of paragraph (5) in any action arising under subsection (d). Nothing in this section shall be construed to treat interactive computer services as common carriers or telecommunications carriers.

"(f)(1) No cause of action may be brought in any court or administrative agency against any person on account of any activity that is not in violation of any law punishable by criminal or civil penalty, and that the person has taken in good faith to implement a defense authorized under this section or otherwise to restrict or prevent the transmission of, or access to, a communication specified in this section.

"(2) No State or local government may impose any liability for commercial activities or actions by commercial entities, nonprofit libraries, or institutions of higher education in connection with an activity or action described in subsection (a)(2) or (d) that is inconsistent with the treatment of those activities or actions under this section: *Provided, however,* That nothing herein shall preclude any State or local government from enacting and enforcing complementary oversight, liability, and regulatory systems, procedures, and requirements, so long as such systems, procedures, and requirements govern only intrastate services and do not result in the imposition of inconsistent rights, duties or obligations on the provision of interstate services. Nothing in this subsection shall preclude any State or local government from governing conduct not covered by this section.

"(g) Nothing in subsection (a), (d), (e), or (f) or in the defenses to prosecution under subsection (a) or (d) shall be construed to affect or limit the application or enforcement of any other Federal law.

"(h) For purposes of this section—

"(1) The use of the term "telecommunications device" in this section—

"(A) shall not impose new obligations on broadcasting station licensees and cable operators covered by obscenity and indecency provisions elsewhere in this Act; and

"(B) does not include an interactive computer service.

"(2) The term "interactive computer service" has the meaning provided in section 230(e)(2).

"(3) The term "access software" means software (including client or server software) or enabling tools that do not create or provide the content of the communication but that allow a user to do any one or more of the following:

"(A) filter, screen, allow, or disallow content;

"(B) pick, choose, analyze, or digest content; or

"(C) transmit, receive, display, forward, cache, search, subset, organize, reorganize, or translate content.

"(4) The term "institution of higher education" has the meaning provided in section 1201 of the Higher Education Act of 1965 (20 U.S.C. 1141).

"(5) The term "library" means a library eligible for participation in State-based plans for funds under title III of the Library Services and Construction Act (20 U.S.C. 355e et seq.).".

SEC. 503. OBSCENE PROGRAMMING ON CABLE TELEVISION.
Section 639 (47 U.S.C. 559) is amended by striking "not more than $10,000" and inserting "under title 18, United States Code,".

SEC. 504. SCRAMBLING OF CABLE CHANNELS FOR NONSUBSCRIBERS.
Part IV of title VI (47 U.S.C. 551 et seq.) is amended by adding at the end the following:

"SEC. 640. SCRAMBLING OF CABLE CHANNELS FOR NONSUBSCRIBERS.

"(a) SUBSCRIBER REQUEST- Upon request by a cable service subscriber, a cable operator shall, without charge, fully scramble or otherwise fully block the audio and video programming of each channel carrying such programming so that one not a subscriber does not receive it.

"(b) DEFINITION- As used in this section, the term "scramble" means to rearrange the content of the signal of the programming so that the programming cannot be viewed or heard in an understandable manner.".

SEC. 505. SCRAMBLING OF SEXUALLY EXPLICIT ADULT VIDEO SERVICE PROGRAMMING.
(a) REQUIREMENT- Part IV of title VI (47 U.S.C. 551 et seq.), as amended by this Act, is further amended by adding at the end the following:

"SEC. 641. SCRAMBLING OF SEXUALLY EXPLICIT ADULT VIDEO SERVICE PROGRAMMING.
"(a) REQUIREMENT- In providing sexually explicit adult programming or other programming that is indecent on any channel of its service primarily dedicated to sexually-oriented programming, a multichannel video programming distributor shall fully scramble or otherwise fully block the video and audio portion of such channel so that one not a subscriber to such channel or programming does not receive it.
"(b) IMPLEMENTATION- Until a multichannel video programming distributor complies with the requirement set forth in subsection (a), the distributor shall limit the access of children to the programming referred to in that subsection by not providing such programming during the hours of the day (as determined by the Commission) when a significant number of children are likely to view it.
"(c) DEFINITION- As used in this section, the term "scramble" means to rearrange the content of the signal of the programming so that the programming cannot be viewed or heard in an understandable manner.".
(b) EFFECTIVE DATE- The amendment made by subsection (a) shall take effect 30 days after the date of enactment of this Act.

SEC. 506. CABLE OPERATOR REFUSAL TO CARRY CERTAIN PROGRAMS.
(a) PUBLIC, EDUCATIONAL, AND GOVERNMENTAL CHANNELS- Section 611(e) (47 U.S.C. 531(e)) is amended by inserting before the period the following: ", except a cable operator may refuse to transmit any public access program or portion of a public access program which contains obscenity, indecency, or nudity".
(b) CABLE CHANNELS FOR COMMERCIAL USE- Section 612(c)(2) (47 U.S.C. 532(c)(2)) is amended by striking "an operator" and inserting "a cable operator may refuse to transmit any leased access

program or portion of a leased access program which contains obscenity, indecency, or nudity and".

SEC. 507. CLARIFICATION OF CURRENT LAWS REGARDING COMMUNICATION OF OBSCENE MATERIALS THROUGH THE USE OF COMPUTERS.

(a) IMPORTATION OR TRANSPORTATION- Section 1462 of title 18, United States Code, is amended—

(1) in the first undesignated paragraph, by inserting "or interactive computer service (as defined in section 230(e)(2) of the Communications Act of 1934)" after "carrier"; and

(2) in the second undesignated paragraph—

(A) by inserting "or receives," after "takes";

(B) by inserting "or interactive computer service (as defined in section 230(e)(2) of the Communications Act of 1934)" after "common carrier"; and

(C) by inserting "or importation" after "carriage".

(b) TRANSPORTATION FOR PURPOSES OF SALE OR DISTRI-BUTION- The first undesignated paragraph of section 1465 of title 18, United States Code, is amended—

(1) by striking "transports in" and inserting "transports or travels in, or uses a facility or means of,";

(2) by inserting "or an interactive computer service (as defined in section 230(e)(2) of the Communications Act of 1934) in or affecting such commerce" after "foreign commerce" the first place it appears;

(3) by striking ", or knowingly travels in" and all that follows through "obscene material in interstate or foreign commerce," and inserting "of".

(c) INTERPRETATION- The amendments made by this section are clarifying and shall not be interpreted to limit or repeal any prohibition contained in sections 1462 and 1465 of title 18, United States Code, before such amendment, under the rule established in *United States v. Alpers*, 338 U.S. 680 (1950).

SEC. 508. COERCION AND ENTICEMENT OF MINORS.

Section 2422 of title 18, United States Code, is amended—

(1) by inserting "(a)" before "Whoever knowingly"; and

(2) by adding at the end the following:

"(b) Whoever, using any facility or means of interstate or foreign commerce, including the mail, or within the special maritime and territorial jurisdiction of the United States, knowingly persuades, induces, entices, or coerces any individual who has not attained the age of 18 years to engage in prostitution or any sexual act for which any person may be criminally prosecuted, or attempts to do so, shall be fined under this title or imprisoned not more than 10 years, or both.".

SEC. 509. ONLINE FAMILY EMPOWERMENT.

Title II of the Communications Act of 1934 (47 U.S.C. 201 et seq.) is amended by adding at the end the following new section:

"SEC. 230. PROTECTION FOR PRIVATE BLOCKING AND SCREENING OF OFFENSIVE MATERIAL.

"(a) FINDINGS- The Congress finds the following:

"(1) The rapidly developing array of Internet and other interactive computer services available to individual Americans represent an extraordinary advance in the availability of educational and informational resources to our citizens.

"(2) These services offer users a great degree of control over the information that they receive, as well as the potential for even greater control in the future as technology develops.

"(3) The Internet and other interactive computer services offer a forum for a true diversity of political discourse, unique opportunities for cultural development, and myriad avenues for intellectual activity.

"(4) The Internet and other interactive computer services have flourished, to the benefit of all Americans, with a minimum of government regulation.

"(5) Increasingly Americans are relying on interactive media for a variety of political, educational, cultural, and entertainment services.

"(b) POLICY- It is the policy of the United States—

"(1) to promote the continued development of the Internet and other interactive computer services and other interactive media;

"(2) to preserve the vibrant and competitive free market that presently exists for the Internet and other interactive computer services, unfettered by Federal or State regulation;

"(3) to encourage the development of technologies which maximize user control over what information is received by individuals, families, and schools who use the Internet and other interactive computer services;

"(4) to remove disincentives for the development and utilization of blocking and filtering technologies that empower parents to restrict their children's access to objectionable or inappropriate online material; and

"(5) to ensure vigorous enforcement of Federal criminal laws to deter and punish trafficking in obscenity, stalking, and harassment by means of computer.

"(c) PROTECTION FOR "GOOD SAMARITAN" BLOCKING AND SCREENING OF OFFENSIVE MATERIAL-

"(1) TREATMENT OF PUBLISHER OR SPEAKER- No provider or user of an interactive computer service shall be treated as the publisher or speaker of any information provided by another information content provider.

"(2) CIVIL LIABILITY- No provider or user of an interactive computer service shall be held liable on account of—

"(A) any action voluntarily taken in good faith to restrict access to or availability of material that the provider or user considers to be obscene, lewd, lascivious, filthy, excessively violent, harassing, or otherwise objectionable, whether or not such material is constitutionally protected; or

"(B) any action taken to enable or make available to information content providers or others the technical means to restrict access to material described in paragraph (1).

"(d) EFFECT ON OTHER LAWS-

"(1) NO EFFECT ON CRIMINAL LAW- Nothing in this section shall be construed to impair the enforcement of section 223 of this Act, chapter 71 (relating to obscenity) or 110 (relating to sexual exploitation of children) of title 18, United States Code, or any other Federal criminal statute.

"(2) NO EFFECT ON INTELLECTUAL PROPERTY LAW- Nothing in this section shall be construed to limit or expand any law pertaining to intellectual property.

"(3) STATE LAW- Nothing in this section shall be construed to prevent any State from enforcing any State law that is consistent with this section. No cause of action may be brought and no liability may be imposed under any State or local law that is inconsistent with this section.

"(4) NO EFFECT ON COMMUNICATIONS PRIVACY LAW- Nothing in this section shall be construed to limit the application of the Electronic Communications Privacy Act of 1986 or any of the amendments made by such Act, or any similar State law.

"(e) DEFINITIONS- As used in this section:

"(1) INTERNET- The term "Internet" means the international computer network of both Federal and non-Federal interoperable packet switched data networks.

"(2) INTERACTIVE COMPUTER SERVICE- The term "interactive computer service" means any information service, system, or access software provider that provides or enables computer access by multiple users to a computer server, including specifically a service or system that provides access to the Internet and such systems operated or services offered by libraries or educational institutions.

"(3) INFORMATION CONTENT PROVIDER- The term "information content provider" means any person or entity that is responsible, in whole or in part, for the creation or development of information provided through the Internet or any other interactive computer service.

"(4) ACCESS SOFTWARE PROVIDER- The term "access software provider" means a provider of software (including client or server software), or enabling tools that do any one or more of the following:

"(A) filter, screen, allow, or disallow content;

"(B) pick, choose, analyze, or digest content; or

"(C) transmit, receive, display, forward, cache, search, subset, organize, reorganize, or translate content.".

SUBTITLE B—VIOLENCE

SEC. 551. PARENTAL CHOICE IN TELEVISION PROGRAMMING.

(a) FINDINGS- The Congress makes the following findings:

(1) Television influences children's perception of the values and behavior that are common and acceptable in society.

(2) Television station operators, cable television system operators, and video programmers should follow practices in connection with video programming that take into consideration that television broadcast and cable programming has established a uniquely pervasive presence in the lives of American children.

(3) The average American child is exposed to 25 hours of television each week and some children are exposed to as much as 11 hours of television a day.

(4) Studies have shown that children exposed to violent video programming at a young age have a higher tendency for violent and aggressive behavior later in life than children not so exposed, and that children exposed to violent video programming are prone to assume that acts of violence are acceptable behavior.

(5) Children in the United States are, on average, exposed to an estimated 8,000 murders and 100,000 acts of violence on television by the time the child completes elementary school.

(6) Studies indicate that children are affected by the pervasiveness and casual treatment of sexual material on television, eroding the ability of parents to develop responsible attitudes and behavior in their children.

(7) Parents express grave concern over violent and sexual video program-

ming and strongly support technology that would give them greater control to block video programming in the home that they consider harmful to their children.

(8) There is a compelling governmental interest in empowering parents to limit the negative influences of video programming that is harmful to children.

(9) Providing parents with timely information about the nature of upcoming video programming and with the technological tools that allow them easily to block violent, sexual, or other programming that they believe harmful to their children is a nonintrusive and narrowly tailored means of achieving that compelling governmental interest.

(b) ESTABLISHMENT OF TELEVISION RATING CODE-

(1) AMENDMENT- Section 303 (47 U.S.C. 303) is amended by adding at the end the following:

"(w) Prescribe—

"(1) on the basis of recommendations from an advisory committee established by the Commission in accordance with section 551(b)(2) of the Telecommunications Act of 1996, guidelines and recommended procedures for the identification and rating of video programming that contains sexual, violent, or other indecent material about which parents should be informed before it is displayed to children: *Provided*, That nothing in this paragraph shall be construed to authorize any rating of video programming on the basis of its political or religious content; and

"(2) with respect to any video programming that has been rated, and in consultation with the television industry, rules requiring distributors of such video programming to transmit such rating to permit parents to block the display of video programming that they have determined is inappropriate for their children.".

(2) ADVISORY COMMITTEE REQUIREMENTS- In establishing an advisory committee for purposes of the amendment made by paragraph (1) of this subsection, the Commission shall—

(A) ensure that such committee is composed of parents, television broadcasters, television programming producers, cable operators, appropriate public interest groups, and other interested individuals from the private sector and is fairly balanced in terms of political affiliation, the points of view represented, and the functions to be performed by the committee;

(B) provide to the committee such staff and resources as may be necessary to permit it to perform its functions efficiently and promptly; and

(C) require the committee to submit a final report of its recommendations within one year after the date of the appointment of the initial members.

(c) REQUIREMENT FOR MANUFACTURE OF TELEVISIONS THAT BLOCK PROGRAMS- Section 303 (47 U.S.C. 303), as amended by subsection

(a), is further amended by adding at the end the following:

"(x) Require, in the case of an apparatus designed to receive television signals that are shipped in interstate commerce or manufactured in the United States and that have a picture screen 13 inches or greater in size (measured diagonally), that such apparatus be equipped with a feature designed to enable viewers to block display of all programs with a common rating, except as otherwise permitted by regulations pursuant to section 330(c)(4).".

(d) SHIPPING OF TELEVISIONS THAT BLOCK PROGRAMS-

(1) REGULATIONS- Section 330 (47 U.S.C. 330) is amended—

(A) by redesignating subsection (c) as subsection (d); and

(B) by adding after subsection (b) the following new subsection (c):

"(c)(1) Except as provided in paragraph (2), no person shall ship in interstate commerce or manufacture in the United States any apparatus described in section 303(x) of this Act except in accordance with rules prescribed by the Commission pursuant to the authority granted by that section.

"(2) This subsection shall not apply to carriers transporting apparatus referred to in paragraph (1) without trading in it.

"(3) The rules prescribed by the Commission under this subsection shall provide for the oversight by the Commission of the adoption of standards by industry for blocking technology. Such rules shall require that all such apparatus be able to receive the rating signals which have been transmitted by way of line 21 of the vertical blanking interval and which conform to the signal and blocking specifications established by industry under the supervision of the Commission.

"(4) As new video technology is developed, the Commission shall take such action as the Commission determines appropriate to ensure that blocking service continues to be available to consumers. If the Commission determines that an alternative blocking technology exists that—

"(A) enables parents to block programming based on identifying programs without ratings,

"(B) is available to consumers at a cost which is comparable to the cost of technology that allows parents to block programming based on common ratings, and

"(C) will allow parents to block a broad range of programs on a multichannel system as effectively and as easily as technology that allows parents to block programming based on common ratings, the Commission shall amend the rules prescribed pursuant to section 303(x) to require that the apparatus described in such section be equipped with either the blocking technology described in such section or the alternative blocking technology described in this paragraph.".

(2) CONFORMING AMENDMENT- Section 330(d), as redesignated by subsection (d)(1)(A), is amended by striking "section 303(s), and section 303(u)" and inserting in lieu thereof "and sections 303(s), 303(u), and 303(x)".

(e) APPLICABILITY AND EFFECTIVE DATES-

(1) APPLICABILITY OF RATING PROVISION- The amendment made by subsection (b) of this section shall take effect 1 year after the date of enactment of this Act, but only if the Commission determines, in consultation with appropriate public interest groups and interested individuals from the private sector, that distributors of video programming have not, by such date—

(A) established voluntary rules for rating video programming that contains sexual, violent, or other indecent material about which parents should be informed before it is displayed to children, and such rules are acceptable to the Commission; and

(B) agreed voluntarily to broadcast signals that contain ratings of such programming.

(2) EFFECTIVE DATE OF MANUFACTURING PROVISION- In prescribing regulations to implement the amendment made by subsection (c), the Federal Communications Commission shall, after consultation with the television manufacturing industry, specify the effective date for the applicability of the requirement to the apparatus covered by such amendment, which date shall not be less than two years after the date of enactment of this Act.

SEC. 552. TECHNOLOGY FUND.

It is the policy of the United States to encourage broadcast television, cable, satellite, syndication, other video programming distributors, and relevant related industries (in consultation with appropriate public interest groups and interested individuals from the private sector) to—

(1) establish a technology fund to encourage television and electronics equipment manufacturers to facilitate the development of technology which would empower parents to block programming they deem inappropriate for their children and to encourage the availability thereof to low income parents;

(2) report to the viewing public on the status of the development of affordable, easy to use blocking technology; and

(3) establish and promote effective procedures, standards, systems, advisories, or other mechanisms for ensuring that users have easy and complete access to the information necessary to effectively utilize blocking technology and to encourage the availability thereof to low income parents.

SUBTITLE C—JUDICIAL REVIEW
SEC. 561. EXPEDITED REVIEW.
(a) THREE-JUDGE DISTRICT COURT HEARING- Notwithstanding any other provision of law, any civil action challenging the constitutionality, on its face, of this title or any amendment made by this title, or any provision thereof, shall be heard by a district court of 3 judges convened pursuant to the provisions of section 2284 of title 28, United States Code.
(b) APPELLATE REVIEW- Notwithstanding any other provision of law, an interlocutory or final judgment, decree, or order of the court of 3 judges in an action under subsection (a) holding this title or an amendment made by this title, or any provision thereof, unconstitutional shall be reviewable as a matter of right by direct appeal to the Supreme Court. Any such appeal shall be filed not more than 20 days after entry of such judgment, decree, or order.

TITLE VI—EFFECT ON OTHER LAWS
SEC. 601. APPLICABILITY OF CONSENT DECREES AND
OTHER LAW.
(a) APPLICABILITY OF AMENDMENTS TO FUTURE
CONDUCT-
(1) AT&T CONSENT DECREE- Any conduct or activity that was, before the date of enactment of this Act, subject to any restriction or obligation imposed by the AT&T Consent Decree shall, on and after such date, be subject to the restrictions and obligations imposed by the Communications Act of 1934 as amended by this Act and shall not be subject to the restrictions and the obligations imposed by such Consent Decree.
(2) GTE CONSENT DECREE- Any conduct or activity that was, before the date of enactment of this Act, subject to any restriction or obligation imposed by the GTE Consent Decree shall, on and after such date, be subject to the restrictions and obligations imposed by the Communications Act of 1934 as amended by this Act and shall not be subject to the restrictions and the obligations imposed by such Consent Decree.
(3) MCCAW CONSENT DECREE- Any conduct or activity that was,

before the date of enactment of this Act, subject to any restriction or obligation imposed by the McCaw Consent Decree shall, on and after such date, be subject to the restrictions and obligations imposed by the Communications Act of 1934 as amended by this Act and subsection (d) of this section and shall not be subject to the restrictions and the obligations imposed by such Consent Decree.

(b) ANTITRUST LAWS-

(1) SAVINGS CLAUSE- Except as provided in paragraphs (2) and (3), nothing in this Act or the amendments made by this Act shall be construed to modify, impair, or supersede the applicability of any of the antitrust laws.

(2) REPEAL- Subsection (a) of section 221 (47 U.S.C. 221(a)) is repealed.

(3) CLAYTON ACT- Section 7 of the Clayton Act (15 U.S.C. 18) is amended in the last paragraph by striking "Federal Communications Commission,".

(c) FEDERAL, STATE, AND LOCAL LAW-

(1) NO IMPLIED EFFECT- This Act and the amendments made by this Act shall not be construed to modify, impair, or supersede Federal, State, or local law unless expressly so provided in such Act or amendments.

(2) STATE TAX SAVINGS PROVISION- Notwithstanding paragraph (1), nothing in this Act or the amendments made by this Act shall be construed to modify, impair, or supersede, or authorize the modification, impairment, or supersession of, any State or local law pertaining to taxation, except as provided in sections 622 and 653(c) of the Communications Act of 1934 and section 602 of this Act.

(d) COMMERCIAL MOBILE SERVICE JOINT MARKETING- Notwithstanding section 22.903 of the Commission's regulations (47 C.F.R. 22.903) or any other Commission regulation, a Bell operating company or any other company may, except as provided in sections 271(e)(1) and 272 of the Communications Act of 1934 as amended by this Act as they relate to wireline service, jointly market and sell commercial mobile services in conjunction with telephone exchange service, exchange access, intraLATA telecommunications service, interLATA telecommunications service, and information services.

(e) DEFINITIONS- As used in this section:

(1) AT&T CONSENT DECREE- The term "AT&T Consent Decree" means the order entered August 24, 1982, in the antitrust action styled *United States v. Western Electric*, Civil Action No. 82-0192, in the United States District Court for the District of Columbia, and includes any judg-

ment or order with respect to such action entered on or after August 24, 1982.

(2) GTE CONSENT DECREE- The term "GTE Consent Decree" means the order entered December 21, 1984, as restated January 11, 1985, in the action styled *United States v. GTE Corp.*, Civil Action No. 83-1298, in the United States District Court for the District of Columbia, and any judgment or order with respect to such action entered on or after December 21, 1984.

(3) MCCAW CONSENT DECREE- The term "McCaw Consent Decree" means the proposed consent decree filed on July 15, 1994, in the antitrust action styled *United States v. AT&T Corp. and McCaw Cellular Communications, Inc.*, Civil Action No. 94-01555, in the United States District Court for the District of Columbia. Such term includes any stipulation that the parties will abide by the terms of such proposed consent decree until it is entered and any order entering such proposed consent decree.

(4) ANTITRUST LAWS- The term "antitrust laws" has the meaning given it in subsection (a) of the first section of the Clayton Act (15 U.S.C. 12(a)), except that such term includes the Act of June 19, 1936 (49 Stat. 1526; 15 U.S.C. 13 et seq.), commonly known as the Robinson-Patman Act, and section 5 of the Federal Trade Commission Act (15 U.S.C. 45) to the extent that such section 5 applies to unfair methods of competition.

SEC. 602. PREEMPTION OF LOCAL TAXATION WITH RESPECT TO DIRECT-TO-HOME SERVICES.

(a) PREEMPTION- A provider of direct-to-home satellite service shall be exempt from the collection or remittance, or both, of any tax or fee imposed by any local taxing jurisdiction on direct-to-home satellite service.

(b) DEFINITIONS- For the purposes of this section—

(1) DIRECT-TO-HOME SATELLITE SERVICE- The term "direct-to-home satellite service" means only programming transmitted or broadcast by satellite directly to the subscribers' premises without the use of ground receiving or distribution equipment, except at the subscribers' premises or in the uplink process to the satellite.

(2) PROVIDER OF DIRECT-TO-HOME SATELLITE SERVICE- For purposes of this section, a "provider of direct-to-home satellite service" means a person who transmits, broadcasts, sells, or distributes direct-to-home satellite service.

(3) LOCAL TAXING JURISDICTION- The term "local taxing juris-

diction" means any municipality, city, county, township, parish, transportation district, or assessment jurisdiction, or any other local jurisdiction in the territorial jurisdiction of the United States with the authority to impose a tax or fee, but does not include a State.

(4) STATE- The term "State" means any of the several States, the District of Columbia, or any territory or possession of the United States.

(5) TAX OR FEE- The terms "tax" and "fee" mean any local sales tax, local use tax, local intangible tax, local income tax, business license tax, utility tax, privilege tax, gross receipts tax, excise tax, franchise fees, local telecommunications tax, or any other tax, license, or fee that is imposed for the privilege of doing business, regulating, or raising revenue for a local taxing jurisdiction.

(c) PRESERVATION OF STATE AUTHORITY- This section shall not be construed to prevent taxation of a provider of direct-to-home satellite service by a State or to prevent a local taxing jurisdiction from receiving revenue derived from a tax or fee imposed and collected by a State.

TITLE VII—MISCELLANEOUS PROVISIONS
SEC. 701. PREVENTION OF UNFAIR BILLING PRACTICES FOR INFORMATION OR SERVICES PROVIDED OVER TOLL-FREE TELEPHONE CALLS.

(a) PREVENTION OF UNFAIR BILLING PRACTICES-

(1) IN GENERAL- Section 228(c) (47 U.S.C. 228(c)) is amended—

(A) by striking out subparagraph (C) of paragraph (7) and inserting in lieu thereof the following:

"(C) the calling party being charged for information conveyed during the call unless—

"(i) the calling party has a written agreement (including an agreement transmitted through electronic medium) that meets the requirements of paragraph (8); or

"(ii) the calling party is charged for the information in accordance with paragraph (9); or";

(B)(i) by striking "or" at the end of subparagraph (C) of such paragraph;

(ii) by striking the period at the end of subparagraph (D) of such paragraph and inserting a semicolon and "or"; and

(iii) by adding at the end thereof the following:

"(E) the calling party being assessed, by virtue of being asked to connect or otherwise transfer to a pay-per-call service, a charge for the call."; and

(C) by adding at the end the following new paragraphs:

"(8) SUBSCRIPTION AGREEMENTS FOR BILLING FOR INFORMATION PROVIDED VIA TOLL-FREE CALLS-

"(A) IN GENERAL- For purposes of paragraph (7)(C)(i), a written subscription does not meet the requirements of this paragraph unless the agreement specifies the material terms and conditions under which the information is offered and includes—

"(i) the rate at which charges are assessed for the information;

"(ii) the information provider's name;

"(iii) the information provider's business address;

"(iv) the information provider's regular business telephone number;

"(v) the information provider's agreement to notify the subscriber at least one billing cycle in advance of all future changes in the rates charged for the information; and

"(vi) the subscriber's choice of payment method, which may be by direct remit, debit, prepaid account, phone bill, or credit or calling card.

"(B) BILLING ARRANGEMENTS- If a subscriber elects, pursuant to subparagraph (A)(vi), to pay by means of a phone bill—

"(i) the agreement shall clearly explain that the subscriber will be assessed for calls made to the information service from the subscriber's phone line;

"(ii) the phone bill shall include, in prominent type, the following disclaimer:

"Common carriers may not disconnect local or long distance telephone service for failure to pay disputed charges for information services."; and

"(iii) the phone bill shall clearly list the 800 number dialed.

"(C) USE OF PINS TO PREVENT UNAUTHORIZED USE- A written agreement does not meet the requirements of this paragraph unless it—

"(i) includes a unique personal identification number or other subscriber-specific identifier and requires a subscriber to use this number or identifier to obtain access to the information provided and includes instructions on its use; and

"(ii) assures that any charges for services accessed by use of the subscriber's personal identification number or subscriber-specific identifier be assessed to subscriber's source of payment elected pursuant to subparagraph (A)(vi).

"(D) EXCEPTIONS- Notwithstanding paragraph (7)(C), a written agreement that meets the requirements of this paragraph is not required—

"(i) for calls utilizing telecommunications devices for the deaf;

"(ii) for directory services provided by a common carrier or its affiliate or by a local exchange carrier or its affiliate; or

"(iii) for any purchase of goods or of services that are not information services.

"(E) TERMINATION OF SERVICE- On receipt by a common carrier of a complaint by any person that an information provider is in violation of the provisions of this section, a carrier shall—

"(i) promptly investigate the complaint; and

"(ii) if the carrier reasonably determines that the complaint is valid, it may terminate the provision of service to an information provider unless the provider supplies evidence of a written agreement that meets the requirements of this section.

"(F) TREATMENT OF REMEDIES- The remedies provided in this paragraph are in addition to any other remedies that are available under title V of this Act.

"(9) CHARGES BY CREDIT, PREPAID, DEBIT, CHARGE, OR CALLING CARD IN ABSENCE OF AGREEMENT- For purposes of paragraph (7)(C)(ii), a calling party is not charged in accordance with this paragraph unless the calling party is charged by means of a credit, prepaid, debit, charge, or calling card and the information service provider includes in response to each call an introductory disclosure message that—

"(A) clearly states that there is a charge for the call;

"(B) clearly states the service's total cost per minute and any other fees for the service or for any service to which the caller may be transferred;

"(C) explains that the charges must be billed on either a credit, prepaid, debit, charge, or calling card;

"(D) asks the caller for the card number;

"(E) clearly states that charges for the call begin at the end of the introductory message; and

"(F) clearly states that the caller can hang up at or before the end of the introductory message without incurring any charge whatsoever.

"(10) BYPASS OF INTRODUCTORY DISCLOSURE MESSAGE- The requirements of paragraph (9) shall not apply to calls from repeat callers using a bypass mechanism to avoid listening to the introductory message: *Provided,* That information providers shall disable such a bypass mechanism after the institution of any price increase and for a period of time determined to be sufficient by the Federal Trade Commission to give callers adequate and sufficient notice of a price increase.

"(11) DEFINITION OF CALLING CARD- As used in this subsection, the term "calling card" means an identifying number or code unique to the individual, that is issued to the individual by a common carrier and enables the individual to be charged by means of a phone bill for charges incurred independent of where the call originates.".

(2) REGULATIONS- The Federal Communications Commission shall

revise its regulations to comply with the amendment made by paragraph (1) not later than 180 days after the date of enactment of this Act.
(3) EFFECTIVE DATE- The amendments made by paragraph (1) shall take effect on the date of enactment of this Act.
(b) CLARIFICATION OF "PAY-PER-CALL SERVICES"-
(1) TELEPHONE DISCLOSURE AND DISPUTE RESOLUTION ACT- Section 204(1) of the Telephone Disclosure and Dispute Resolution Act (15 U.S.C. 5714(1)) is amended to read as follows:
"(1) The term "pay-per-call services" has the meaning provided in section 228(i) of the Communications Act of 1934, except that the Commission by rule may, notwithstanding subparagraphs (B) and (C) of section 228(i)(1) of such Act, extend such definition to other similar services providing audio information or audio entertainment if the Commission determines that such services are susceptible to the unfair and deceptive practices that are prohibited by the rules prescribed pursuant to section 201(a).".
(2) COMMUNICATIONS ACT- Section 228(i)(2) (47 U.S.C. 228(i)(2)) is amended by striking "or any service the charge for which is tariffed,".

SEC. 702. PRIVACY OF CUSTOMER INFORMATION.
Title II is amended by inserting after section 221 (47 U.S.C. 221) the following new section:

"SEC. 222. PRIVACY OF CUSTOMER INFORMATION.
"(a) IN GENERAL- Every telecommunications carrier has a duty to protect the confidentiality of proprietary information of, and relating to, other telecommunication carriers, equipment manufacturers, and customers, including telecommunication carriers reselling telecommunications services provided by a telecommunications carrier.
"(b) CONFIDENTIALITY OF CARRIER INFORMATION- A telecommunications carrier that receives or obtains proprietary information from another carrier for purposes of providing any telecommunications service shall use such information only for such purpose, and shall not use such information for its own marketing efforts.
"(c) CONFIDENTIALITY OF CUSTOMER PROPRIETARY NETWORK INFORMATION-
"(1) PRIVACY REQUIREMENTS FOR TELECOMMUNICATIONS CARRIERS- Except as required by law or with the approval of the customer, a telecommunications carrier that receives or obtains customer proprietary network information by virtue of its provision of a telecom-

munications service shall only use, disclose, or permit access to individually identifiable customer proprietary network information in its provision of (A) the telecommunications service from which such information is derived, or (B) services necessary to, or used in, the provision of such telecommunications service, including the publishing of directories.

"(2) DISCLOSURE ON REQUEST BY CUSTOMERS- A telecommunications carrier shall disclose customer proprietary network information, upon affirmative written request by the customer, to any person designated by the customer.

"(3) AGGREGATE CUSTOMER INFORMATION- A telecommunications carrier that receives or obtains customer proprietary network information by virtue of its provision of a telecommunications service may use, disclose, or permit access to aggregate customer information other than for the purposes described in paragraph (1). A local exchange carrier may use, disclose, or permit access to aggregate customer information other than for purposes described in paragraph (1) only if it provides such aggregate information to other carriers or persons on reasonable and nondiscriminatory terms and conditions upon reasonable request therefor.

"(d) EXCEPTIONS- Nothing in this section prohibits a telecommunications carrier from using, disclosing, or permitting access to customer proprietary network information obtained from its customers, either directly or indirectly through its agents—

"(1) to initiate, render, bill, and collect for telecommunications services;

"(2) to protect the rights or property of the carrier, or to protect users of those services and other carriers from fraudulent, abusive, or unlawful use of, or subscription to, such services; or

"(3) to provide any inbound telemarketing, referral, or administrative services to the customer for the duration of the call, if such call was initiated by the customer and the customer approves of the use of such information to provide such service.

"(e) SUBSCRIBER LIST INFORMATION- Notwithstanding subsections (b), (c), and (d), a telecommunications carrier that provides telephone exchange service shall provide subscriber list information gathered in its capacity as a provider of such service on a timely and unbundled basis, under nondiscriminatory and reasonable rates, terms, and conditions, to any person upon request for the purpose of publishing directories in any format.

"(f) DEFINITIONS- As used in this section:

"(1) CUSTOMER PROPRIETARY NETWORK INFORMATION- The term "customer proprietary network information" means—

"(A) information that relates to the quantity, technical configuration, type, destination, and amount of use of a telecommunications service subscribed to by any customer of a telecommunications carrier, and that is made available to the carrier by the customer solely by virtue of the carrier-customer relationship; and

"(B) information contained in the bills pertaining to telephone exchange service or telephone toll service received by a customer of a carrier; except that such term does not include subscriber list information.

"(2) AGGREGATE INFORMATION- The term "aggregate customer information" means collective data that relates to a group or category of services or customers, from which individual customer identities and characteristics have been removed.

"(3) SUBSCRIBER LIST INFORMATION- The term "subscriber list information" means any information—

"(A) identifying the listed names of subscribers of a carrier and such subscribers" telephone numbers, addresses, or primary advertising classifications (as such classifications are assigned at the time of the establishment of such service), or any combination of such listed names, numbers, addresses, or classifications; and"(B) that the carrier or an affiliate has published, caused to be published, or accepted for publication in any directory format.".

SEC. 703. POLE ATTACHMENTS.

Section 224 (47 U.S.C. 224) is amended—

(1) in subsection (a)(1), by striking the first sentence and inserting the following: "The term "utility" means any person who is a local exchange carrier or an electric, gas, water, steam, or other public utility, and who owns or controls poles, ducts, conduits, or rights-of-way used, in whole or in part, for any wire communications.";

(2) in subsection (a)(4), by inserting after "system" the following: "or provider of telecommunications service";

(3) by inserting after subsection (a)(4) the following:

"(5) For purposes of this section, the term "telecommunications carrier" (as defined in section 3 of this Act) does not include any incumbent local exchange carrier as defined in section 251(h).";

(4) by inserting after "conditions" in subsection (c)(1) a comma and the following: "or access to poles, ducts, conduits, and rights-of-way as provided in subsection (f),";

(5) in subsection (c)(2)(B), by striking "cable television services" and inserting "the services offered via such attachments";

(6) by inserting after subsection (d)(2) the following:

"(3) This subsection shall apply to the rate for any pole attachment used by a cable television system solely to provide cable service. Until the effective date of the regulations required under subsection (e), this subsection shall also apply to the rate for any pole attachment used by a cable system or any telecommunications carrier (to the extent such carrier is not a party to a pole attachment agreement) to provide any telecommunications service."; and

(7) by adding at the end thereof the following:

"(e)(1) The Commission shall, no later than 2 years after the date of enactment of the Telecommunications Act of 1996, prescribe regulations in accordance with this subsection to govern the charges for pole attachments used by telecommunications carriers to provide telecommunications services, when the parties fail to resolve a dispute over such charges. Such regulations shall ensure that a utility charges just, reasonable, and nondiscriminatory rates for pole attachments.

"(2) A utility shall apportion the cost of providing space on a pole, duct, conduit, or right-of-way other than the usable space among entities so that such apportionment equals two-thirds of the costs of providing space other than the usable space that would be allocated to such entity under an equal apportionment of such costs among all attaching entities.

"(3) A utility shall apportion the cost of providing usable space among all entities according to the percentage of usable space required for each entity.

"(4) The regulations required under paragraph (1) shall become effective 5 years after the date of enactment of the Telecommunications Act of 1996. Any increase in the rates for pole attachments that result from the adoption of the regulations required by this subsection shall be phased in equal annual increments over a period of 5 years beginning on the effective date of such regulations.

"(f)(1) A utility shall provide a cable television system or any telecommunications carrier with nondiscriminatory access to any pole, duct, conduit, or right-of-way owned or controlled by it.

"(2) Notwithstanding paragraph (1), a utility providing electric service may deny a cable television system or any telecommunications carrier access to its poles, ducts, conduits, or rights-of-way, on a non-discriminatory basis where there is insufficient capacity and for reasons of safety, reliability and generally applicable engineering purposes.

"(g) A utility that engages in the provision of telecommunications services or cable services shall impute to its costs of providing such services (and charge any affiliate, subsidiary, or associate company engaged in the provision of such services) an equal amount to the pole attachment rate for which such company would be liable under this section.

"(h) Whenever the owner of a pole, duct, conduit, or right-of-way intends to modify or alter such pole, duct, conduit, or right-of-way, the owner shall provide written notification of such action to any entity that has obtained an attachment to such conduit or right-of-way so that such entity may have a reasonable opportunity to add to or modify its existing attachment. Any entity that adds to or modifies its existing attachment after receiving such notification shall bear a proportionate share of the costs incurred by the owner in making such pole, duct, conduit, or right-of-way accessible.

"(i) An entity that obtains an attachment to a pole, conduit, or right-of-way shall not be required to bear any of the costs of rearranging or replacing its attachment, if such rearrangement or replacement is required as a result of an additional attachment or the modification of an existing attachment sought by any other entity (including the owner of such pole, duct, conduit, or right-of-way).".

SEC. 704. FACILITIES SITING; RADIO FREQUENCY EMISSION STANDARDS.

(a) NATIONAL WIRELESS TELECOMMUNICATIONS SITING POLICY- Section 332(c) (47 U.S.C. 332(c)) is amended by adding at the end the following new paragraph:

"(7) PRESERVATION OF LOCAL ZONING AUTHORITY-

"(A) GENERAL AUTHORITY- Except as provided in this paragraph, nothing in this Act shall limit or affect the authority of a State or local government or instrumentality thereof over decisions regarding the placement, construction, and modification of personal wireless service facilities.

"(B) LIMITATIONS-

"(i) The regulation of the placement, construction, and modification of personal wireless service facilities by any State or local government or instrumentality thereof—

"(I) shall not unreasonably discriminate among providers of functionally equivalent services; and

"(II) shall not prohibit or have the effect of prohibiting the provision of personal wireless services.

"(ii) A State or local government or instrumentality thereof shall act on any request for authorization to place, construct, or modify personal wireless service facilities within a reasonable period of time after the request is duly filed with such government or instrumentality, taking into account the nature and scope of such request.

"(iii) Any decision by a State or local government or instrumentality thereof to deny a request to place, construct, or modify personal wireless service facilities shall be in writing and supported by substantial evidence contained in a written record.

"(iv) No State or local government or instrumentality thereof may regulate the placement, construction, and modification of personal wireless service facilities on the basis of the environmental effects of radio frequency emissions to the extent that such facilities comply with the Commission's regulations concerning such emissions.

"(v) Any person adversely affected by any final action or failure to act by a State or local government or any instrumentality thereof that is inconsistent with this subparagraph may, within 30 days after such action or failure to act, commence an action in any court of competent jurisdiction. The court shall hear and decide such action on an expedited basis. Any person adversely affected by an act or failure to act by a State or local government or any instrumentality thereof that is inconsistent with clause (iv) may petition the Commission for relief.

"(C) DEFINITIONS- For purposes of this paragraph—

"(i) the term "personal wireless services" means commercial mobile services, unlicensed wireless services, and common carrier wireless exchange access services;

"(ii) the term "personal wireless service facilities" means facilities for the provision of personal wireless services; and

"(iii) the term "unlicensed wireless service" means the offering of telecommunications services using duly authorized devices which do not require individual licenses, but does not mean the provision of direct-to-home satellite services (as defined in section 303(v)).".

(b) RADIO FREQUENCY EMISSIONS- Within 180 days after the enactment of this Act, the Commission shall complete action in ET Docket 93-62 to prescribe and make effective rules regarding the environmental effects of radio frequency emissions.

(c) AVAILABILITY OF PROPERTY- Within 180 days of the enactment of this Act, the President or his designee shall prescribe procedures by which Federal departments and agencies may make available on a fair, reasonable, and nondiscriminatory basis, property, rights-of-way, and easements under their control for the placement of new telecommunications services that are dependent, in whole or in part, upon the utilization of Federal spectrum rights for the transmission or reception of such services. These procedures may establish a presumption that requests for the use of property, rights-of-way, and easements by duly authorized providers should be granted absent unavoidable direct conflict with the department or agency's mission, or the current or planned use of the property, rights-of-way, and easements in question. Reasonable fees may be charged to providers of such telecommunications services for use of property, rights-of-way, and easements. The Commission shall provide technical support to States to encourage them to make property, rights-of-way, and easements under their jurisdiction available for such purposes.

SEC. 705. MOBILE SERVICES DIRECT ACCESS TO LONG DISTANCE CARRIERS.

Section 332(c) (47 U.S.C. 332(c)) is amended by adding at the end the following new paragraph:

"(8) MOBILE SERVICES ACCESS- A person engaged in the provision of commercial mobile services, insofar as such person is so engaged, shall not be required to provide equal access to common carriers for the provision of telephone toll services. If the Commission determines that subscribers to such services are denied access to the provider of telephone toll services of the subscribers' choice, and that such denial is contrary to the public interest, convenience, and necessity, then the Commission shall prescribe regulations to afford subscribers unblocked access to the provider of telephone toll services of the subscribers' choice through the use of a carrier identification code assigned to such provider or other mechanism. The requirements for unblocking shall not apply to mobile satellite services unless the Commission finds it to be in the public interest to apply such requirements to such services."

SEC. 706. ADVANCED TELECOMMUNICATIONS INCENTIVES.

(a) IN GENERAL- The Commission and each State commission with regulatory jurisdiction over telecommunications services shall encourage the deployment on a reasonable and timely basis of advanced telecommunications capability to all Americans (including, in particular, elementary and secondary schools and classrooms) by utilizing, in a manner consistent with the public interest, convenience, and necessity, price cap regulation, regulatory forbearance, measures that promote competition in the local telecommunications market, or other regulating methods that remove barriers to infrastructure investment.

(b) INQUIRY- The Commission shall, within 30 months after the date of enactment of this Act, and regularly thereafter, initiate a notice of inquiry concerning the availability of advanced telecommunications capability to all Americans (including, in particular, elementary and sec-

ondary schools and classrooms) and shall complete the inquiry within
180 days after its initiation. In the inquiry, the Commission shall determine whether advanced telecommunications capability is being deployed
to all Americans in a reasonable and timely fashion. If the Commission's
determination is negative, it shall take immediate action to accelerate
deployment of such capability by removing barriers to infrastructure investment and by promoting competition in the telecommunications
market.

(c) DEFINITIONS- For purposes of this subsection:

(1) ADVANCED TELECOMMUNICATIONS CAPABILITY- The
term "advanced telecommunications capability" is defined, without regard to any transmission media or technology, as high-speed, switched,
broadband telecommunications capability that enables users to originate
and receive high-quality voice, data, graphics, and video telecommunications using any technology.

(2) ELEMENTARY AND SECONDARY SCHOOLS- The term "elementary and secondary schools" means elementary and secondary
schools, as defined in paragraphs (14) and (25), respectively, of section
14101 of the Elementary and Secondary Education Act of 1965 (20
U.S.C. 8801).

SEC. 707. TELECOMMUNICATIONS DEVELOPMENT FUND.

(a) DEPOSIT AND USE OF AUCTION ESCROW ACCOUNTS-
Section 309(j)(8) (47 U.S.C. 309(j)(8)) is amended by adding at the end
the following new subparagraph:

"(C) DEPOSIT AND USE OF AUCTION ESCROW ACCOUNTS-
Any deposits the Commission may require for the qualification of any
person to bid in a system of competitive bidding pursuant to this subsection shall be deposited in an interest bearing account at a financial institution designated for purposes of this subsection by the Commission (after
consultation with the Secretary of the Treasury). Within 45 days following the conclusion of the competitive bidding—

"(i) the deposits of successful bidders shall be paid to the Treasury;

"(ii) the deposits of unsuccessful bidders shall be returned to such bidders; and

"(iii) the interest accrued to the account shall be transferred to the Telecommunications Development Fund established pursuant to section 714
of this Act.".

(b) ESTABLISHMENT AND OPERATION OF FUND- Title VII is
amended by inserting after section 713 (as added by section 305) the
following new section:

"SEC. 714. TELECOMMUNICATIONS DEVELOPMENT FUND.

"(a) PURPOSE OF SECTION- It is the purpose of this section—

"(1) to promote access to capital for small businesses in order to enhance competition in the telecommunications industry;

"(2) to stimulate new technology development, and promote employment and training; and

"(3) to support universal service and promote delivery of telecommunications services to underserved rural and urban areas.

"(b) ESTABLISHMENT OF FUND- There is hereby established a body corporate to be known as the Telecommunications Development Fund, which shall have succession until dissolved. The Fund shall maintain its principal office in the District of Columbia and shall be deemed, for purposes of venue and jurisdiction in civil actions, to be a resident and citizen thereof.

"(c) BOARD OF DIRECTORS-

"(1) COMPOSITION OF BOARD; CHAIRMAN- The Fund shall have a Board of Directors which shall consist of 7 persons appointed by the Chairman of the Commission. Four of such directors shall be representative of the private sector and three of such directors shall be representative of the Commission, the Small Business Administration, and the Department of the Treasury, respectively. The Chairman of the Commission shall appoint one of the representatives of the private sector to serve as chairman of the Fund within 30 days after the date of enactment of this section, in order to facilitate rapid creation and implementation of the Fund. The directors shall include members with experience in a number of the following areas: finance, investment banking, government banking, communications law and administrative practice, and public policy.

"(2) TERMS OF APPOINTED AND ELECTED MEMBERS- The directors shall be eligible to serve for terms of 5 years, except of the initial members, as designated at the time of their appointment—

"(A) 1 shall be eligible to service for a term of 1 year;

"(B) 1 shall be eligible to service for a term of 2 years;

"(C) 1 shall be eligible to service for a term of 3 years;

"(D) 2 shall be eligible to service for a term of 4 years; and

"(E) 2 shall be eligible to service for a term of 5 years (1 of whom shall be the Chairman). Directors may continue to serve until their successors have been appointed and have qualified.

"(3) MEETINGS AND FUNCTIONS OF THE BOARD- The Board of Directors shall meet at the call of its Chairman, but at least quarterly. The Board shall determine the general policies which shall govern the operations of the Fund. The Chairman of the Board shall, with the approval of the Board, select, appoint, and compensate qualified persons to fill the offices as may be provided for in the bylaws, with such functions, powers, and duties as may be prescribed by the bylaws or by the Board of Directors, and such persons shall be the officers of the Fund and shall discharge all such functions, powers, and duties.

"(d) ACCOUNTS OF THE FUND- The Fund shall maintain its accounts at a financial institution designated for purposes of this section by the Chairman of the Board (after consultation with the Commission and the Secretary of the Treasury). The accounts of the Fund shall consist of—

"(1) interest transferred pursuant to section 309(j)(8)(C) of this Act;

"(2) such sums as may be appropriated to the Commission for advances to the Fund;

"(3) any contributions or donations to the Fund that are accepted by the Fund; and

"(4) any repayment of, or other payment made with respect to, loans, equity, or other extensions of credit made from the Fund.

"(e) USE OF THE FUND- All moneys deposited into the accounts of the Fund shall be used solely for—

"(1) the making of loans, investments, or other extensions of credits to eligible small businesses in accordance with subsection (f);

"(2) the provision of financial advice to eligible small businesses;

"(3) expenses for the administration and management of the Fund (including salaries, expenses, and the rental or purchase of office space for the fund);

"(4) preparation of research, studies, or financial analyses; and

"(5) other services consistent with the purposes of this section.

"(f) LENDING AND CREDIT OPERATIONS- Loans or other extensions of credit from the Fund shall be made available in accordance with the requirements of the Federal Credit Reform Act of 1990 (2 U.S.C. 661 et seq.) and any other applicable law to an eligible small business on the basis of—

"(1) the analysis of the business plan of the eligible small business;

"(2) the reasonable availability of collateral to secure the loan or credit extension;

"(3) the extent to which the loan or credit extension promotes the purposes of this section; and

"(4) other lending policies as defined by the Board.

"(g) RETURN OF ADVANCES- Any advances appropriated pursuant to subsection (d)(2) shall be disbursed upon such terms and conditions (including conditions relating to the time or times of repayment) as are specified in any appropriations Act providing such advances.

"(h) GENERAL CORPORATE POWERS- The Fund shall have power—

"(1) to sue and be sued, complain and defend, in its corporate name and through its own counsel;

"(2) to adopt, alter, and use the corporate seal, which shall be judicially noticed;

"(3) to adopt, amend, and repeal by its Board of Directors, bylaws, rules, and regulations as may be necessary for the conduct of its business;

"(4) to conduct its business, carry on its operations, and have officers and exercise the power granted by this section in any State without regard to any qualification or similar statute in any State;

"(5) to lease, purchase, or otherwise acquire, own, hold, improve, use, or otherwise deal in and with any property, real, personal, or mixed, or any interest therein, wherever situated, for the purposes of the Fund;

"(6) to accept gifts or donations of services, or of property, real, personal, or mixed, tangible or intangible, in aid of any of the purposes of the Fund;

"(7) to sell, convey, mortgage, pledge, lease, exchange, and otherwise dispose of its property and assets;

"(8) to appoint such officers, attorneys, employees, and agents as may be required, to determine their qualifications, to define their duties, to fix their salaries, require bonds for them, and fix the penalty thereof; and

"(9) to enter into contracts, to execute instruments, to incur liabilities, to make loans and equity investment, and to do all things as are necessary or incidental to the proper management of its affairs and the proper conduct of its business.

"(i) ACCOUNTING, AUDITING, AND REPORTING- The accounts of the Fund shall be audited annually. Such audits shall be conducted in accordance with generally accepted auditing standards by independent certified public accountants. A report of each such audit shall be furnished to the Secretary of the Treasury and the Commission. The representatives of the Secretary and the Commission shall have access to all books, accounts, financial records, reports, files, and all other papers, things, or property belonging to or in use by the Fund and necessary to facilitate the audit.

"(j) REPORT ON AUDITS BY TREASURY- A report of each such audit for a fiscal year shall be made by the Secretary of the Treasury to the President and to the Congress not later than 6 months following the close of such fiscal year. The report shall set forth the scope of the audit and shall include a statement of assets and liabilities, capital and surplus or deficit; a statement of surplus or deficit analysis; a statement of income and expense; a statement of sources and application of funds; and such comments and information as may be deemed necessary to keep the President and the Congress informed of the operations and financial condition of the Fund, together with such recommendations with respect thereto as the Secretary may deem advisable.

"(k) DEFINITIONS- As used in this section:

"(1) ELIGIBLE SMALL BUSINESS- The term "eligible small business" means business enterprises engaged in the telecommunications industry that have $50,000,000 or less in annual revenues, on average over the past 3 years prior to submitting the application under this section.

"(2) FUND- The term "Fund" means the Telecommunications Development Fund established pursuant to this section.

"(3) TELECOMMUNICATIONS INDUSTRY- The term "telecommunications industry" means communications businesses using regulated or unregulated facilities or services and includes broadcasting, telecommunications, cable, computer, data transmission, software, programming, advanced messaging, and electronics businesses.".

SEC. 708. NATIONAL EDUCATION TECHNOLOGY FUNDING CORPORATION.

(a) FINDINGS; PURPOSE-

(1) FINDINGS- The Congress finds as follows:

(A) CORPORATION- There has been established in the District of Columbia a private, nonprofit corporation known as the National Education Technology Funding Corporation which is not an agency or independent establishment of the Federal Government.

(B) BOARD OF DIRECTORS- The Corporation is governed by a Board of Directors, as prescribed in the Corporation's articles of incorporation, consisting of 15 members, of which—

(i) five members are representative of public agencies representative of schools and public libraries;

(ii) five members are representative of State government, including persons knowledgeable about State finance, technology and education; and

(iii) five members are representative of the private sector, with expertise in network technology, finance and management.

(C) CORPORATE PURPOSES- The purposes of the Corporation, as set forth in its articles of incorporation, are—

(i) to leverage resources and stimulate private investment in education technology infrastructure;

(ii) to designate State education technology agencies to receive loans, grants or other forms of assistance from the Corporation;

(iii) to establish criteria for encouraging States to—

(I) create, maintain, utilize and upgrade interactive high capacity networks capable of providing audio, visual and data communications for elementary schools, secondary schools and public libraries;

(II) distribute resources to assure equitable aid to all elementary schools and secondary schools in the State and achieve universal access to network technology; and

(III) upgrade the delivery and development of learning through innovative technology-based instructional tools and applications;

(iv) to provide loans, grants and other forms of assistance to State education technology agencies, with due regard for providing a fair balance among types of school districts and public libraries assisted and the disparate needs of such districts and libraries;

(v) to leverage resources to provide maximum aid to elementary schools, secondary schools and public libraries; and

(vi) to encourage the development of education telecommunications and information technologies through public-private ventures, by serving as a clearinghouse for information on new education technologies, and by providing technical assistance, including assistance to States, if needed, to establish State education technology agencies.

(2) PURPOSE- The purpose of this section is to recognize the Corporation as a nonprofit corporation operating under the laws of the District of Columbia, and to provide authority for Federal departments and agencies to provide assistance to the Corporation.

(b) DEFINITIONS- For the purpose of this section—

(1) the term "Corporation" means the National Education Technology Funding Corporation described in subsection (a)(1)(A);

(2) the terms "elementary school" and "secondary school" have the same meanings given such terms in section 14101 of the Elementary and Secondary Education Act of 1965; and

(3) the term "public library" has the same meaning given such term in section 3 of the Library Services and Construction Act.

(c) ASSISTANCE FOR EDUCATION TECHNOLOGY PURPOSES-

(1) RECEIPT BY CORPORATION- Notwithstanding any other provision of law, in order to carry out the corporate purposes described in subsection (a)(1)(C), the Corporation shall be eligible to receive discretionary grants, contracts, gifts, contributions, or technical assistance from any Federal department or agency, to the extent otherwise permitted by law.

(2) AGREEMENT- In order to receive any assistance described in paragraph (1) the Corporation shall enter into an agreement with the Federal department or agency providing such assistance, under which the Corporation agrees—

(A) to use such assistance to provide funding and technical assistance only for activities which the Board of Directors of the Corporation determines are consistent with the corporate purposes described in subsection (a)(1)(C);

(B) to review the activities of State education technology agencies and other entities receiving assistance from the Corporation to assure that the corporate purposes described in subsection (a)(1)(C) are carried out;

(C) that no part of the assets of the Corporation shall accrue to the benefit of any member of the Board of Directors of the Corporation, any officer or employee of the Corporation, or any other individual, except as salary or reasonable compensation for services;

(D) that the Board of Directors of the Corporation will adopt policies and procedures to prevent conflicts of interest;

(E) to maintain a Board of Directors of the Corporation consistent with subsection (a)(1)(B);

(F) that the Corporation, and any entity receiving the assistance from the Corporation, are subject to the appropriate oversight procedures of the Congress; and

(G) to comply with—

(i) the audit requirements described in subsection

(d); and

(ii) the reporting and testimony requirements described in subsection (e).

(3) CONSTRUCTION- Nothing in this section shall be construed to establish the Corporation as an agency or independent establishment of the Federal Government, or to establish the members of the Board of Directors of the Corporation, or the officers and employees of the Corporation, as officers or employees of the Federal Government.

(d) AUDITS-

(1) AUDITS BY INDEPENDENT CERTIFIED PUBLIC ACCOUNTANTS-

(A) IN GENERAL- The Corporation's financial statements shall be audited annually in accordance with generally accepted auditing standards by independent certified public accountants who are certified by a regulatory authority of a State or other political subdivision of the United States. The audits shall be conducted at the place or places where the accounts of the Corporation are normally kept. All books, accounts, financial records, reports, files, and all other papers, things, or property belonging to or in use by the Corporation and necessary to facilitate the audit shall be made available to the person or persons conducting the audits, and full facilities for verifying transactions with the balances or securities held by depositories, fiscal agents, and custodians shall be afforded to such person or persons.

(B) REPORTING REQUIREMENTS- The report of each annual audit described in subparagraph (A) shall be included in the annual report required by subsection (e)(1).

(2) RECORDKEEPING REQUIREMENTS; AUDIT AND EXAMINATION OF BOOKS-

(A) RECORDKEEPING REQUIREMENTS- The Corporation shall ensure that each recipient of assistance from the Corporation keeps—

(i) separate accounts with respect to such assistance;

(ii) such records as may be reasonably necessary to fully disclose—

(I) the amount and the disposition by such recipient of the proceeds of such assistance;

(II) the total cost of the project or undertaking in connection with which such assistance is given or used; and

(III) the amount and nature of that portion of the cost of the project or undertaking supplied by other sources; and

(iii) such other records as will facilitate an effective audit.

(B) AUDIT AND EXAMINATION OF BOOKS- The Corporation shall ensure that the Corporation, or any of the Corporation's duly authorized representatives, shall have access for the purpose of audit and examination to any books, documents, papers, and records of any recipient of assistance from the Corporation that are pertinent to such assistance. Representatives of the Comptroller General shall also have such access for such purpose.

(e) ANNUAL REPORT; TESTIMONY TO THE CONGRESS-

(1) ANNUAL REPORT- Not later than April 30 of each year, the Corporation shall publish an annual report for the preceding fiscal year and submit that report to the President and the Congress. The report shall include a comprehensive and detailed evaluation of the Corporation's operations, activities, financial condition, and accomplishments under this section and may include such recommendations as the Corporation deems appropriate.

(2) TESTIMONY BEFORE CONGRESS- The members of the Board of Directors, and officers, of the Corporation shall be available to testify before appropriate committees of the Congress with respect to the report described in paragraph (1), the report of any audit made by the Comp-

troller General pursuant to this section, or any other matter which any such committee may determine appropriate.

SEC. 709. REPORT ON THE USE OF ADVANCED TELECOMMUNICATIONS SERVICES FOR MEDICAL PURPOSES.

The Secretary of Commerce, in consultation with the Secretary of Health and Human Services and other appropriate departments and agencies, shall submit a report to the Committee on Commerce of the House of Representatives and the Committee on Commerce, Science, and Transportation of the Senate concerning the activities of the Joint Working Group on Telemedicine, together with any findings reached in the studies and demonstrations on telemedicine funded by the Public Health Service or other Federal agencies. The report shall examine questions related to patient safety, the efficacy and quality of the services provided, and other legal, medical, and economic issues related to the utilization of advanced telecommunications services for medical purposes. The report shall be submitted to the respective committees by January 31, 1997.

SEC. 710. AUTHORIZATION OF APPROPRIATIONS.

(a) IN GENERAL- In addition to any other sums authorized by law, there are authorized to be appropriated to the Federal Communications Commission such sums as may be necessary to carry out this Act and the amendments made by this Act.

(b) EFFECT ON FEES- For the purposes of section 9(b)(2) (47 U.S.C. 159(b)(2)), additional amounts appropriated pursuant to subsection (a) shall be construed to be changes in the amounts appropriated for the performance of activities described in section 9(a) of the Communications Act of 1934.

(c) FUNDING AVAILABILITY- Section 309(j)(8)(B) (47 U.S.C. 309(j)(8)(B)) is amended by adding at the end the following new sentence: "Such offsetting collections are authorized to remain available until expended.". Speaker of the House of Representatives. Vice President of the United States and President of the Senate.

Appendix A2:
Communications Act of 1934
(condensed business applications version)

AN ACT To provide for the regulation of interstate and foreign communication by wire or radio, and for other purposes.

Title I—General Provisions

Sec. 1. [47 U.S.C. 151] Purposes of Act, Creation of Federal Communications Commission.

For the purpose of regulating interstate and foreign commerce in communication by wire and radio so as to make available, so far as possible, to all the people of the United States, without discrimination on the basis of race, color, religion, national origin, or sex, a rapid, efficient, Nation-wide, and world-wide wire and radio communication service with adequate facilities at reasonable charges, for the purpose of the national defense, for the purpose of promoting safety of life and property through the use of wire and radio communication, and for the purpose of securing a more effective execution of this policy by centralizing authority heretofore granted by law to several agencies and by granting additional authority with respect to interstate and foreign commerce in wire and radio communication, there is hereby created a commission to be known as the "Federal Communications Commission," which shall be constituted as hereinafter provided, and which shall execute and enforce the provisions of this Act.

Sec. 2. [47 U.S.C. 152] Application of Act.

(a) The provisions of this act shall apply to all interstate and foreign communication by wire or radio and all interstate and foreign transmission of energy by radio, which originates and/or is received within the United States, and to all persons engaged within the United States in such communication or such transmission of energy by radio, and to the licensing and regulating of all radio stations as hereinafter provided; but it shall not apply to persons engaged in wire or radio communication or transmission in the Canal Zone, or to wire or radio communication or transmission wholly within the Canal Zone. The provisions of this Act shall apply with respect to cable service, to all persons engaged within the United States in providing such service, and to the facilities of cable operators which relate to such service, as provided in title VI.

(b) Except as provided in sections 223 through 227, inclusive, and section 332, and subject to the provisions of section 301 and title VI, nothing in this Act shall be construed to apply or to give the Commission jurisdiction with respect to (1) charges, classifications, practices, services, facilities, or regulations for or in connection with intrastate communication service by wire or radio of any carrier, or (2) any carrier engaged in interstate or foreign communication solely through physical connection with the facilities of another carrier not directly or indirectly controlling or controlled by, or under direct or indirect common control with such carrier, or (3) any carrier engaged in interstate or foreign communication solely through connection by radio, or by wire and radio, with facilities, located in an adjoining State or in Canada or Mexico (where they adjoin the State in which the carrier is doing business), of another carrier not directly or indirectly controlling or controlled by, or under direct or indirect common control with such carrier, or (4) any carrier to which clause (2) or clause (3) would be applicable except for furnishing interstate mobile radio communication service or radio communication service to mobile stations on land vehicles in Canada or Mexico; except that sections 201 through 205 of this Act, both inclusive, shall, except as otherwise provided therein, apply to carriers described in clauses (2), (3), and (4).

Sec. 3. [47 U.S.C. 153] Definitions.

For the purposes of this Act, unless the context otherwise requires—

(1) Affiliate.—The term "affiliate" means a person that (directly or indirectly) owns or controls, is owned or controlled by, or is under common ownership or control with, another person. For purposes of this paragraph, the term "own" means to own an equity interest (or the equivalent thereof) of more than 10 percent.

(2) Amateur station.—The term "amateur station" means a radio station operated by a duly authorized person interested in radio technique solely with a personal aim and without pecuniary interest.

(3) AT&T consent decree.—The term "AT&T Consent Decree" means the order entered August 24, 1982, in the antitrust action styled United States v. Western Electric, Civil Action No. 82-0192, in the United States District Court for the District of Columbia, and includes any judgment or order with respect to such action entered on or after August 24, 1982.

(4) Bell operating company.—The term "Bell operating company"—

(A) means any of the following companies: Bell Telephone Company of Nevada, Illinois Bell Telephone Company, Indiana Bell Telephone Company, Incorporated, Michigan Bell Telephone Company, New England Telephone and Telegraph Company, New Jersey Bell Telephone Company, New York Telephone Company, U S West Communications Company, South Central Bell Telephone Company, Southern Bell Telephone and Telegraph Company, Southwestern Bell Telephone Company, The Bell Telephone Company of Pennsylvania, The Chesapeake and Potomac Telephone Company, The Chesapeake and Potomac Telephone Company of Maryland, The Chesapeake and Potomac Telephone Company of Virginia, The Chesapeake and Potomac Telephone Company of West Virginia, The Diamond State Telephone Company, The Ohio Bell Telephone Company, The Pacific Telephone and Telegraph Company, or Wisconsin Telephone Company; and

(B) includes any successor or assign of any such company that provides wireline telephone exchange service; but

(C) does not include an affiliate of any such company, other than an affiliate described in subparagraph (A) or (B).

(5) Broadcast station.—The term "broadcast station," "broadcasting station," or "radio broadcast station" means a radio station equipped to engage in broadcasting as herein defined.

(6) Broadcasting.—The term "broadcasting" means the dissemination of radio communications intended to be received by the public, directly or by the intermediary of relay stations.

(7) Cable service.—The term "cable service" has the meaning given such term in section 602.

(8) Cable system.—The term "cable system" has the meaning given such term in section 602.

(9) Chain broadcasting.—The term "chain broadcasting" means simultaneous broadcasting of an identical program by two or more connected stations.

(10) Common carrier.—The term "common carrier" or "carrier" means any person engaged as a common carrier for hire, in interstate or foreign communication by wire or radio or in interstate or foreign radio transmission of energy, except where reference is made to common carriers not subject to this Act; but a person engaged in radio broadcasting shall not, insofar as such person is so engaged, be deemed a common carrier.

(11) Connecting carrier.—The term "connecting carrier" means a carrier described in clauses (2), (3), or (4) of section 2(b).

(12) Construction permit.—The term "construction permit" or "permit for construction" means that instrument of authorization required by this Act or the rules and regulations of the Commission made pursuant to this Act for the construction of a station, or the installation of apparatus, for the transmission of energy, or communications, or signals by radio, by whatever name the instrument may be designated by the Commission.

(13) Corporation.—The term "corporation" includes any corporation, joint-stock company, or association.

(14) Customer premises equipment.—The term "customer premises equipment" means equipment employed on the premises of a person (other than a carrier) to originate, route, or terminate telecommunications.

(15) Dialing parity.—The term "dialing parity" means that a person that is not an affiliate of a local exchange carrier is able to provide telecommunications services in such a manner that customers have the ability to route automatically, without the use of any access code, their telecommunications to the telecommunications services provider of the customer's designation from among 2 or more telecommunications services providers (including such local exchange carrier).

(16) Exchange access.—The term "exchange access" means the offering of access to telephone exchange services or facilities for the purpose of the origination or termination of telephone toll services.

(17) Foreign communication.—The term "foreign communication" or "foreign transmission" means communication or transmission from or to any place in the United States to or from a foreign country, or between a station in the United States and a mobile station located outside the United States.

(18) Great lakes agreement.—The term "Great Lakes Agreement" means the Agreement for the Promotion of Safety on the Great Lakes by Means of Radio in force and the regulations referred to therein.

(19) Harbor.—The term "harbor" or "port" means any place to which ships may resort for shelter or to load or unload passengers or goods, or to obtain fuel, water, or supplies. This term shall apply to such places whether proclaimed public or not and whether natural or artificial.

(20) Information service.—The term "information service" means the offering of a capability for generating, acquiring, storing, transforming, processing, retrieving, utilizing, or making available information via telecommunications, and includes electronic publishing, but does not include any use of any such capability for the management, control, or operation of a telecommunications system or the management of a telecommunications service.

(21) Interlata service.—The term "interLATA service" means telecommunications between a point located in a local access and transport area and a point located outside such area.

(22) Interstate communication.—The term "interstate communication" or "interstate transmission" means communication or transmission (A) from any State, Territory, or possession of the United States (other than the Canal Zone), or the District of Columbia, to any other State, Territory, or possession of the United States (other than the Canal Zone), or the District of Columbia, (B) from or to the United States to or from the Canal Zone, insofar as such communication or transmission takes place within the United States, or (C) between points within the United States but through a foreign country; but shall not, with respect to the provisions of title II of this Act (other than section 223 thereof), include wire or radio communication between points in the same State, Territory, or possession of the United States, or the District of Columbia, through any place outside thereof, if such communication is regulated by a State commission.

(23) Land station.—The term "land station" means a station, other than a mobile station, used for radio communication with mobile stations.

(24) Licensee.—The term "licensee" means the holder of a radio station license granted or continued in force under authority of this Act.

(25) Local access and transport area.—The term "local access and transport area" or "LATA" means a contiguous geographic area—

(A) established before the date of enactment of the Telecommunications Act of 1996 by a Bell operating company such that no exchange area includes points within more than 1 metropolitan statistical area, consolidated metropolitan statistical area, or State, except as expressly permitted under the AT&T Consent Decree; or

(B) established or modified by a Bell operating company after such date of enactment and approved by the Commission.

(26) Local exchange carrier.—The term "local exchange carrier" means any person that is engaged in the provision of telephone exchange service or exchange access. Such term does not include a person insofar as such person is engaged in the provision of a commercial mobile service under section 332(c), except to the extent that the Commission finds that such service should be included in the definition of such term.

(27) Mobile service.—The term "mobile service" means a radio communication service carried on between mobile stations or receivers and land stations, and by mobile stations communicating among themselves, and includes (A) both one-way and two-way radio communication services, (B) a mobile service which provides a regularly interacting group of base, mobile, portable, and associated control and relay stations (whether licensed on an individual, cooperative, or multiple basis) for private one-way or two-way land mobile radio communications by eligible users over designated areas of operation, and (C) any service for which a license is required in a personal communications service established pursuant to the proceeding entitled "Amendment to the Commission's Rules to Establish New Personal Communications Services" (GEN Docket No. 90-314; ET Docket No. 92-100), or any successor proceeding.

(28) Mobile station.—The term "mobile station" means a radio-communication station capable of being moved and which ordinarily does move.

(29) Network element.—The term "network element" means a facility or equipment used in the provision of a telecommunications service. Such term also includes features, functions, and capabilities that are provided by means of such facility or equipment, including subscriber numbers, databases, signaling systems, and information sufficient for billing and collection or used in the transmission, routing, or other provision of a telecommunications service.

(30) Number portability.—The term "number portability" means the ability of users of telecommunications services to retain, at the same location, existing telecommunications numbers without impairment of quality, reliability, or convenience when switching from one telecommunications carrier to another.

(31)(A) Operator.—The term "operator" on a ship of the United States means, for the purpose of parts II and III of title III of this Act, a person holding a radio operator's license of the proper class as prescribed and issued by the Commission.

(B) "Operator" on a foreign ship means, for the purpose of part II of title III of this Act, a person holding a certificate as such of the proper class complying with the provision of the radio regulations annexed to the International Telecommunication Convention in force, or complying with an agreement or treaty between the United States and the country in which the ship is registered.

(32) Person.—The term "person" includes an individual, partnership, association, joint-stock company, trust, or corporation.

(33) Radio communication.—The term "radio communication" or "communication by radio" means the transmission by radio of writing, signs, signals, pictures, and sounds of all kinds, including all instrumentalities, facilities, apparatus, and services (among other things, the receipt, forwarding, and delivery of communications) incidental to such transmission.

(34)(A) Radio officer.—The term "radio officer" on a ship of the United States means, for the purpose of part II of title III of this Act, a person holding at least a first or second class radiotelegraph operator's license as prescribed and issued by the Commission. When such person is employed to operate a radiotelegraph station aboard a ship of the United States, he is also required to be licensed as a "radio officer" in accordance with the Act of May 12, 1948 (46 U.S.C. 229a-h).

(B) "Radio officer" on a foreign ship means, for the purpose of part II of title III of this Act, a person holding at least a first or second class radiotelegraph operator's certificate complying with the provisions of the radio regulations annexed to the International Telecommunication Convention in force.

(35) Radio station.—The term "radio station" or "station" means a station equipped to engage in radio communication or radio transmission of energy.

(36) Radiotelegraph auto alarm.—The term "radiotelegraph auto alarm" on a ship of the United States subject to the provisions of part II of title III of this Act means an automatic alarm receiving apparatus which responds to the radiotelegraph alarm signal and has been approved by the Commission. "Radiotelegraph auto alarm" on a foreign ship means an automatic alarm receiving apparatus which responds to the radiotelegraph alarm signal and has been approved by the government of the country in which the ship is registered : Provide, That the United States and the country in which the ship is registered are parties to the same treaty, convention, or agreement prescribing the requirements for

such apparatus. Nothing in this Act or in any other provision of law shall be construed to require the recognition of a radiotelegraph auto alarm as complying with part II of title III of this Act, on a foreign ship subject to such part, where the country in which the ship is registered and the United States are not parties to the same treaty, convention, or agreements prescribing the requirements for such apparatus.

(37) Rural telephone company.—The term "rural telephone company" means a local exchange carrier operating entity to the extent that such entity—

(A) provides common carrier service to any local exchange carrier study area that does not include either—

(i) any incorporated place of 10,000 inhabitants or more, or any part thereof, based on the most recently available population statistics of the Bureau of the Census; or

(ii) any territory, incorporated or unincorporated, included in an urbanized area, as defined by the Bureau of the Census as of August 10, 1993;

(B) provides telephone exchange service, including exchange access, to fewer than 50,000 access lines;

(C) provides telephone exchange service to any local exchange carrier study area with fewer than 100,000 access lines; or

(D) has less than 15 percent of its access lines in communities of more than 50,000 on the date of enactment of the Telecommunications Act of 1996.

(38) Safety convention.—The term "safety convention" means the International Convention for the Safety of Life at Sea in force and the regulations referred to therein.

(39)(A) Ship.—The term "ship" or "vessel" includes every description of watercraft or other artificial contrivance, except aircraft, used or capable of being used as a means of transportation on water, whether or not it is actually afloat.

(B) A ship shall be considered a passenger ship if it carries or is licensed or certificated to carry more than twelve passengers.

(C) A cargo ship means any ship not a passenger ship.

(D) A passenger is any person carried on board a ship or vessel except (1) the officers and crew actually employed to man and operate the ship, (2) persons employed to carry on the business of the ship, and (3) persons on board a ship when they are carried, either because of the obligation laid upon the master to carry shipwrecked, distressed, or other persons in like or similar situations or by reason of any circumstance over which neither the master, the owner, nor the charterer (if any) has control.

(E) "Nuclear ship" means a ship provided with a nuclear power-plant.

(40) State.—The term "State" includes the District of Columbia and the Territories and possessions.

(41) State commission.—The term "State commission" means the commission, board, or official (by whatever name designated) which under the laws of any State has regulatory jurisdiction with respect to intrastate operations of carriers.

(42) Station license.—The term "station license," "radio station license," or "license" means that instrument of authorization required by this Act or the rules and regulations of the Commission made pursuant to this Act, for the use or operation of apparatus for transmission of energy, or communications, or signals by radio by whatever name the instrument may be designated by the Commission.

(43) Telecommunications.—The term "telecommunications" means the transmission, between or among points specified by the user, of information of the user's choosing, without change in the form or content of the information as sent and received.

(44) Telecommunications carrier.—The term "telecommunications carrier" means any provider of telecommunications services, except that such term does not include aggregators of telecommunications services (as defined in section 226). A telecommunications carrier shall be treated as a common carrier under this Act only to the extent that it is engaged in providing telecommunications services, except that the Commission shall determine whether the provision of fixed and mobile satellite service shall be treated as common carriage.

(45) Telecommunications equipment.—The term "telecommunications equipment" means equipment, other than customer premises equipment, used by a carrier to provide telecommunications services, and includes software integral to such equipment (including upgrades).

(46) Telecommunications service.—The term "telecommunications service" means the offering of telecommunications for a fee directly to the public, or to such classes of users as to be effectively available directly to the public, regardless of the facilities used.

(47) Telephone exchange service.—The term "telephone exchange service" means (A) service within a telephone exchange, or within a connected system of telephone exchanges within the same exchange area operated to furnish to subscribers intercommunicating service of the character ordinarily furnished by a single exchange, and which is covered

by the exchange service charge, or (B) comparable service provided through a system of switches, transmission equipment, or other facilities (or combination thereof) by which a subscriber can originate and terminate a telecommunications service.

(48) Telephone toll service.—The term "telephone toll service" means telephone service between stations in different exchange areas for which there is made a separate charge not included in contracts with subscribers for exchange service.

(49) Transmission of energy by radio.—The term "transmission of energy by radio" or "radio transmission of energy" includes both such transmission and all instrumentalities, facilities, and services incidental to such transmission.

(50) United states.—The term "United States" means the several States and Territories, the District of Columbia, and the possessions of the United States, but does not include the Canal Zone.

(51) Wire communication.—The term "wire communication" or "communication by wire" means the transmission of writing, signs, signals, pictures, and sounds of all kinds by aid of wire, cable, or other like connection between the points of origin and reception of such transmission, including all instrumentalities, facilities, apparatus, and services (among other things, the receipt, forwarding, and delivery of communications) incidental to such transmission.

Sec. 4. [47 U.S.C. 154] Provisions Relating to the Commission.

(a) The Federal Communications Commission (in this Act referred to as the "Commission") shall be composed of five Commissioners appointed by the President, by and with the advice and consent of the Senate, one of whom the President shall designate as chairman.

(b)(1) Each member of the Commission shall be a citizen of the United States.

(2)(A) No member of the Commission or person employed by the Commission shall—

(i) be financially interested in any company or other entity engaged in the manufacture or sale of telecommunications equipment which is subject to regulation by the Commission;

(ii) be financially interested in any company or other entity engaged in the business of communication by wire or radio or in the use of the electromagnetic spectrum;

(iii) be financially interested in any company or other entity which controls any company or other entity specified in clause (i) or clause (ii), or which derives a significant portion of its total income from ownership of stocks, bonds, or other securities of any such company or other entity; or

(iv) be employed by, hold any official relation to, or own any stocks, bonds, or other securities of, any person significantly regulated by the Commission under this Act;

except that the prohibitions established in this subparagraph shall apply only to financial interests in any company or other entity which has a significant interest in communications, manufacturing, or sales activities which are subject to regulation by the Commission.

(B)(i) The Commission shall have authority to waive, from time to time, the application of the prohibitions established in subparagraph (A) to persons employed by the Commission if the Commission determines that the financial interests of a person which are involved in a particular case are minimal, except that such waiver authority shall be subject to the provisions of section 208 of title 18, United States Code. The waiver authority established in this subparagraph shall not apply with respect to members of the Commission.

(ii) In any case in which the Commission exercises the waiver authority established in this subparagraph, the Commission shall publish notice of such action in the Federal Register and shall furnish notice of such action to the appropriate committees of each House of the Congress. Each such notice shall include information regarding the identity of the person receiving the waiver, the position held by such person, and the nature of the financial interests which are the subject of the waiver.

(3) The Commission, in determining whether a company or other entity has a significant interest in communications, manufacturing, or sales activities which are subject to regulation by the Commission, shall consider (without excluding other relevant factors)—

(A) the revenues, investments, profits, and managerial efforts directed to the related communications, manufacturing, or sales activities of the company or other entity involved, as compared to the other aspects of the business of such company or other entity;

(B) the extent to which the Commission regulates and oversees the activities of such company or other entity;

(C) the degree to which the economic interests of such company or other entity may be affected by any action of the Commission; and

(D) the perceptions held by the public regarding the business activities of such company or other entity.

(4) Members of the Commission shall not engage in any other business, vocation, profession, or employment while serving as such members.

(5) The maximum number of commissioners who may be members of the same political party shall be a number equal to the least number of commissioners which constitutes a majority of the full membership of the Commission.

(c) Commissioners shall be appointed for terms of five years and until their successors are appointed and have been confirmed and taken the oath of office, except that they shall not continue to serve beyond the expiration of the next session of Congress subsequent to the expiration of said fixed term of office; except that any person chosen to fill a vacancy shall be appointed only for the unexpired term of the Commissioner whom he succeeds. No vacancy in the Commission shall impair the right of the remaining commissioners to exercise all the powers of the Commission.

(d) Each Commissioner shall receive an annual salary at the annual rate payable from time to time for level IV of the Executive Schedule, payable in monthly installments. The Chairman of the Commission, during the period of his service as Chairman, shall receive an annual salary at the annual rate payable from time to time for level III of the Executive Schedule.

(e) The principal office of the Commission shall be in the District of Columbia, where its general sessions shall be held; but whenever the convenience of the public or of the parties may be promoted or delay or expense prevented thereby, the Commission may hold special sessions in any part of the United States.

(f)(1) The Commission shall have authority, subject to the provisions of the civil-service laws and the Classification Act of 1949, as amended, to appoint such officers, engineers, accountants, attorneys, inspectors, examiners, and other employees as are necessary in the exercise of its functions.

(2) Without regard to the civil-service laws, but subject to the Classification Act of 1949, each commissioner may appoint three professional assistants and a secretary, each of whom shall perform such duties as such commissioner shall direct. In addition, the chairman of the Commission may appoint, without regard to the civil-service laws, but subject to the Classification Act of 1949, an administrative assistant who shall perform such duties as the chairman shall direct.

(3) The Commission shall fix a reasonable rate of extra compensation for overtime services of engineers in charge and radio engineers of the Field Engineering and Monitoring Bureau of the Federal Communications Commission, who may be required to remain on duty between the hours of 5 o'clock postmeridian and 8 o'clock antemeridian or on Sundays or holidays to perform services in connection with the inspection of ship radio equipment and apparatus for the purposes of part II of title III of this Act or the Great Lakes Agreement, on the basis of one-half day's additional pay for each two

hours or fraction thereof of at least one hour that the overtime extends beyond 5 o'clock postmeridian (but not to exceed two and one-half days' pay for the full period from 5 o'clock postmeridian to 8 o'clock antemeridian) and two additional days' pay for Sunday or holiday duty. The said extra compensation for overtime services shall be paid by the master, owner, or agent of such vessel to the local United States collector of customs or his representative, who shall deposit such collection into the Treasury of the United States to an appropriately designated receipt account: *Provided,* That the amounts of such collections received by the said collector of customs or his representatives shall be covered into the Treasury as miscellaneous receipts; and the payments of such extra compensation to the several employees entitled thereto shall be made from the annual appropriations for salaries and expenses of the Commission: *Provided further,* That to the extent that the annual appropriations which are hereby authorized to be made from the general fund of the Treasury are insufficient, there are hereby authorized to be appropriated from the general fund of the Treasury such additional amounts as may be necessary to the extent that the amounts of such receipts are in excess of the amounts appropriated: *Provided further,* That such extra compensation shall be paid if such field employees have been ordered to report for duty and have so reported whether the actual inspection of the radio equipment or apparatus takes place or not: And provided further, That in those ports where customary working hours are other than those hereinabove mentioned, the engineers in charge are vested with authority to regulate the hours of such employees so as to agree with prevailing working hours in said ports where inspections are to be made, but nothing contained in this proviso shall be construed in any manner to alter the length of a working day for the engineers in charge and radio engineers or the overtime pay herein fixed: and *Provided further,* That, in the alternative, an entity designated by the Commission may make the inspections referred to in this paragraph.

(4)(A) The Commission, for purposes of preparing or administering any examination for an amateur station operator license, may accept and employ the voluntary and uncompensated services of any individual who holds an amateur station operator license of a higher class than the class of license for which the examination is being prepared or administered. In the case of examinations for the highest class of amateur station operator license, the Commission may accept and employ such services of any individual who holds such class of license.

(B)(i) The Commission, for purposes of monitoring violations of any provision of this Act (and of any regulation prescribed by the Commission under this Act) relating to the amateur radio service, may—

(I) recruit and train any individual licensed by the Commission to operate an amateur station; and

(II) accept and employ the voluntary and uncompensated services of such individual.

(ii) The Commission, for purposes of recruiting and training individuals under clause (i) and for purposes of screening, annotating, and summarizing violation reports referred under clause (i), may accept and employ the voluntary and uncompensated services of any amateur station operator organization.

(iii) The functions of individuals recruited and trained under this subparagraph shall be limited to—

(I) the detection of improper amateur radio transmissions;

(II) the conveyance to Commission personnel of information which is essential to the enforcement of this Act (or regulations prescribed by the Commission under this Act) relating to the amateur radio service; and

(III) issuing advisory notices, under the general direction of the Commission, to persons who apparently have violated any provision of this Act (or regulations prescribed by the Commission under this Act) relating to the amateur radio service.

Nothing in this clause shall be construed to grant individuals recruited and trained under this subparagraph any authority to issue sanctions to violators or to take any enforcement action other than any action which the Commission may prescribe by rule.

(C)(i) The Commission, for purposes of monitoring violations of any provision of this Act (and of any regulation prescribed by the Commission under this Act) relating to the citizens band radio service, may—

(I) recruit and train any citizens band radio operator; and

(II) accept and employ the voluntary and uncompensated services of such operator.

(ii) The Commission, for purposes of recruiting and training individuals under clause (i) and for purposes of screening, annotating, and summarizing violation reports referred under clause (i), may accept and employ the voluntary and uncompensated services of any citizens band radio operator organization. The Commission, in accepting and employing services of individuals under this subparagraph, shall seek to achieve a broad representation of individuals and organizations interested in citizens band radio operation.

(iii) The functions of individuals recruited and trained under this subparagraph shall be limited to—

(I) the detection of improper citizens band radio transmissions;

(II) the conveyance to Commission personnel of information which is essential to the enforcement of this Act (or regulations prescribed by the Commission under this Act) relating to the citizens band radio service; and

(III) issuing advisory notices, under the general direction of the Commission, to persons who apparently have violated any provision of this Act (or regulations prescribed by the Commission under this Act) relating to the citizens band radio service.

Nothing in this clause shall be construed to grant individuals recruited and trained under this subparagraph any authority to issue sanctions to violators or to take any enforcement action other than any action which the Commission may prescribe by rule.

(D) The Commission shall have the authority to endorse certification of individuals to perform transmitter installation, operation, maintenance, and repair duties in the private land mobile services and fixed services (as defined by the Commission by rule) if such certification programs are conducted by organizations or committees which are representative of the users in those services and which consist of individuals who are not officers or employees of the Federal Government.

(E) The authority of the Commission established in this paragraph shall not be subject to or affected by the provisions of part III of title 5, United States Code, or section 3679(b) of the Revised Statutes (31 U.S.C. 665(b)).

(F) Any person who provides services under this paragraph shall not be considered, by reason of having provided such services, a Federal employee.

(G) The Commission, in accepting and employing services of individuals under subparagraphs (A) and (B), shall seek to achieve a broad representation of individuals and organizations interested in amateur station operation.

(H) The Commission may establish rules of conduct and other regulations governing the service of individuals under this paragraph.

(I) With respect to the acceptance of voluntary uncompensated services for the preparation, processing, or administration of examinations for amateur station operator licenses, pursuant to subparagraph (A) of this paragraph, individuals, or organizations which provide or coordinate such authorized volunteer services may recover from examinees reimbursement for out-of-pocket costs. The total amount of allowable cost reimbursement per examinee shall not exceed $4, adjusted annually every January 1 for changes in the Department of Labor Consumer Price Index.

(5)(A) The Commission, for purposes of preparing and administering any examination for a commercial radio operator license or endorsement, may accept and employ the services of persons that the Commission determines to be qualified. Any person so employed may not receive compensation for such services, but may recover from examinees such fees as the Commission permits, considering such factors as public service and cost estimates submitted by such person.

(B) The Commission may prescribe regulations to select, oversee, sanction, and dismiss any person authorized under this paragraph to be employed by the Commission.

(C) Any person who provides services under this paragraph or who provides goods in connection with such services shall not, by reason of having provided such service or goods, be considered a Federal or special government employee.

(g)(1) The Commission may make such expenditures (including expenditures for rent and personal services at the seat of government and elsewhere, for office supplies, lawbooks, periodicals, and books of reference, for printing and binding, for land for use as sites for radio monitoring stations and related facilities, including living quarters where necessary in remote areas, for the construction of such stations and facilities, and for the improvement, furnishing, equipping, and repairing of such stations and facilities and of laboratories and other related facilities (including construction of minor subsidiary buildings and structures not exceeding $25,000 in any one instance) used in connection with technical research activities), as may be necessary for the execution of the functions vested in the Commission and as may be appropriated for by the Congress in accordance with the authorizations of appropriations established in section 6. All expenditures of the Commission, including all necessary expenses for transportation incurred by the commissioners or by their employees, under their orders, in making any investigation or upon any official business in any other places than in the city of Washington, shall be allowed and paid on the presentation of itemized vouchers therefor approved by the chairman of the Commission or by such other members or officer thereof as may be designated by the Commission for that purpose.

(2)(A) If—

(i) the necessary expenses specified in the last sentence of paragraph (1) have been incurred for the purpose of enabling commissioners or employees of the Commission to attend and participate in any convention, conference, or meeting;

(ii) such attendance and participation are in furtherance of the functions of the Commission; and

(iii) such attendance and participation are requested by the person sponsoring such convention, conference, or meeting;

then the Commission shall have authority to accept direct reimbursement from such sponsor for such necessary expenses.

(B) The total amount of unreimbursed expenditures made by the Commission for travel for any fiscal year, together with the total amount of reimbursements which the Commission accepts under subparagraph (A) for such fiscal year, shall not exceed the level of travel expenses appropriated to the Commission for such fiscal year.

(C) The Commission shall submit to the appropriate committees of the Congress, and publish in the Federal Register, quarterly reports specifying reimbursements which the Commission has accepted under this paragraph.

(D) The provisions of this paragraph shall cease to have any force or effect at the end of fiscal year 1994.

(E) Funds which are received by the Commission as reimbursements under the provisions of this paragraph after the close of a fiscal year shall remain available for obligation.

(3)(A) Notwithstanding any other provision of law, in furtherance of its functions the Commission is authorized to accept, hold, administer, and use unconditional gifts, donations, and bequests of real, personal, and other property (including voluntary and uncompensated services, as authorized by section 3109 of title 5, United States Code).

(B) The Commission, for purposes of providing radio club and military-recreational call signs, may utilize the voluntary, uncompensated, and unreimbursed services of amateur radio organizations authorized by the Commission that have tax-exempt status under section 501(c)(3) of the Internal Revenue Code of 1986.

(C) For the purpose of Federal law on income taxes, estate taxes, and gift taxes, property or services accepted under the authority of subparagraph (A) shall be deemed to be a gift, bequest, or devise to the United States.

(D) The Commission shall promulgate regulations to carry out the provisions of this paragraph. Such regulations shall include provisions to preclude the acceptance of any gift, bequest, or donation that would create a conflict of interest or the appearance of a conflict of interest.

(h) Three members of the Commission shall constitute a quorum thereof. The Commission shall have an official seal which shall be judicially noticed.

(i) The Commission may perform any and all acts, make such rules and regulations, and issue such orders, not inconsistent with this Act, as may be necessary in the execution of its functions.

(j) The Commission may conduct its proceedings in such manner as will best conduce to the proper dispatch of business and to the ends of justice. No commissioner shall participate in any hearing or proceeding in which he has a pecuniary interest. Any party may appear before the Commission and be heard in person or by attorney. Every vote and official act the Commission shall be entered of record, and its proceedings shall be public upon the request of any party interested. The Commission is authorized to withhold publication of records or proceedings containing secret information affecting the national defense.

(k) The Commission shall make an annual report to Congress, copies of which shall be distributed as are other reports transmitted to Congress. Such reports shall contain—

(1) such information and data collected by the Commission as may be considered of value in the determination of questions connected with the regulation of interstate and foreign wire and radio communication and radio transmission of energy;

(2) such information and data concerning the functioning of the Commission as will be of value to Congress in appraising the amount and character of the work and accomplishments of the Commission and the adequacy of its staff and equipment;

(3) an itemized statement of all funds expended during the preceding year by the Commission, of the sources of such funds, and of the authority in this Act or elsewhere under which such expenditures were made; and

(4) specific recommendations to Congress as to additional legislation which the Commission deems necessary or desirable, including all legislative proposals submitted for approval to the Director of the Office of Management and Budget.

(l) All reports of investigations made by the Commission shall be entered of record, and a copy thereof shall be furnished to the party who may have complained, and to any common carrier or licensee that may have been complained of.

(m) The Commission shall provide for the publication of its reports and decisions in such form and manner as may be best adapted for public information and use, and such authorized publications shall be competent evidence of the reports and decisions of the Commission therein contained in all courts of the United States and of the several States without any further proof or authentication thereof.

(n) Rates of compensation of persons appointed under this section shall be subject to the reduction applicable to officers and employees of the Federal Government generally.

(o) For the purpose of obtaining maximum effectiveness from the use of radio and wire communications in connection with safety of life and property, the Commission shall investigate and study all phases of the problem and the best methods of obtaining the cooperation and coordination of these systems.

Sec. 5. [47 U.S.C. 155] Organization and Functioning of the Commission.

(a) The member of the Commission designated by the President as chairman shall be the chief executive officer of the Commission. It shall be his duty to preside at all meetings and sessions of the Commission, to represent the

Commission in all matters relating to legislation and legislative reports, except that any commissioner may present his own or minority views or supplemental reports, to represent the Commission in all matters requiring conferences or communications with other governmental officers, departments or agencies, and generally to coordinate and organize the work of the Commission in such manner as to promote prompt and efficient disposition of all matters within the jurisdiction of the Commission. In the case of a vacancy in the office of the chairman of the Commission, or the absence or inability of the chairman to serve, the Commission may temporarily designate one of its members to act as chairman until the cause or circumstance requiring such designation shall have been eliminated or corrected.

(b) From time to time as the Commission may find necessary, the Commission shall organize its staff into (1) integrated bureaus, to function on the basis of the Commission's principal workload operations, and (2) such other divisional organizations as the Commission may deem necessary. Each such integrated bureau shall include such legal, engineering, accounting, administrative, clerical, and other personnel as the Commission may determine to be necessary to perform its functions.

(c)(1) When necessary to the proper functioning of the Commission and the prompt and orderly conduct of its business, the Commission may, by published rule or by order, delegate any of its functions (except functions granted to the Commission by this paragraph and by paragraphs (4), (5), and (6) of this subsection and except any action referred to in sections 204(a)(2), 208(b), and 405(b)) to a panel of commissioners, an individual commissioner, an employee board, or an individual employee, including functions with respect to hearing, determining, ordering, certifying, reporting, or otherwise acting as to any work, business, or matter; except that in delegating review functions to employees in cases of adjudication (as defined in the Administrative Procedure Act), the delegation in any such case may be made only to an employee board consisting of two or more employees referred to in paragraph (8). Any such rule or order may be adopted, amended, or rescinded only by a vote of a majority of the members of the Commission then holding office. Except for cases involving the authorization of service in the instructional television fixed service, or as otherwise provided in this Act, nothing in this paragraph shall authorize the Commission to provide for the conduct, by any person or persons other than persons referred to in paragraph (2) or (3) of section 556(b) of title 5, United States Code, of any hearing to which such section applies.

(2) As used in this subsection (d) the term "order, decision, report, or action" does not include an initial, tentative, or recommended decision to which exceptions may be filed as provided in section 409(b).

(3) Any order, decision, report, or action made or taken pursuant to any such delegation, unless reviewed as provided in paragraph (4), shall have the same force and effect, and shall be made, evidenced, and enforced in the same manner, as orders, decisions, reports, or other actions of the Commission.

(4) Any person aggrieved by any such order, decision, report or action may file an application for review by the Commission within such time and in such manner as the Commission shall prescribe, and every such application shall be passed upon by the Commission. The Commission, on its own initiative, may review in whole or in part, at such time and in such manner as it shall determine, any order, decision, report, or action made or taken pursuant to any delegation under paragraph (1).

(5) In passing upon applications for review, the Commission may grant, in whole or in part, or deny such applications without specifying any reasons therefore. No such application for review shall rely on questions of fact or law upon which the panel of commissioners, individual commissioner, employee board, or individual employee has been afforded no opportunity to pass.

(6) If the Commission grants the application for review, it may affirm, modify, or set aside the order, decision, report, or action, or it may order a rehearing upon such order, decision, report, or action in accordance with section 405.

(7) The filing of an application for review under this subsection shall be a condition precedent to judicial review of any order, decision, report, or action made or taken pursuant to a delegation under paragraph (1). The time within which a petition for review must be filed in a proceeding to which section 402(a) applies, or within which an appeal must be taken under section 402(b), shall be computed from the date upon which public notice is given of orders disposing of all applications for review filed in any case.

(8) The employees to whom the Commission may delegate review functions in any case of adjudication (as defined in the Administrative Procedure Act) shall be qualified, by reason of their training, experience, and competence, to perform such review functions, and shall perform no duties inconsistent with such review functions. Such employees shall be in a grade classification or salary level commensurate with their important duties, and in no event less than the grade classification or salary level of the employee or employees whose actions are to be reviewed. In the performance of such review functions such employees shall be assigned to cases in rotation so far as practicable and shall not be responsible to or subject to the supervision or direction of any officer, employee, or agent engaged in the performance of investigative or prosecuting functions for any agency.

(9) The secretary and seal of the Commission shall be the secretary and seal of each panel of the Commission, each individual commissioner, and each

employee board or individual employee exercising functions delegated pursuant to paragraph (1) of this subsection.

(d) Meetings of the Commission shall be held at regular intervals, not less frequently than once each calendar month, at which times the functioning of the Commission and the handling of its work load shall be reviewed and such orders shall be entered and other action taken as may be necessary or appropriate to expedite the prompt and orderly conduct of the business of the Commission with the objective of rendering a final decision (1) within three months from the date of filing in all original application, renewal, and transfer cases in which it will not be necessary to hold a hearing, and (2) within six months from the final date of the hearing in all hearing cases.

(e) The Commission shall have a Managing Director who shall be appointed by the Chairman subject to the approval of the Commission. The Managing Director, under the supervision and direction of the Chairman, shall perform such administrative and executive functions as the Chairman shall delegate. The Managing Director shall be paid at a rate equal to the rate then payable for level V of the Executive Schedule.

Sec. 7. [47 U.S.C. 157] New Technologies and Services.

(a) It shall be the policy of the United States to encourage the provision of new technologies and services to the public. Any person or party (other than the Commission) who opposes a new technology or service proposed to be permitted under this Act shall have the burden to demonstrate that such proposal is inconsistent with the public interest.

(b) The Commission shall determine whether any new technology or service proposed in a petition or application is in the public interest within one year after such petition or application is filed. If the Commission initiates its own proceeding for a new technology or service, such proceeding shall be completed within 12 months after it is initiated.

Sec. 10. [47 U.S.C. 160] Competition in Provision of Telecommunications Service.

(a) Regulatory flexibility.—Notwithstanding section 332(c)(1)(A) of this Act, the Commission shall forbear from applying any regulation or any provision of this Act to a telecommunications carrier or telecommunications service, or class of telecommunications carriers or telecommunications services, in any or some of its or their geographic markets, if the Commission determines that—

(1) enforcement of such regulation or provision is not necessary to ensure that the charges, practices, classifications, or regulations by, for, or in connection with that telecommunications carrier or telecommunications service are just and reasonable and are not unjustly or unreasonably discriminatory;

(2) enforcement of such regulation or provision is not necessary for the protection of consumers; and

(3) forbearance from applying such provision or regulation is consistent with the public interest.

(b) Competitive Effect To Be Weighed.—In making the determination under subsection (a)(3), the Commission shall consider whether forbearance from enforcing the provision or regulation will promote competitive market conditions, including the extent to which such forbearance will enhance competition among providers of telecommunications services. If the Commission determines that such forbearance will promote competition among providers of telecommunications services, that determination may be the basis for a Commission finding that forbearance is in the public interest.

(c) Petition for Forbearance.—Any telecommunications carrier, or class of telecommunications carriers, may submit a petition to the Commission requesting that the Commission exercise the authority granted under this section with respect to that carrier or those carriers, or any service offered by that carrier or carriers. Any such petition shall be deemed granted if the Commission does not deny the petition for failure to meet the requirements for forbearance under subsection (a) within one year after the Commission receives it, unless the one-year period is extended by the Commission. The Commission may extend the initial one-year period by an additional 90 days if the Commission finds that an extension is necessary to meet the requirements of subsection (a). The Commission may grant or deny a petition in whole or in part and shall explain its decision in writing.

(d) Limitation.—Except as provided in section 251(f), the Commission may not forbear from applying the requirements of section 251(c) or 271 under subsection (a) of this section until it determines that those requirements have been fully implemented.

(e) State Enforcement After Commission Forbearance.—A State commission may not continue to apply or enforce any provision of this Act that the Commission has determined to forbear from applying under subsection (a).

Sec. 11. [47 U.S.C. 161] Regulatory Reform.

(a) Biennial Review of Regulations.—In every even-numbered year (beginning with 1998), the Commission—

(1) shall review all regulations issued under this Act in effect at the time of the review that apply to the operations or activities of any provider of telecommunications service; and

(2) shall determine whether any such regulation is no longer necessary in the public interest as the result of meaningful economic competition between providers of such service.

(b) Effect of Determination.—The Commission shall repeal or modify any regulation it determines to be no longer necessary in the public interest.

Title II—Common Carriers
Part I—Common Carrier Regulation

Sec. 201. [47 U.S.C. 201] Service and Charges.

(a) It shall be the duty of every common carrier engaged in interstate or foreign communication by wire or radio to furnish such communication service upon reasonable request therefor; and, in accordance with the orders of the Commission, in cases where the Commission, after opportunity for hearing, finds such action necessary or desirable in the public interest, to establish physical connections with other carriers, to establish through routes and charges applicable thereto and the divisions of such charges, and to establish and provide facilities and regulations for operating such through routes.

(b) All charges, practices, classifications, and regulations for and in connection with such communication service, shall be just and reasonable, and any such charge, practice, classification, or regulation that is unjust or unreasonable is hereby declared to be unlawful: *Provided*, That communications by wire or radio subject to this Act may be classified into day, night, repeated, unrepeated, letter, commercial, press, Government and such other classes as the Commission may decide to be just and reasonable, and different charges may be made for the different classes of communications: *Provided further*, That nothing in this Act or in any other provision of law shall be construed to prevent a common carrier subject to this Act from entering into or operating under any contract with any common carrier not subject to this Act, for the exchange of their services, if the Commission is of the opinion that such contract is not contrary to the public interest: *Provided further*, That nothing in this Act or in any other provision of law shall prevent a common carrier subject to this Act from furnishing reports of positions of ships at sea to newspapers of general circulation, either at a nominal charge or without charge, provided the name of such common carrier is displayed along with such ship position reports. The Commissioner may prescribe such rules and regulations as may be necessary in the public interest to carry out the provisions of this Act.

Sec. 203. [47 U.S.C. 203] Schedules of Charges.

(a) Every common carrier, except connecting carriers, shall, within such reasonable time as the Commission shall designate, file with the Commission and print and keep open for public inspection schedules showing all charges for itself and its connecting carriers for interstate and foreign wire or radio communication between the different points on its own system, and between points on its own system and points on the system of its connecting carriers or points on the system of any other carrier subject to this Act when a through route has been established, whether such charges are joint or separate, and showing the classifications, practices, and regulations affecting such charges. Such schedules shall contain such other information, and be printed in such form, and be posted and kept open for public inspection in such places, as the Commission may by regulation require, and each such schedule shall give notice of its effective date; and such common carrier shall furnish such schedules to each of its connecting carriers, and such connecting carriers shall keep such schedules open for inspection in such public places as the Commission may require.

(b)(1) No change shall be made in the charges, classifications, regulations, or practices which have been so filed and published except after one hundred and twenty days notice to the Commission and to the public, which shall be published in such form and contain such information as the Commission may by regulations prescribe.

(2) The Commission may, in its discretion and for good cause shown, modify any requirement made by or under the authority of this section either in particular instances or by general order applicable to special circumstances or conditions except that the Commission may not require the notice period specified in paragraph (1) to be more than one hundred and twenty days.

(c) No carrier, unless otherwise provided by or under authority of this Act, shall engage or participate in such communication unless schedules have been filed and published in accordance with the provisions of this Act and with the regulations made thereunder; and no carrier shall (1) charge, demand, collect, or receive a greater or less or different compensation, for such communication, or for any service in connection therewith, between the points named in any such schedule than the charges specified in the schedule then in effect, or (2) refund or remit by any means or device any portion of the charges so specified, or (3) extend to any person any privileges or facilities, in such communication, or employ or enforce any classifications, regulations, or practices affecting such charges, except as specified in such schedule.

(d) The Commission may reject and refuse to file any schedule entered for filing which does not provide and give lawful notice of its effective date. Any schedule so rejected by the Commission shall be void and its use shall be unlawful.

(e) In case of failure or refusal on the part of any carrier to comply with the provisions of this section or of any regulation or order made by the Commission thereunder, such carrier shall forfeit to the United States the sum of $6,000 for each such offense, and $300 for each and every day of the continuance of such offense.

Sec. 204. [47 U.S.C. 204] Hearing as to Lawfulness of New Charges; Suspension.

(a)(1) Whenever there is filed with the Commission any new or revised charge, classification, regulation, or practice, the Commission may either upon complaint or upon its own initiative without complaint, upon reasonable notice, enter upon a hearing concerning the lawfulness thereof; and pending such hearing and the decision thereon the Commission, upon delivering to the carrier or carriers affected thereby a statement in writing of its reasons for such suspension, may suspend the operation of such charge, classification, regulation, or practice, in whole or in part but not for a longer period than five months beyond the time when it would otherwise go into effect; and after full hearing the Commission may make such order with reference thereto as would be proper in a proceeding initiated after such charge, classification, regulation, or practice had become effective. If the proceeding has not been concluded and an order made within the period of the suspension, the proposed new or revised charge, classification, regulation, or practice shall go into effect at the end of such period; but in case of a proposed charge for a new service or a revised charge, the Commission may by order require the interested carrier or carriers to keep accurate account of all amounts received by reason of such charge for a new service or revised charge, specifying by whom and in whose behalf such amounts are paid, and upon completion of the hearing and decision may by further order require the interested carrier or carriers to refund, with interest, to the persons in whose behalf such amounts were paid, such portion of such charge for a new service or revised charges as by its decision shall be found not justified. At any hearing involving a new or revised charge, or a proposed new or revised charge, the burden of proof to show that the new or revised charge, or proposed charge, is just and reasonable shall be upon the carrier, and the Commission shall give to the hearing and decision of such questions preference over all other questions pending before it and decide the same as speedily as possible.

(2)(A) Except as provided in subparagraph (B), the Commission shall, with respect to any hearing under this section, issue an order concluding such hearing within 5 months after the date that the charge, classification, regulation, or practice subject to the hearing becomes effective.

(B) The Commission shall, with respect to any such hearing initiated prior to the date of enactment of this paragraph, issue an order concluding the hearing not later than 12 months after such date of enactment.

(C) Any order concluding a hearing under this section shall be a final order and may be appealed under section 402(a).

(3) A local exchange carrier may file with the Commission a new or revised charge, classification, regulation, or practice on a streamlined basis. Any such charge, classification, regulation, or practice shall be deemed lawful and shall be effective 7 days (in the case of a reduction in rates) or 15 days (in the case of an increase in rates) after the date on which it is filed with the Commission unless the Commission takes action under paragraph (1) before the end of that 7-day or 15-day period, as is appropriate.

(b) Notwithstanding the provisions of subsection (a) of this section, the Commission may allow part of a charge, classification, regulation, or practice to go into effect, based upon a written showing by the carrier or carriers affected, and an opportunity for written comment thereon by affected persons, that such partial authorization is just, fair, and reasonable. Additionally, or in combination with a partial authorization, the Commission, upon a similar showing, may allow all or part of a charge, classification, regulation, or practice to go into effect on a temporary basis pending further order of the Commission. Authorizations of temporary new or increased charges may include an accounting order of the type provided for in subsection (a).

Sec. 208. [47 U.S.C. 208] Complaints to the Commission.

(a) Any person, any body politic or municipal organization, or State commission, complaining of anything done or omitted to be done by any common carrier subject to this Act, in contravention of the provisions thereof, may apply to said Commission by petition which shall briefly state the facts, whereupon a statement of the complaint thus made shall be forwarded by the Commission to such common carrier, who shall be called upon to satisfy the complaint or to answer the same in writing within a reasonable time to be specified by the Commission. If such common carrier within the time specified shall make reparation for the injury alleged to have been caused, the common carrier shall be relieved of liability to the complainant only for the particular violation of law thus complained of. If such carrier or carriers shall not satisfy the complaint within the time specified or there shall appear to be any reasonable ground for investigating said complaint, it shall be the duty of the Commission to investigate the matters complained of in such manner and by such means as it shall deem proper. No complaint shall at any time be dismissed because of the absence of direct damage to the complainant.

(b)(1) Except as provided in paragraph (2), the Commission shall, with respect to any investigation under this section of the lawfulness of a charge, classification, regulation, or practice, issue an order concluding such investigation within 5 months after the date on which the complaint was filed.

(2) The Commission shall, with respect to any such investigation initiated prior to the date of enactment of this subsection, issue an order concluding the investigation not later than 12 months after such date of enactment.

(3) Any order concluding an investigation under paragraph (1) or (2) shall be a final order and may be appealed under section 402(a).

Sec. 213. [47 U.S.C. 213] Valuation of Carrier Property.

(a) The Commission may from time to time, as may be necessary for the proper administration of this Act, and after opportunity for hearing, make a valuation of all or of any part of the property owned or used by any carrier subject to this Act, as of such date as the Commission may fix.

(b) The Commission may at any time require any such carrier to file with the Commission an inventory of all or of any part of the property owned or used by said carrier, which inventory shall show the units of said property classified in such detail, and in such manner, as the Commission shall direct, and shall show the estimated cost of reproduction new of said units, and their reproduction cost new less depreciation, as of such date as the Commission may direct; and such carrier shall file such inventory within such reasonable time as the Commission by order shall require.

(c) The Commission may at any time require any such carrier to file with the Commission a statement showing the original cost at the time of dedication to the public use of all or of any part of the property owned or used by said carrier. For the showing of such original cost said property shall be classified, and the original cost shall be defined, in such manner as the Commission may prescribe; and if any part of such cost cannot be determined from accounting or other records, the portion of the property for which such cost cannot be determined shall be reported to the Commission; and if the Commission shall so direct, the original cost thereof shall be estimated in such manner as the Commission may prescribe. If the carrier owning the property at the time such original cost is reported shall have paid more or less than the original cost to acquire the same, the amount of such cost of acquisition, and any facts which the Commission may require in connection therewith, shall be reported with such original cost. The report made by a carrier under this paragraph shall show the source or sources from which the original cost reported was obtained, and such other information as to the manner in which the report was prepared, as the Commission shall require.

(d) Nothing shall be included in the original cost reported for the property of any carrier under paragraph (c) of this section on account of any easement, license, or franchise granted by the United States or by any State or political subdivision thereof, beyond the reasonable necessary expense lawfully incurred in obtaining such easement, license, or franchise from the public authority aforesaid, which expense shall be reported separately from all other costs in such detail as the Commission may require; and nothing shall be included in any valuation of the property of any carrier made by the Commission on account of any such easement, license, or franchise, beyond such reasonable necessary expense lawfully incurred as aforesaid.

(e) The Commission shall keep itself informed of all new construction, extensions, improvements, retirements, or other changes in the condition, quantity, use, and classification of the property of common carriers, and of the cost of all additions and betterments thereto and of all changes in the investment therein, and may keep itself informed of current changes in costs and values of carrier properties.

(f) For the purpose of enabling the Commission to make a valuation of any of the property of any such carrier, or to find the original cost of such property, or to find any other facts concerning the same which are required for use by the Commission, it shall be the duty of each such carrier to furnish to the Commission, within such reasonable time as the Commission may order, any information with respect thereto which the Commission may by order require, including copies of maps, contracts, reports of engineers, and other data, records, and papers, and to grant to all agents of the Commission free access to its property and its accounts, records, and memoranda whenever and wherever requested by any such duly authorized agent, and to cooperate with and aid the Commission in the work of making any such valuation of finding in such manner and to such extent as the Commission may require and direct, and all rules and regulations made by the Commission for the purpose of administering this section shall have the full force and effect of law. Unless otherwise ordered by the Commission, with the reasons therefor, the records and data of the Commission shall be open to the inspection and examination of the public. The Commission, in making any such valuation, shall be free to adopt any method of valuation which shall be lawful.

(g) Nothing in this section shall impair or diminish the powers of any State commission.

Part II—Development of Competitive Markets

Sec. 251. [47 U.S.C. 251] Interconnection.

(a) General Duty of Telecommunications Carriers.—Each telecommunications carrier has the duty—

(1) to interconnect directly or indirectly with the facilities and equipment of other telecommunications carriers; and

(2) not to install network features, functions, or capabilities that do not comply with the guidelines and standards established pursuant to section 255 or 256.

(b) Obligations of All Local Exchange Carriers.—Each local exchange carrier has the following duties:

(1) Resale.—The duty not to prohibit, and not to impose unreasonable or discriminatory conditions or limitations on, the resale of its telecommunications services.

(2) Number portability.—The duty to provide, to the extent technically feasible, number portability in accordance with requirements prescribed by the Commission.

(3) Dialing parity.—The duty to provide dialing parity to competing providers of telephone exchange service and telephone toll service, and the duty to permit all such providers to have nondiscriminatory access to telephone numbers, operator services, directory assistance, and directory listing, with no unreasonable dialing delays.

(4) Access to rights-of-way.—The duty to afford access to the poles, ducts, conduits, and rights-of-way of such carrier to competing providers of telecommunications services on rates, terms, and conditions that are consistent with section 224.

(5) Reciprocal compensation.—The duty to establish reciprocal compensation arrangements for the transport and termination of telecommunications.

(c) Additional Obligations of Incumbent Local Exchange Carriers.—In addition to the duties contained in subsection (b), each incumbent local exchange carrier has the following duties:

(1) Duty to negotiate.—The duty to negotiate in good faith in accordance with section 252 the particular terms and conditions of agreements to fulfill the duties described in paragraphs (1) through (5) of

subsection (b) and this subsection. The requesting telecommunications carrier also has the duty to negotiate in good faith the terms and conditions of such agreements.

(2) Interconnection.—The duty to provide, for the facilities and equipment of any requesting telecommunications carrier, interconnection with the local exchange carrier's network—

(A) for the transmission and routing of telephone exchange service and exchange access;

(B) at any technically feasible point within the carrier's network;

(C) that is at least equal in quality to that provided by the local exchange carrier to itself or to any subsidiary, affiliate, or any other party to which the carrier provides interconnection; and

(D) on rates, terms, and conditions that are just, reasonable, and nondiscriminatory, in accordance with the terms and conditions of the agreement and the requirements of this section and section 252.

(3) Unbundled access.—The duty to provide, to any requesting telecommunications carrier for the provision of a telecommunications service, nondiscriminatory access to network elements on an unbundled basis at any technically feasible point on rates, terms, and conditions that are just, reasonable, and nondiscriminatory in accordance with the terms and conditions of the agreement and the requirements of this section and section 252. An incumbent local exchange carrier shall provide such unbundled network elements in a manner that allows requesting carriers to combine such elements in order to provide such telecommunications service.

(4) Resale.—The duty—

(A) to offer for resale at wholesale rates any telecommunications service that the carrier provides at retail to subscribers who are not telecommunications carriers; and

(B) not to prohibit, and not to impose unreasonable or discriminatory conditions or limitations on, the resale of such telecommunications service, except that a State commission may, consistent with regulations prescribed by the Commission under this section, prohibit a reseller that obtains at wholesale rates a telecommunications service that is available at retail only to a category of subscribers from offering such service to a different category of subscribers.

(5) Notice of changes.—The duty to provide reasonable public notice of changes in the information necessary for the transmission and routing of services using that local exchange carrier's facilities or networks, as well as of any other changes that would affect the interoperability of those facilities and networks.

(6) Collocation.—The duty to provide, on rates, terms, and conditions that are just, reasonable, and nondiscriminatory, for physical collocation of equipment necessary for interconnection or access to unbundled network elements at the premises of the local exchange carrier, except that the carrier may provide for virtual collocation if the local exchange carrier demonstrates to the State commission that physical collocation is not practical for technical reasons or because of space limitations.

(d) Implementation.—

(1) In general.—Within 6 months after the date of enactment of the Telecommunications Act of 1996, the Commission shall complete all actions necessary to establish regulations to implement the requirements of this section.

(2) Access standards.—In determining what network elements should be made available for purposes of subsection (c)(3), the Commission shall consider, at a minimum, whether—

(A) access to such network elements as are proprietary in nature is necessary; and

(B) the failure to provide access to such network elements would impair the ability of the telecommunications carrier seeking access to provide the services that it seeks to offer.

(3) Preservation of state access regulations.—In prescribing and enforcing regulations to implement the requirements of this section, the Commission shall not preclude the enforcement of any regulation, order, or policy of a State commission that—

(A) establishes access and interconnection obligations of local exchange carriers;

(B) is consistent with the requirements of this section; and

(C) does not substantially prevent implementation of the requirements of this section and the purposes of this part.

(e) Numbering Administration.—

(1) Commission authority and jurisdiction.—The Commission shall create or designate one or more impartial entities to administer telecommunications numbering and to make such numbers available on an equitable basis. The Commission shall have exclusive jurisdiction over those portions of the North American Numbering Plan that pertain to the United States. Nothing in this paragraph shall preclude the Commission from delegating to State commissions or other entities all or any portion of such jurisdiction.

(2) Costs.—The cost of establishing telecommunications numbering administration arrangements and number portability shall be borne

by all telecommunications carriers on a competitively neutral basis as determined by the Commission.

(f) Exemptions, Suspensions, and Modifications.—

(1) Exemption for certain rural telephone companies.—

(A) Exemption.—Subsection (c) of this section shall not apply to a rural telephone company until (i) such company has received a bona fide request for interconnection, services, or network elements, and (ii) the State commission determines (under subparagraph (B)) that such request is not unduly economically burdensome, is technically feasible, and is consistent with section 254 (other than subsections (b)(7) and (c)(1)(D) thereof).

(B) State termination of exemption and implementation schedule.—The party making a bona fide request of a rural telephone company for interconnection, services, or network elements shall submit a notice of its request to the State commission. The State commission shall conduct an inquiry for the purpose of determining whether to terminate the exemption under subparagraph (A). Within 120 days after the State commission receives notice of the request, the State commission shall terminate the exemption if the request is not unduly economically burdensome, is technically feasible, and is consistent with section 254 (other than subsections (b)(7) and (c)(1)(D) thereof). Upon termination of the exemption, a State commission shall establish an implementation schedule for compliance with the request that is consistent in time and manner with Commission regulations.

(C) Limitation on exemption.—The exemption provided by this paragraph shall not apply with respect to a request under subsection (c) from a cable operator providing video programming, and seeking to provide any telecommunications service, in the area in which the rural telephone company provides video programming. The limitation contained in this subparagraph shall not apply to a rural telephone company that is providing video programming on the date of enactment of the Telecommunications Act of 1996.

(2) Suspensions and modifications for rural carriers.—A local exchange carrier with fewer than 2 percent of the Nation's subscriber lines installed in the aggregate nationwide may petition a State commission for a suspension or modification of the application of a requirement or requirements of subsection (b) or (c) to telephone exchange service facilities specified in such petition. The State commission shall grant such petition to the extent that, and for such duration as, the State commission determines that such suspension or modification—

(A) is necessary—

(i) to avoid a significant adverse economic impact on users of telecommunications services generally;

(ii) to avoid imposing a requirement that is unduly economically burdensome; or

(iii) to avoid imposing a requirement that is technically infeasible; and

(B) is consistent with the public interest, convenience, and necessity.

The State commission shall act upon any petition filed under this paragraph within 180 days after receiving such petition. Pending such action, the State commission may suspend enforcement of the requirement or requirements to which the petition applies with respect to the petitioning carrier or carriers.

(g) Continued Enforcement of Exchange Access and Interconnection Requirements.—On and after the date of enactment of the Telecommunications Act of 1996, each local exchange carrier, to the extent that it provides wireline services, shall provide exchange access, information access, and exchange services for such access to interexchange carriers and information service providers in accordance with the same equal access and nondiscriminatory interconnection restrictions and obligations (including receipt of compensation) that apply to such carrier on the date immediately preceding the date of enactment of the Telecommunications Act of 1996 under any court order, consent decree, or regulation, order, or policy of the Commission, until such restrictions and obligations are explicitly superseded by regulations prescribed by the Commission after such date of enactment. During the period beginning on such date of enactment and until such restrictions and obligations are so superseded, such restrictions and obligations shall be enforceable in the same manner as regulations of the Commission.

(h) Definition of Incumbent Local Exchange Carrier.—

(1) Definition.—For purposes of this section, the term "incumbent local exchange carrier" means, with respect to an area, the local exchange carrier that—

(A) on the date of enactment of the Telecommunications Act of 1996, provided telephone exchange service in such area; and

(B)(i) on such date of enactment, was deemed to be a member of the exchange carrier association pursuant to section 69.601(b) of the Commission's regulations (47 C.F.R. 69.601(b)); or

(ii) is a person or entity that, on or after such date of enactment, became a successor or assign of a member described in clause (i).

(2) Treatment of comparable carriers as incumbents.—The Commission may, by rule, provide for the treatment of a local exchange carrier (or class or category thereof) as an incumbent local exchange carrier for purposes of this section if—

(A) such carrier occupies a position in the market for telephone exchange service within an area that is comparable to the position occupied by a carrier described in paragraph (1);

(B) such carrier has substantially replaced an incumbent local exchange carrier described in paragraph (1); and

(C) such treatment is consistent with the public interest, convenience, and necessity and the purposes of this section.

(i) Savings Provision.—Nothing in this section shall be construed to limit or otherwise affect the Commission's authority under section 201.

Sec. 252. [47 U.S.C. 252] Procedures for Negotiation, Arbitration, and Approval of Agreements.

(a) Agreements Arrived at Through Negotiation.—

(1) Voluntary negotiations.—Upon receiving a request for interconnection, services, or network elements pursuant to section 251, an incumbent local exchange carrier may negotiate and enter into a binding agreement with the requesting telecommunications carrier or carriers without regard to the standards set forth in subsections (b) and (c) of section 251. The agreement shall include a detailed schedule of itemized charges for interconnection and each service or network element included in the agreement. The agreement, including any interconnection agreement negotiated before the date of enactment of the Telecommunications Act of 1996, shall be submitted to the State commission under subsection (e) of this section.

(2) Mediation.—Any party negotiating an agreement under this section may, at any point in the negotiation, ask a State commission to participate in the negotiation and to mediate any differences arising in the course of the negotiation.

(b) Agreements Arrived at Through Compulsory Arbitration.—

(1) Arbitration.—During the period from the 135th to the 160th day (inclusive) after the date on which an incumbent local exchange carrier receives a request for negotiation under this section, the carrier or any other party to the negotiation may petition a State commission to arbitrate any open issues.

(2) Duty of petitioner.—

(A) A party that petitions a State commission under paragraph (1) shall, at the same time as it submits the petition, provide the State commission all relevant documentation concerning-

(i) the unresolved issues;

(ii) the position of each of the parties with respect to those issues; and

(iii) any other issue discussed and resolved by the parties.

(B) A party petitioning a State commission under paragraph (1) shall provide a copy of the petition and any documentation to the other party or parties not later than the day on which the State commission receives the petition.

(3) Opportunity to respond.—A non-petitioning party to a negotiation under this section may respond to the other party's petition and provide such additional information as it wishes within 25 days after the State commission receives the petition.

(4) Action by state commission.—

(A) The State commission shall limit its consideration of any petition under paragraph (1) (and any response thereto) to the issues set forth in the petition and in the response, if any, filed under paragraph (3).

(B) The State commission may require the petitioning party and the responding party to provide such information as may be necessary for the State commission to reach a decision on the unresolved issues. If any party refuses or fails unreasonably to respond on a timely basis to any reasonable request from the State commission, then the State commission may proceed on the basis of the best information available to it from whatever source derived.

(C) The State commission shall resolve each issue set forth in the petition and the response, if any, by imposing appropriate conditions as required to implement subsection (c) upon the parties to the agreement, and shall conclude the resolution of any unresolved issues not later than 9 months after the date on which the local exchange carrier received the request under this section.

(5) Refusal to negotiate.—The refusal of any other party to the negotiation to participate further in the negotiations, to cooperate with the State commission in carrying out its function as an arbitrator, or to continue to negotiate in good faith in the presence, or with the assistance, of the State commission shall be considered a failure to negotiate in good faith.

(c) Standards for Arbitration.—In resolving by arbitration under subsection (b) any open issues and imposing conditions upon the parties to the agreement, a State commission shall—

(1) ensure that such resolution and conditions meet the requirements of section 251, including the regulations prescribed by the Commission pursuant to section 251;

(2) establish any rates for interconnection, services, or network elements according to subsection (d); and

(3) provide a schedule for implementation of the terms and conditions by the parties to the agreement.

(d) Pricing Standards.—

(1) Interconnection and network element charges.—Determinations by a State commission of the just and reasonable rate for the interconnection of facilities and equipment for purposes of subsection (c)(2)
of section 251, and the just and reasonable rate for network elements for
purposes of subsection (c)(3) of such section—

(A) shall be—

(i) based on the cost (determined without reference to a
rate-of-return or other rate-based proceeding) of providing
the interconnection or network element (whichever is applicable), and

(ii) nondiscriminatory, and

(B) may include a reasonable profit.

(2) Charges for transport and termination of traffic.—

(A) In general.—For the purposes of compliance by an incumbent local exchange carrier with section 251(b)(5), a State commission shall not consider the terms and conditions for reciprocal
compensation to be just and reasonable unless—

(i) such terms and conditions provide for the mutual
and reciprocal recovery by each carrier of costs associated with
the transport and termination on each carrier's network facilities of calls that originate on the network facilities of the other
carrier; and

(ii) such terms and conditions determine such costs on
the basis of a reasonable approximation of the additional costs
of terminating such calls.

(B) Rules of construction.—This paragraph shall not be construed—

(i) to preclude arrangements that afford the mutual recovery of costs through the offsetting of reciprocal obligations,
including arrangements that waive mutual recovery (such as
bill-and-keep arrangements); or

(ii) to authorize the Commission or any State commission to engage in any rate regulation proceeding to establish

with particularity the additional costs of transporting or terminating calls, or to require carriers to maintain records with respect to the additional costs of such calls.

(3) Wholesale prices for telecommunications services.—For the purposes of section 251(c)(4), a State commission shall determine wholesale rates on the basis of retail rates charged to subscribers for the telecommunications service requested, excluding the portion thereof attributable to any marketing, billing, collection, and other costs that will be avoided by the local exchange carrier.

(e) Approval by State Commission.—

(1) Approval required.—Any interconnection agreement adopted by negotiation or arbitration shall be submitted for approval to the State commission. A State commission to which an agreement is submitted shall approve or reject the agreement, with written findings as to any deficiencies.

(2) Grounds for rejection.—The State commission may only reject—

(A) an agreement (or any portion thereof) adopted by negotiation under subsection (a) if it finds that—

(i) the agreement (or portion thereof) discriminates against a telecommunications carrier not a party to the agreement; or

(ii) the implementation of such agreement or portion is not consistent with the public interest, convenience, and necessity; or

(B) an agreement (or any portion thereof) adopted by arbitration under subsection (b) if it finds that the agreement does not meet the requirements of section 251, including the regulations prescribed by the Commission pursuant to section 251, or the standards set forth in subsection (d) of this section.

(3) Preservation of authority.—Notwithstanding paragraph (2), but subject to section 253, nothing in this section shall prohibit a State commission from establishing or enforcing other requirements of State law in its review of an agreement, including requiring compliance with intrastate telecommunications service quality standards or requirements.

(4) Schedule for decision.—If the State commission does not act to approve or reject the agreement within 90 days after submission by the parties of an agreement adopted by negotiation under subsection (a), or within 30 days after submission by the parties of an agreement adopted by arbitration under subsection (b), the agreement shall be deemed approved. No State court shall have jurisdiction to review the action of

a State commission in approving or rejecting an agreement under this section.

(5) Commission to act if state will not act.—If a State commission fails to act to carry out its responsibility under this section in any proceeding or other matter under this section, then the Commission shall issue an order preempting the State commission's jurisdiction of that proceeding or matter within 90 days after being notified (or taking notice) of such failure, and shall assume the responsibility of the State commission under this section with respect to the proceeding or matter and act for the State commission.

(6) Review of state commission actions.—In a case in which a State fails to act as described in paragraph (5), the proceeding by the Commission under such paragraph and any judicial review of the Commission's actions shall be the exclusive remedies for a State commission's failure to act. In any case in which a State commission makes a determination under this section, any party aggrieved by such determination may bring an action in an appropriate Federal district court to determine whether the agreement or statement meets the requirements of section 251 and this section.

(f) Statements of Generally Available Terms.—

(1) In general.—A Bell operating company may prepare and file with a State commission a statement of the terms and conditions that such company generally offers within that State to comply with the requirements of section 251 and the regulations thereunder and the standards applicable under this section.

(2) State commission review.—A State commission may not approve such statement unless such statement complies with subsection (d) of this section and section 251 and the regulations thereunder. Except as provided in section 253, nothing in this section shall prohibit a State commission from establishing or enforcing other requirements of State law in its review of such statement, including requiring compliance with intrastate telecommunications service quality standards or requirements.

(3) Schedule for review.—The State commission to which a statement is submitted shall, not later than 60 days after the date of such submission—

(A) complete the review of such statement under paragraph

(2) (including any reconsideration thereof), unless the submitting carrier agrees to an extension of the period for such review; or

(B) permit such statement to take effect.

(4) Authority to continue review.—Paragraph (3) shall not preclude the State commission from continuing to review a statement that has

been permitted to take effect under subparagraph (B) of such paragraph or from approving or disapproving such statement under paragraph (2).

(5) Duty to negotiate not affected.—The submission or approval of a statement under this subsection shall not relieve a Bell operating company of its duty to negotiate the terms and conditions of an agreement under section 251.

(g) Consolidation of State Proceedings.—Where not inconsistent with the requirements of this Act, a State commission may, to the extent practical, consolidate proceedings under sections 214(e), 251(f), 253, and this section in order to reduce administrative burdens on telecommunications carriers, other parties to the proceedings, and the State commission in carrying out its responsibilities under this Act.

(h) Filing Required.—A State commission shall make a copy of each agreement approved under subsection (e) and each statement approved under subsection (f) available for public inspection and copying within 10 days after the agreement or statement is approved. The State commission may charge a reasonable and nondiscriminatory fee to the parties to the agreement or to the party filing the statement to cover the costs of approving and filing such agreement or statement.

(i) Availability to Other Telecommunications Carriers.—A local exchange carrier shall make available any interconnection, service, or network element provided under an agreement approved under this section to which it is a party to any other requesting telecommunications carrier upon the same terms and conditions as those provided in the agreement.

(j) Definition of Incumbent Local Exchange Carrier.—For purposes of this section, the term "incumbent local exchange carrier" has the meaning provided in section 251(h).

Sec. 253. [47 U.S.C. 253] Removal of Barriers to Entry.

(a) In General.—No State or local statute or regulation, or other State or local legal requirement, may prohibit or have the effect of prohibiting the ability of any entity to provide any interstate or intrastate telecommunications service.

(b) State Regulatory Authority.—Nothing in this section shall affect the ability of a State to impose, on a competitively neutral basis and consistent with section 254, requirements necessary to preserve and advance universal service, protect the public safety and welfare, ensure the continued quality of telecommunications services, and safeguard the rights of consumers.

(c) State and Local Government Authority.—Nothing in this section affects the authority of a State or local government to manage the public rights-of-way or to require fair and reasonable compensation from telecommunications providers, on a competitively neutral and nondiscriminatory basis, for use of public rights-of-way on a nondiscriminatory basis, if the compensation required is publicly disclosed by such government.

(d) Preemption.—If, after notice and an opportunity for public comment, the Commission determines that a State or local government has permitted or imposed any statute, regulation, or legal requirement that violates subsection (a) or (b), the Commission shall preempt the enforcement of such statute, regulation, or legal requirement to the extent necessary to correct such violation or inconsistency.

(e) Commercial Mobile Service Providers.—Nothing in this section shall affect the application of section 332(c)(3) to commercial mobile service providers.

(f) Rural Markets.—It shall not be a violation of this section for a State to require a telecommunications carrier that seeks to provide telephone exchange service or exchange access in a service area served by a rural telephone company to meet the requirements in section 214(e)(1) for designation as an eligible telecommunications carrier for that area before being permitted to provide such service. This subsection shall not apply—

(1) to a service area served by a rural telephone company that has obtained an exemption, suspension, or modification of section 251(c)(4) that effectively prevents a competitor from meeting the requirements of section 214(e)(1); and

(2) to a provider of commercial mobile services.

Appendix B:
Glossary

Absolute advantage Capacity of a country to produce a specific good with fewer resources (per unit of output) than other countries.

Acquisition A purchase of one firm by another; in rare instances, a strategic alliance among multiple firms may result in acquisition of one or more companies.

Administrative law That body of public law that governs bureaucratic management and regulatory policy.

Affiliation Association between two or more firms through nonequity agreements, the principal objective of which concerns strengthening distribution channels for each affiliate.

Aggregation The process of attracting customers so as to expand an existing consumer base.

Antitrust law Federal statutes preventing, modifying, or restricting proposed mergers or acquisitions that result in reduction of market competition. The Sherman, Clayton, FTC, and Robinson-Patman Acts (passed in 1890, 1914, 1915, and 1936, respectively) define contemporary antitrust law.

Baby Boom Generation That generation born between 1946 and 1964; some demographers extend this definition to include births beginning in 1940 through 1965. This generation, comprising 80 million, is the largest American demographic group in history.

BOP Business opportunity paradigm: a model, designated in this volume, and validated through recent empirical research, which suggests that strategic planning in many telecommunications enterprises should focus on the development of tertiary markets.

Brand A symbol, image, term, name, design, or label that differentiates a firm from its competitors.

Brand awareness A measure, statistic, or other evaluative tool that isolates a consumer's knowledge of a brand's image.

Brand extension The extension of a firm's reputation to the development of other products and markets.

Business cycle Variations in supply and demand in the macroeconomy that influence prospective consumption for all goods and services. Recent studies on the business cycle reveal that post–World War II prosperities are lengthening, while recessions are contracting in length.

Business process reengineering The restructuring of a firm's functions based on the introduction of advanced communications and information technologies.

Cherry picking A tactic typically applied by a firm bringing a new product to market; the term is intended to convey that high-income consumers become the initial target of entrepreneurs seeking a market presence for the product.

Classic strategic planning A simple six-step procedure for strategic planning, embracing mission definition, elaboration of company goals, an assessment of the firm's competitive environment, an internal audit of the firm's resources, identifiable action steps, and a plan to implement tactics designed to achieve strategic goals. The inability of classic planning methodology to contend with volatile competitive environments has inspired a host of alternative planning models in recent years.

Communications Act of 1934 The fundamental federal statute guiding communications policy through 1996.

Comparative advantage The capacity of a country to produce a specific good at a lower opportunity cost than its trading partner.

Competitive advantage An economic expression used to define the inherent benefits, gains, or other favorable circumstances that a firm can exploit relative to its competition; the expression sustains particular efficacy in the field of industry analysis.

Competitive analysis The procedure whereby the competitive environment of a firm is dissected.

Competitive environment A reference in modern strategic management that emphasizes the intensification, volatility, and uncertainty associated with non-incremental change in all sectors of the economy.

Competitive research Also described as *surveillance research*, in which a firm defines strategic threats likely to emerge from the competition.

Consumer decision making The process by which consumers assimilate, retain, store, and retrieve product characteristics prior to purchase.

Consumer primary An organizational perspective that holds that consumers, during the impending age of customization, represent the exclusive focus of market research.

Continuous learning markets The great long-term areas of consumer demand likely to be the strategic objective of major firms within the next decade; these markets are characterized by the need to certify education and training, as well as to link collaborative networks of businesspeople, communities, and other groups of common interest.

Convergence A word of dual meanings: may be referred to as the growing interaction and influence of telephony, computer, and broadcasting technologies designated to create new products and services; may also be applied to describe the recent spate of partnerships, alliances, affiliations, and other business formations uniting former competitors in introducing new products to market.

Coopetition A business model that conveys the desire of two or more firms to cooperate, rather than directly compete, in producing or distributing a product or service.

Corporate consolidation Reduction in the number of firms serving a market through mergers and acquisitions.

Current account A term used to refer to a country's balance-of-payments position in international trade. American balance-of-payments (or net trade position) has been measurably enhanced in recent years through the export of communications products, equipment, and services, despite the fact that the United States has not experienced a balanced-trade position since 1980.

Customization The design of products and services geared to customer-directed specifications.

Demand   Consumption arising in response to pricing alternatives.

Demographics   The statistical characteristics or variations assigned to a group or population.

Deregulation   Reduction, relaxation, or modification of regulatory powers as applied to an agency of federal, state, or local government; the American experiment with deregulation across a variety of industries began in 1978 and continues today.

Determinants of demand   Those factors responsible for consumption, including income, taste, competitive alternatives, demographic variables, and future expectations of consumers, households, and organizations.

Determinants of supply   Those variables responsible for production, including technology, cost, managerial innovation, number of sellers, and future expectations of firms regarding price and profitability.

Digitization   A term of multiple meanings, but typically used to refer to the generation, storage, distribution, or assimilation of information in nonmaterial form; distilled to its strategic significance, the term confers economic advantage to those companies able to leverage information technology or communications equipment in response to a customer base.

Disintermediation   The elimination of labor (reduction or elimination of intermediaries) involved in the distribution of goods and services to market, typically precipitated by advances in communications and information technology.

Duopoly   A market served by two producers.

Economic profit   The amount of profit a firm earns beyond the cost of production; this term is often distinguished from *accounting profit*, which designates that economic costs can exceed explicit costs in certain instances.

Economies of scale   The decrease in cost per unit of a product or service, instigated by volume buying, division of labor, and managerial efficiencies.

Economies of scope   The expertise that a firm can apply in utilizing complementary skills of its personnel to produce two or more goods or services.

Electronic commerce   Also referred to as *e-commerce*, the label applied to networked markets of consumers and producers; electronic commerce is growing exponentially relative to growth rates exhibited in both the domestic and global economies, implying it is supplanting traditional models of commerce and thus reshaping traditional strategic management models.

Externalities   Third-party consequences resulting from a transaction between buyers and sellers.

Five-Force Model   A method of competitive analysis, originated by Michael Porter at Harvard Business School, which integrates five analytical elements to discern optimal strategic plans for a firm.

Fixed costs   A firm's charges that remain constant regardless of increases or decreases in sales activity.

GATT   General Agreement on Tariffs and Trade: devised in 1948 as an international body of government officials whose principal objective has been the promotion of global economic growth through reduced governmental trade restrictions and lower tariffs.

GDP   Gross Domestic Product: a statistical measure gauging annual national wealth through the unduplicated count of all goods and services.

Generation X   That portion of the population born between 1965 and 1982.

Generation Y   That fraction of the population born between 1983 and 2002; sometimes referred to as the *millennial generation*.

Geodemographics   The application of demographic variables to specify the relationship between consumer behavior and regional affiliation.

Geographic Information Systems (GIS)   Databases containing geodemographic information (*see* geodemographics).

Globalization   The increasing interaction of nations, firms, and individuals in pursuing common economic objectives.

Government lag   Often applied to bureaucratic regulators who take an inordinate amount of time to supply licenses, respond to corporate requests for information, or who are otherwise slow to respond to business concerns.

Herfindahl-Hirshman Index   A method of economic analysis validating the presence of emerging monopolies through the accumulation of market share; increasingly, the index is decisive in launching antitrust investigations in those firms or industries that exhibit a strategic predilection toward monopoly.

Horizontal integration   Absorption of competitors for purposes of enlarging market share; also defined as the acquisition of competitors at the same level in the distribution chain.

IBOP   International Business Opportunity Paradigm: a model implying that long-term communications demand will be fueled to a considerable degree by tertiary markets.

Income elasticity   The relationship between income and consumption, as measured in percentage changes.

Indifference curve   Utility derived from a choice between two alternative options for consumption; the curve measures the degree to which a consumer is able to purchase one product relative to the proportionate consumption of another.

Innovation   The capacity to improve or refine an existing product line, as distinguished from *creativity*, which implies the power to promulgate birth of new products. Viewed from a strategic perspective, most firms can generally be categorized as either innovative or creative with respect to the products they bring to market.

Intellectual capital   The value of a firm's labor force, as measured in knowledge, intelligence, creativity, or capacity to innovate; increasingly, the economic significance of labor is being recognized by strategists as mobility of the labor force defines future strategic advantage in the information technology and communications industries.

Intermediation   The creation or evolution of intermediaries necessary to produce, distribute, market, or resell goods or services; after experiencing a century of growing economic intermediation, the U.S. economy has recently exhibited definite signs of disintermediation in the production and distribution of goods and services to market.

International convergence   The interaction and evolving relationship between the computer, broadcasting, and telephony industries on a cross-national or cross-regional basis.

Internetworked enterprise   A firm able to leverage communications technologies so as to establish and strengthen relationships with suppliers, distributors, and consumers.

Intrapreneurship   Creation of a new firm within an existing corporate structure, in which entrepreneurs are able to exploit the financial, marketing, and managerial resources of the parent to facilitate production and distribution of a new product line.

Just-in-time (JIT) inventory/purchasing   Refers to advances in information and manufacturing technology that facilitate congruence between production and demand; the telecommunications sectors have become critical in refining JIT management methods for all other industries.

Liberalization   In regulatory applications, the relaxation of existing government sanctions or constraints; in international trade, the relaxation of quotas, tariffs, or domestic content provisions.

Market   May be dichotomized as real versus theoretical; a real market consists of buyers and sellers interacting in the economy at any given time, while the theoretical (or maximum) market is one in which all potential buyers and sellers can simultaneously interact.

Market acceptance   A term applied to a product that has gained foothold in a market, or has attained a minimum of 10% market share. Also used to differentiate strategic management philosophy among high-technology firms: is it more advantageous to be first-to-market with new product, or first-to-market acceptance with an existing or veteran product line?

Market equilibrium   The quantity of a product or service that people purchase that is simultaneously equivalent to the quantity produced by sellers at any given time.

Market research   The surveying of demographic, economic, and other data in support of strategic plans and sales objectives.

Market structure   A label used to describe market characteristics of a given market sector. Market structures include monopoly, oligopoly, competitive, purely competitive, monopolistic competitive, and duopoly forms.

Mass customization   The process of customizing a product line for mass consumption; the capacity of a firm to design or personalize a product or service according to the unique needs of every customer within a large and expanding customer base.

Merger   Integration of two firms that typically relinquish their individual identities in favor of a new title and complementary activities.

Monopolistic competition   A market served by many producers, but distinguished from perfect competition through brand, image, or reputation.

Monopoly   In theory, a market served by a single producer. An effective monopoly can operate holding as little as two-thirds market share.

NAFTA   Trade agreement negotiated and ratified by the United States, Canada, and Mexico in 1994. The agreement contains various provisions affecting telecommunications trade between these nations. NAFTA stands as a prototype for liberalized trade in multinational negotiations and is likely to instigate eventual complete free trade throughout the Western Hemisphere.

Opportunity cost   The amount that is foregone by an individual or firm in pursuing one objective as against other available options.

Panel consensus   An organizational technique designed to produce a forecast such that members concur in its underlying reasoning.

Pareto Optimality   An economic postulate which holds that self-interest and social interest are optimized in perfectly competitive markets; as applied to the emerging field of customer relationship management (CRM), one feature of this economic law implies that 80% of a firm's revenue, within parameters, results from the transactions generated by 20% of its consumer base.

Particle marketing   Market research aimed at the customization of products and services; also referred to as *one-to-one marketing*, the logic of this tactic holds that each customer is treated as a market unto him- or herself.

Post-industrial society   Also referred to as post-industrialization, this expression has become a synonym for the *information society*, and identifies: (1) the value of information to knowledge-based economies; (2) nations, including the United States, Western Europe, Canada, and Japan, whose highest percentage of the labor force works in service or knowledge-based industries.

Price elasticity   Degree of responsiveness of buyers to changes in product price.

Price implosion   A theoretical model projecting a market collapse as profits through "price wars" depress an industry; this disequilibrium results in eventual equilibrium, as failing firms are supplanted by survivors.

Primary demand   Initial consumer demand satisfied by the introduction of a product or service.

Productivity   Ordinarily referred to as *labor productivity*, and measured in a firm's output relative to resources expended in capital and technology; may also be measured in labor output per hour. Since 1997, studies reveal that American labor productivity has been rising relative to the level of organizational expenditures invested in new information technologies and communications equipment.

Reengineering   *See* Business process reengineering.

Regulation   Government rules imposed on the pricing, quality, distribution, or other characteristics of products and services.

Regulatory overview   An expression indicating the breadth of influence of authority held by government regulators over their prescribed duties (also taken to mean regulators' own interpretation of their responsibilities as defined by statutory law).

Reintermediation The procedure by which individuals and firms reinstall themselves into the distribution chain by providing value (in pricing, quality, or service) to consumers that did not formerly exist.

Relational marketing A term used to define and evaluate a firm's relationship to its best (most profitable) customers.

S-curve A model that holds that emerging consumer demand rises at an exponential rate, assuming certain positive conditions, after 10% market penetration is attained.

Secondary demand Ancillary demand, with resultant intermediate businesses, occurring after a new technology is introduced.

Service explosion An economic model intended to convey the intent of the framers of the 1996 deregulation bill; proliferation of providers whose competitive interaction results in price reduction and service improvement for consumers.

Statutory law That body of law devised and passed by Congress (i.e., statutes are the province of elected officials as opposed to common or other areas of law). Of particular consequence for telecommunications strategists are statutes governing administrative and antitrust law.

Strategic alliance Relationship between two or more firms seeking entry to market not otherwise attainable through individual action; this relationship is not based on an equity interest, as is the case in strategic partnerships.

Strategic choice Identification of the optimal strategy among a variety of available options.

Strategic management Often used interchangeably with strategic planning, the expression is typically applied to the ongoing development, monitoring, and management of strategy within the organization.

Strategic partnership A relationship between two or more firms, the purpose of which is to bring to market a product requiring long-term capital, labor, and R&D commitment.

Strategic variations Isolation of the most desirable strategic options.

Strategy Plans that encompass organizational decision making, allocation of resources, specification of goals and objectives, and the firm's response to changes in the competitive landscape.

Systems analysis   As applied to economics, a tool used to identify all inputs, processes, outputs, and feedback so as to predict the future demand for newly introduced products or services.

TCN   Technology Capital Network: an organization dedicated to connecting inventors and entrepreneurs with venture capitalists.

Telecommunications Act of 1996   Sometimes referred to as the Telecommunications Reform Act, this statute, passed by Congress in February 1996, stipulated, among other items, the elimination of sanctions preventing commercial interaction between the cable, telephony, and other communications industries. The economic and strategic significance of the law lies in its long-term influence over other sectors of the economy as the next century unfolds.

Tertiary analysis   An economic tool, based on systems theory, which forecasts emerging market opportunities precipitated by externalities associated with new product introduction.

Undifferentiated marketing   A marketing strategy aimed at establishing a general reputation and brand image in the market without regard to segmentation or niche definition.

Value-added   An economic term applied to the contribution of a firm's components to the production, sales, or service of a product.

Value chain   The collection of all activities that allow a firm to exploit its competitive advantage in the market.

Variable cost   Those costs incurred as the level of output adjusts to changes in and availability of labor, capital, and other resources.

Venture capital   Funds invested by individuals or firms in the creation of business start-ups. Venture capitalists seek annual returns in excess of 25% or more before committing funds, given the level of risk.

Vertical integration   The acquisition of firms whose complementary expertise results in reduced costs of production and greater strategic control.

Volume segmentation   The dichotomization of consumer and business markets on the basis of comparative usage; a telecommunications provider might differentiate, for instance, its consumer base according to minimal, average, or excessive users.

WTO   World Trade Organization: a multilateral negotiating body of particular significance to telecommunications global trade. Since 1997, this group has negotiated a reduction in the number of regulatory sanctions and quotas governing international trade in communications equipment and services.

Appendix C:
Web Site Citations for
Telecommunications Analysts

Indicated below are Web site citations useful for telecommunications analysts. These references are divided according to the primary subfields of specialization in business management. Fields of specialization include finance, management, marketing, electronic commerce, economic analysis, and regulatory policy and law. Strategic planners seeking to exploit information technologies require conduits into each of these areas, both in terms of securing primary data as well as establishing methods for communicating with colleagues across disciplinary boundaries.

Also in this section are research institutes, news forums, associations, government directories, and other sources of data of continuing interest to strategists. It is strongly recommended that readers refer to www.umich.edu, www.brint.com, www.infoseek.go/titles, www.planware.org/strategy, www. burrelles.com, and www.kiwiclub.edu for hyperlinked information relative to the following database categories: World Wide Web–based tools; function-specific tools; news forums; associations; government directories; research institutes; and specialized research on strategic management in telecommunications.

C.1 World Wide Web–Based Tools

Unless otherwise noted, each of the following references is presumed to follow the standard URL (Uniform Resource Locator) code: http://www.name.com.

For general articles and commercial data pertaining to subjects in telecommunications, strategic planning, competitive analysis, business management, marketing, finance, regulatory, or emerging legal issues, see:

tele.com	architel.com
Internettelephony.com	rpcp.mit.edu/diig/other sites
futuris.net	millenniumtel.com/ref-gov
dataquest.com	otcog.comgfifax.com
ntq.com	businessmonitor.co.uk
teledot.com	ntsika.org
telecoms-mag.com	fcc.gov.hotissue
informationweek.com	icanet.com
mmm.com	iec.org
ameritech.com	computer.org
att.com	mmta.org
pacbell.com	nab.org
sbc.com	tiaonline.org
ucop.slis.edu	tfi.com
regulation.org	imd.ch/way
cnet.comnews.com	mktg.inc.com
wired.com	telecommagzine.com
techwire.com	burrelles.com
zdnet.com	dryden.com/mgmt/topics
my.yahoo.com	hbsp.harvard.edu/mgmt
lib.depaul.edu	planware.org/strategy
vive.com/www/virtual/research/tel	infoseek.go.com/titles
main.org/computer	ieee.org

Additional Internet business and economics references directly pertaining to the communications industry can be located through Gopher at the following listing: gopher://Niord.SHSU.edu:70/11gopher_root%3A%5B_data.economics%5D; also refer to: http://www.wugate.wustl.edu.

The use of these tools will provide access to these important references (refer to the University of Michigan [www.u-mich.edu] Web site for URL citations).

Economics and Finance

Economics Resources from NetEcArchive
Economics Working Paper Archive
Bureau of Labor Statistics
Bell Atlantic Gopher

Business, Economics, Marketing
ECONsult International Consultancy and Forecasting
EconData
Economic Bulletin Board
Economic Research Service
Internet Business Journal
National Information Infrastructure Information
National Information Infrastructure Task Force
Resources for Economists on the Internet
World Bank Public Information Center

The following Web pages can be accessed initially by tapping into www.
dowjones.com and hyperlinking accordingly via the University of Michigan.
Also refer to: www.sims.berkeley.edu for related hyperlinked information.

Macroeconomic Law and Regulation

ARL Federal Relations E-News
Cases, Statutes, & Topical Highlights, Cyberspace Law & Policy (UCLA On-
    Line Institute)
Clearinghouse on Computer Accommodation (COCA)
Computers and Society
Cyberlaw World Wide
Cyberlaw Center: Cybercrime Resources
Cyberspace Law Center: Law Directories and Other Sites
Cyberspace Law Center: Intellectual Property Resources
Cyberspace Law for Non-Lawyers
Digital Future Coalition
Economics of the Internet
Economics of Networks
Information Law Web
Intellectual Property Law Notes for Technology Companies
Legal Links in Electronic Commerce and Interactive Entertainment
Patent Law Web Server
Regulation of the Communications Functions of the Internet
State Public Utility Regulations
Technology Law Server
Telecom Law and Policy Resources
Technology Law Update
Telecommunications Policy Sites
WWW Multimedia Law
Citizen Internet Empowerment Coalition

Electronic Commerce

Committee on the Law of Commerce in Cyberspace
CommerceNet
Commercial Internet Exchange (CIX)
Information Innovation
Information Warfare
CNET.com
Dataquest Interactive
Year 2000
ZDNet E-Business
OECD Electronic Commerce
Internet.com's Electronic Commerce Guide
IBM Institute for Advanced Commerce
Information Economies Project
NetBusiness
The Emerging Digital Economy
Competence Network
CYBER Review Newsletter
Center for Research in Electronic Commerce
A Framework for Global Electronic Commerce
Corporate Knowledge Management
Information Strategy Online
Governments and Electronic Commerce
NetAcademy on Business Media
Journal of Internet Banking and Commerce
Electronic Banking
Inventory of EU and Electronic Commerce Initiatives
OnLine Commerce and Models for the Future
Jupiter Communications
Transaction.Net
CyberSource
SEMPER
Business of the Internet
Internet Business Center
Frontiers of Electronic Commerce
Information Innovation Homepage
Fisher Center for Information Technology Management
Electronic Commerce Resource Guide
Electronic Commerce and the Internet
Association Franaise pour le Commerce et les Echanges Electroniques

Electronic Money in the United States
EFF's Digital Money
Clickshare
Electronic Money
Harvard Information Infrastructure Project
Internet Law and Policy Forum
Information Infrastructure Task Force
NII Virtual Library
Global Information Infrastructure: Agenda for Cooperation
Collaborations on Internet Security
Merit Network, Inc.
Alliance for Competitive Communication
Internet Telephony Consortium
Consumer Project on Technology

C.2　Function-Specific Tools

The following research tools contain periodic information on domestic and global markets and are listed by organizational function.

Strategic Planning

www.ioma.com
www.aom.pace.edu
www.imd.ch
http://future.sri.com
www.lumina.com/DA
www.hoovers.com
www.dnb.com
www.pathfinder.com
www.infoseek.com/business/
　economics
www.fourmilab.ch
www.enews.com
www.clark.net/pub

www.wueconb.wustl/edu/business
www.hbsp.edu/strategy
www.wbs.warwick.ac.uk./omindex
www.wst.com/library
www.apics.org/SIGs
www.skyenet.net/~leg/tqm
http://deming.eng.clemson.edu
http://deming.org/deminghtml/
　wedi
www.informs.org/jobs/jobs
www.itu.ch
www.wow-com.com
www.mpt.go.jp

Management

www.sbaon-line.gov
www.imd.ch/

www.teleport.com
www.wp.com

Marketing

www.census.gov
www.marketingtools.com
www.advert.com
www.mci.com

www.esri.com
maps.esri.com
www.geo.edu
www.gis.umn.edu

Finance

www.actnet.org
www.moneypages.com
www.interaccess.com
www.investorweb.com
www.pcquote.com
www.hoovers.com
www.corpfinet.com

www.gnn.com
www.dbisna.com
www.sec.gov
www.princeton.edu
www.datasecurities.com
www.ventureconomics.com

C.3 News Forums

News forums are professional organizations that regularly report on the status of macroeconomic and business policy, as well as mergers, acquisitions, and partnerships. The following primary news forums provide daily information on developments within the industry.

www.broadcast.com/business
www.prnewswire.com
www.businesswire.com
www.reuters.com
www.dowjones.com
www.cnbc.com
www.pctv.com
www.upside.com
www.economist.com
www.djia.com

www.securitiesdata.com
www.industrystandard.com
www.bloomberg.com
www.businessweek.com
www.cbs.marketwatch.com
www.cnbc.com
www.cnnfn.com
www.forbes.com
www.msnbc.com/news/
    COMFront.asp

C.4 Associations

The following trade or citizen associations represent fertile avenues for the accumulation of data and economic analysis. You may obtain hypertext links via www.dowjones.com and www.search.com.

Alliance for Competitive Communication
Benton Foundation's Communication Policy Project
Center for Civic Networking
Coalition for Networked Information Policy.Net
Counsel Connect
Cyberspace Law Institute
EDUCOM
MCI Public Policy
MCI Regulatory and Public Policy Department
MFJ Task Force
MIT Research Program on Communications Policy
Online Institute for Cyberspace Law and Policy
Society for Electronic Access
Video Information Providers for Non-Discriminatory Access
Benton Foundation
Canadian International Development Research Center (IDRC)
Centre for International Research on Communication and Information Technologies
Columbia University Institute for Tele-Information
Electronic Frontier Foundation
European Telecommunications Standards Institute
European Union Documents
International Maritime Satellite Organization
Intelsat
International Telecommunications Union
Leland Project
Massachusetts Institute of Technology Media Lab
National Computer Board of Singapore
Pacific Telecommunications Council
World Bank: Infodev Project

C.5 Government Directories

The following government agencies and their Web directories provide regular access to economic policy, regulation, demographic, legal, and other data, as indicated. Additional federal agencies are expected to provide expanding Web pages on communications services in the near future.

State and local government regulatory agencies: www.fcc.gov/state&local
Department of Justice: www.doj.gov

National Technology Transfer Center: www.lcweb.loc.gov
U.S. Patent and Trademark Office: www.uspto.gov
Commerce Information Locator Service: www.doc.gov
Federal Trade Commission: www.ftc.gov
U.S. Census Bureau: www.census.gov
Department of Commerce: www.doc.gov
Security and Exchange Commission: www.sec.gov
National Telecommunications Information Administration: www.ntia.gov
Federal Communications Commission: www.fcc.gov
International Trade Administration: www.ita.gov
Congressional Budget Office: www.cbo.gov

C.6   Research Institutes

The following research institutes conduct research on regulation, deregulation, and related matters. Web pages are available through the University of Michigan and www.regulation.org.

Progress and Freedom Foundation
Claremont Institute
Hoover Institute (Stanford University)
Center for the Study of American Business (Washington University)
Competitive Enterprise Institute
American Enterprise Institute
CATO Institute
Heritage Foundation
University of Michigan
Yale Documents Center
The Brookings Institution

C.7   Academic and Trade Journals in Information and Communications Strategic Management

The following journals, covering various topics of relevance to strategic planners, can be hyperlinked through www.analysys.com:

ATM News Digest (daily journal for broadband networking)
BIT (The Spanish Telecommunication Engineers Association Monthly Magazine)

BT WATCH (weekly monitor of British Telecom produced by Apogeion Global Monitoring)

Business Communications Review

Business News Americas (Latin American daily news service, with telecommunications section)

Butterworth-Heinemann (publisher of books and electronic products for students and professionals in the communications and telecommunications industry)

Cable Foreman Magazine (bridging outside plant engineering and construction)

Cable World

Cabling Installation & Maintenance Magazine (fiber optic communications technology and applications)

Call Centre Focus (magazine covering call center management)

Calling (Telia's on-line magazine, in Swedish)

CHAOnline (Cathey, Hutton and Associates: telecommunications news service)

CLEC News (on-line telecommunications magazine)

Communications News Magazine (monthly trade publication for network managers and IT professionals)

Communications Standards Review (publishes journals reporting on formal telecommunications standards in the ITU, ETSI, and TIA)

Communications Week International

Computer-Mediated Communication Magazine (journal covering CMC issues from December Communications)

Connect-World (publications on telecommunications development and liberalization in Latin America, the Caribbean, and India)

Data Communications Magazine

Digital Horizons Telecommunications Newsletter (new product and technology announcements from vendors of telecommunications, cable, wireless, and computer telephony)

Discount Long Distance Digest (an Internet journal for the long-distance industry)

e.biz (guide to electronic business, from Business Week Online)

e-Summit (monthly on-line report on call center strategy published by MW Productions)

Editors' eNeWS (design house for electronic publishing in IT and communications)

E-Journal (The World Wide Web Virtual Library: Electronic Journals)

ETHOS (recent telematics news provided by the European Telematics Horizontal Observatory Service)

exchange (weekly newsletter for the Australian telecommunications industry)

Fiber Optics Online (news, analysis, product information, and virtual market-place)

Global Communications (information center for IT professionals)

GSM Info Online Search Page (information on networks, services, tariffs, roaming, and coverage)

GSMag (magazine on GSM mobile telephony)

GTB (Global Telecoms Business)

HotWired (incorporating *Wired*)

ICB Toll Free News (regulatory and telecommunications industry reporting)

IEEE Computer Society (list of magazines published by IEEE)

infoHIGHWAY

Inside Cable (newsletter dedicated to the U.K. cable TV and telecommunications industry)

Interop Online Networld + Interop annual conference

Institute for International Research Web sites for telecommunications conferences (and associated newsgroups)

IT News (IT news, information on high-tech products, software, books, etc.)

Journal of Computer Mediated Communication at the Hebrew University of Jerusalem and its Annenberg School for Communication site

Lightwave Magazine

McKinsey Quarterly Telecoms Library

MIT Technology Review

Mobile Computing and Communications

Mobile Office Magazine

Mobile Week (journal of technology issues for users and operators of mobile networks)

Mobilis (features interviews, tutorials, reviews, concerning all aspects of PDAs, wireless communications, and mobile peripherals)

Netcomtalk from the Boston University College of Communication

Network Bibliography (keyword search)

NEWSdesk The Online IT News Network (on-line information service covering IT and telecommunications from Hewlett-Packard)

NewsNow (up-to-the-minute news service with technology section)

NewsPage (news on communications, telecommunications, and datacommunications inter alia)

Outside Plant Magazine (cabling, switches, poles, manholes, installation, and maintenance for telephone and CATV voice and data)

Phillips Business Information, Inc. (satellite, cable, and telecommunications information)

Phillips TelecomWeb (news, research, and analysis in the telecommunications industry)

PNEwire (on-line version of *Public Network Europe* magazine)

Point Topic Ltd. (news and analysis on broadband access to the Internet)

Premises Networks Online (news, analysis, product information, and virtual market place)

The Response Source (service that links journalists with IT sources)

"Rural Telecoms in India" (article by Abhinav Taneja)

Siemens Telcom Report (a Siemens publication covering global telecommunications trends, products, solutions, and services)

Satellite Today (news service by Phillips TelecomWeb)

Satellite Week (journal of news and analysis of technology issues in the satellite industry)

Technology Research Institute (reports on telecommunications billing, data warehousing, provisioning, and customer care)

Telecom Business Online (information source for carriers, convergence, competition, and next-generation networks)

Telecom Sources (trade magazine with product information)

tele.com (publication for telecommunications service providers)

Telecom News Magazine (global news on telecommunications, datacommunications, multimedia, Internet, and events)

Telecom Publishing Group (produces a range of titles in telecommunications)

Telecom Today (telecommunications news and information)

Telecom Tribune (news, information, and statistics on Japanese telecommunications)

Telecoms OSS BSS and IN News (news service run by TelcOSS Executive International, Inc.)

Telecommunications Online

Telecommunications Policy

TRI: Telecommunications Reports International, Inc. (news and information services for the communications, multimedia, and electric utility industries)

Telecoms Newsline (monthly telecommunications industry information source sponsored by Hewlett-Packard)

TeleGeography Home Page

Tele-M@rket (news and discussion forum for the call center business)

TeleNews Asia (news on Asian telecommunications)

Telephony

TelephonyWorld.com (on-line resource for the computer telephony and communication industry)

Telirati (archive of telephony newsletters)

Test and Measurement.com (news, analysis, product information, and virtual marketplace)

Total Telecom (daily news and information service for telecommunications professionals)

UCTIR: U.S.-China Telecommunications and Information Research (on-line newsletters on China's telecommunications and information industries)

UK Telecom Frequently Asked Questions

University of Iowa Libraries Gateway to the Internet (Communications Studies section of an Internet directory)

University of Miami Reading Room (comprehensive list of telecommunications journals)

Video Information Provider Newsletter

Virtual Institute of Information (V.I.I.) (on-line telecommunications information from the Communications and Mass Media Research Institute at Columbia University)

Voice & Data (India communications magazine)

Warren Publishing, Inc. (daily telecommunications news)

Web-Vantage (technology guide for the Internet development community)

Wireless Design Online (news and discussion forum for the wireless engineering market)

Wireless Integration Magazine

Wireless Today (news service by Phillips TelecomWeb)

Wireless Week

C.8 Specialized Research on Strategic Management in Telecommunications

Citations listed under this heading may be accessed through www.brint.com and www.sims.berkeley.edu.

The Internet Economy Indicators

Institute for Technology and Enterprise

Internet Economics

Netomics

Brazilian Journal of Economics

Internet 2

National Coordination Office for Computing, Information and Communications

Virtual Institute of Information

Local Switch Congestion

Internet and Telephony

Boardwatch Magazine

Network Economics
Intermediaries and Cybermediaries
Network Externalities
Internet Engineering and Planning Group
Interoperability and the Economics of Information Infrastructure
USA Global Interoperability for Broadband Networks
Interoperability and the National Information Infrastructure
NSF Networking and Communications Research
Telegeography
Internet Architecture Today
Policy Gateway
Berkman Center for Internet and Society
Internet Public Policy Network
Global Internet Project
Internet Law and Policy Forum
FindLaw
New Information Infrastructure: Strategies for U.S. Policy
Taxes and Electronic Commerce
Alliance for Public Technology
LawResearch
Telepathic
Legal Links in Electronic Commerce and Interactive Entertainment

These specific Web sites are recommended for state-of-the-art analysis and benchmarking in corporate strategic planning departments:

www.trac.org
www.trackingcommunications.com
www.ratexchange.com
www.vista-cctv.com
www.wirelessweek.com
www.broadband-guide.com
http://smartone.svi.org
www.telstra.com/au/infocommunications
www.benton.com
http://china.si.umich.edu/telecom/telecom-info.html
www.ukans.edu/cwis/units/coms2/tech.html
www.its.bldrdoc.gov/its.html
www.onepassinfo.com/opteldef.html

These government and industrial organizations produce primary source data on strategic management problems in the telecommunications industry. Refer to www.doc.gov for hyperlinked information to these sites and their assorted publications. Parenthetical commentary is provided by www.analysis.com and www.brint.com.

Arabsat (Arab Satellite Communications Organization)
Artis Software Corporation (OptSim: high-end optical communications systems simulator)
Batelco (Bahrain Telecommunications Company)
CIENA LightWorks (DWDM and intelligent optical networking)
Comviq (Swedish GSM operator)
Cyprus Telecommunications Authority
Etisalat (United Arab Emirates)
Global One (Jordan)
Graphnet, Inc. (datacommunications services)
GTO: General Telecommunications Organization, Sultanate of Oman
IBM Public Affairs (Living in the Information Society)
IDN: Inter Digital Networks (broadband and metropolitan network operator in the United Kingdom)
Itlissalat Al Maghrib (Morocco)
IWE: Institute for Wireless Education (training courses in wireless technology)
Jinny Paging (Lebanon)
Kuwait: Ministry of Communications
McKinsey Quarterly E-commerce Library
McKinsey Quarterly Telecoms Library
McKinsey & Company
MobiNil (Egypt)
MPT Lebanon (Lebanese Republic Ministry of Posts and Telecommunications)
MTC (Kuwait)
New Telecom Policy 1999 (India's policy statement)
Newentrants.com (New Entrants Tracking Service, maintained by Analysys Ltd.)
Noyes Fiber Systems (manufacturer of fiber optic test equipment)
ntl Group Ltd. (U.K. operator)
OECD: Electronic Commerce.url
POLE-TEC INC (line test equipment)
Q-TEL (Qatar Telecom)
Ravi Visvesvaraya Prasad & Associates (telecommunications, datacommunications, networking, and software consultants in India)
SATRA (South African Regulator)
STC: Saudi Telecom Company

Stratecast Partners (communications industry strategic intelligence)

Tele2 Mobil (Swedish GSM operator)

TeleCap Communications (investment information on new telecommunications companies)

Telecom Business Online (information source for carriers, convergence, competition, and next-generation networks)

Telecommunications, Inc. (Comprehensive Communications Consulting)

TeleYemen

Telsim (Turkey)

Thuraya Satellite Telecommunications Company

TSS: Total Software Solutions, Inc. (wireless and wireline billing systems)

Tunisie Telecom (Tunisia)

Turkcell (Turkish GSM operator)

web2FAX (tool for faxing over the Internet, by Graphnet, Inc.)

Butterworth-Heinemann (publisher of books and electronic products for students and professionals in the communications and telecommunications industry)

Comarco, Inc. (wireless communications products, engineering and management)

CommuniTech (headsets, speakerphones, audioconferencing, and videoconferencing systems)

DSD: Distributed Software Development (billing and customer care solutions for paging operators worldwide)

Dwyer Associates (courses in telecommunications policy and regulation)

e.biz (guide to electronic business, from *Business Week Online*)

Emil Kipriotis SA (manufacturer of communications terminals)

EnterSource (enterprise resource management in telecommunications, energy, and utilities)

FirstLink Communications, Inc. (provider of wireless LAN/WAN/MAN, ISP, voice IP, modems, and T1/E1 communications equipment)

Flagship TeleSystems (billing systems and customer care solutions for telecommunications)

FRIAP: French Internet Access Providers list (listing of French ISPs)

Global Cellular and Paging (wireless products, services, and accessories)

Global Light Telecommunications, Inc. (emerging international carrier)

Grayson Wireless (wireless measurement systems)

Highpoint Telecommunications, Inc. (emerging international carrier)

iaxis Corporation (European carrier's carrier)

RIT Distance Learning (on-line courses at the Rochester Institute of Technology)

SAFCO Technolgies, Inc. (data collection and analysis of wireless networks)

Salient 3 (designs, manufactures, and markets equipment for use by communications network operators worldwide)

Studio Cazzani (telecommunications consultancy focusing on Italy)

The Helein Telecommunications Group (management, operational, and technical consulting to the telecommunications industry)

TMNG: The Management Network Group, Inc. (management and operational consulting company for the telecommunications industry)

TONEX (telecommunications, Internet, and communications training and consulting service)

Trend Communications (hand-held test equipment for ISDN, Frame Relay, and ATM, brand name "aurora")

Videoconference.com (forum for videoconferencing professionals)

World Telecom

3rd Millennium (two-way radio, telecommunications, IT, and broadcast equipment)

"A Global Marketplace for SMEs" (paper outlining potential benefits of e-commerce to SMEs)

ACEL: Atlantic Communications Enterprises Limited (turnkey supplier of telecommunications products and services)

ACIF: Australian Communications Industry Forum (standards, industry codes, and industry support services)

Advanced Datacom Systems (network management, monitoring, and performance management solutions)

AITEC: Australian Information Technology Engineering Centre (joint venture between four Australian universities)

Alan Warwick Consulting Ltd. (U.K. consultant in IT and telecommunications procurement)

BCL (international and long-distance services in St. Petersburg and Northwest Russia)

CBSI: Complete Business Solutions, Inc. (IT and telecommunications consulting services)

Cell-Tel International, Inc. (second market cellular and infrastructure equipment)

Cyber Digital (digital data and voice telephone switches, PBXs, class VI and V switches)

Direct Visual Limited (videoconferencing supply and support)

El Homo Comunicatus (Spanish cable operator)

Fiber Optics Online (news, analysis, product information, and virtual marketplace)

GE American Communications (Americom) (satellite service provider)

GFRC Shelters (modular concrete buildings for the telecommunications industry)

Gilat Interactive Distance Learning (IDL systems)

Gilat Satellite Communications, Inc. (international links and VSAT services)

Glocalnet (operates pan-European voice-over-IP network)

"Guaranteed Electronic Markets" (paper proposing new ways of organizing e-commerce transactions)

ICT electronics (test and measurement solutions for telecommunications based on SDH, PDH, ATM, and IP)

InnoWave Wireless Systems (fixed wireless access solutions)

International Chamber of Commerce (Commission on Telecommunications and Information Technologies)

Lynk (remote LAN/WAN and IP telephony solutions)

MapInfo (spatial solutions for telecommunications strategy issues)

Metrodata (high performance ATM and non-ATM access products)

Microtronix (telephony test equipment for testing to ITU-T, IEEE standards, and many country-specific requirements)

NextCall Telecom plc (U.K. residential service provider)

OFTEL: Inquiry into Electronic Commerce (trade and industry select committee report)

Premises Networks Online (news, analysis, product information, and virtual marketplace)

PTT: Providers of Telecommunications Training (computer-based training courses for the telecommunications industry)

Quallaby Corporation (analysis and reporting solutions for network service providers)

RSA Data Security (encryption and authentication technologies)

Taiwan Telecommunication Industry Company Ltd. (coin, card payphone system, digital pair gain system, facsimiles)

Telecommunications Engineering Technology (undergraduate degree at RIT: Rochester Institute of Technology)

TelephonyWorld.com (on-line resource for the computer telephony and communications industry)

Telesoft Design Ltd. (design and manufacture of medium-scale multiprotocol converters, CTI OEM boards, and protocol test solutions)

Telution (customer care and billing solution for competitive communications providers)

Test and Measurement.com (news, analysis, product information, and virtual marketplace)

The Communications Group (forum for communications lawyers, consultants, regulators, policy-makers, and other professionals)

Thomcast Group (multimedia equipment and AM, DAB, FM, MMDS, and TV transmitters)

TLA Communications (international telecommunications access, with a variety of enhanced calling services)

UNCITRAL: United Nations Commission on International Trade Law

VersaTel Telecom International (Benelux service provider)

Webmonkey (e-commerce tutorial)

WinSoft Corporation (custom software solutions for electronic test systems)

Wireless Design Online (news, analysis, product information, and virtual marketplace)

BD Software (customer acquisition, loyalty, credit control, billing, and management software solutions)

Bertsch International (Internet-based learning consultancy specializing in teaching the use of technology in the classroom)

BICC Communications (cable, cable systems, and services)

BSW Telecoms (business and operations software solutions for telecommunications network and service providers)

BTG: Banner Telecom Group plc (mobile telephone accessories)

Clear Communications (intelligent surveillance applications for network quality and performance monitoring)

Comtest Laboratories (Australian telecommunications and electrical safety compliance testing laboratory)

CrossKeys (software solutions to telecommunications service providers for network information and control)

CTIA: Cellular Telecommunications Information Association

David's Sonet/SDH Page (SONET and SDH information, especially on testing)

EAP (Irish provider of data transport services over wireless point-to-multipoint broadband network)

Econet Wireless (GSM operator in Zimbabwe)

e-Summit (monthly, on-line report on call center strategy published by MW Productions)

Foxcom Wireless (fiber optic transmission systems for wireless services)

Galaxy Power, Inc. (intelligent power management solutions)

GTB: Global Telecoms Business

Intec Systems (interconnect billing systems developer)

Integration Technologies (cable telephony upgrades, cable modems, SONET, and WDM networks)

In-Touch Associates (business strategy, marketing, sales, and training through to network planning, engineering, and operations)

INTUG: International Telecommunications User Group

iPass Corporate Access (Internet connectivity for mobile employees)

Israel Ministry of Communications

JSI Capital Advisors (investment banking, industry information, and education services to the independent communications provider industry)

K & M TEL, INC. (refurbished, new and used transmission, central office, switching, power, outside plant equipment, installation, and deinstallation)

Motech Industries Incorporated (general electronic test and measurement equipment)

NETMANSYS (network management software products)

Norcom, Inc. (telecommunications and Internet service provider)

Nursat (licensed carrier of voice, data, video, and Internet services in Kazakhstan, with a satellite network)

OSI: Objective Systems Integrators (service and network management solutions to the worldwide communications industry)

Paralan Corporation (products for the SCSI enhancement marketplace)

Paul Budde Communication (telecommunications and information highways consultancy services and publications)

Payphone Connection (public communications solutions and equipment in Canada)

PCIA: Personal Communications Industry Association

PrimeTEC (European-based carrier's carrier)

Protek (network management specialists who develop Cortex, a network management tool)

Ratekona (Lithuanian broadcasting consultants)

Regulatel (association of regulators in Latin America)

Russell and Partners (telecommunications recruitment consultancy)

Satyam Telecom (customer care and billing, CTI, mediation, clearinghouse, migration services, etc.)

Senao International Co. Ltd. (cellular phone accessories)

Starnet Engineering Ltd. (telehousing and technical support services)

Tait Electronics Ltd. (mobile radio equipment)

TDG: TeleCommons Development Group (international telecommunications and networking consulting firm specializing in solutions for rural areas)

Technology Investment Partners (European corporate finance and project management for growing technology businesses)

Telecom Consulting Group, Inc. (interconnection agreement negotiations and consulting on other telecommunications service and network design issues)

Telecom Solutions (low rate long-distance services, toll-free numbers, international calling)

Telecommunications Management Program (Oklahoma State University)

The GSM Consultancy (consultancy services to the international wireless telecommunications industry)

TSI: Telecom Services International (telecommunications engineers supplying
integrated digital radio link systems for voice and data)

VoiceTrak (voice mail, call accounting, and fax back software)

Wireless 2000 Engineering Advisory Group (professional engineering services
to wireless communications industry)

The following trade and academic journals contain company-specific and in-
dustry-specific articles on telecommunications strategic management (note that
these publications can be accessed through www.analysys.com).

American Telesis, Inc. (wholesaler of private lines or data lines to resellers and
agents)

Arthur Andersen (global multidisciplinary professional services firm)

Band-X: the bandwidth exchange (independent virtual market for international
wholesale telecommunications capacity)

Bloomberg Web site

Calculus Solutions Ltd. (customer care and billing product set for pan-
European emerging and mature telcos)

CalTech International Telecom Corporation (CLEC building VoIP network,
operating in California, Illinois, Colorado, and Arizona)

Cape Saffron Online (trader in international traffic)

CINTEL: Centro de Investigacion de las Telecomunicaciones (nonprofit organization offering consultancy on the strategic and technological aspects
of telecommunications in Colombia and other Latin American countries)

CMMC: Customer Management Multimedia Call Centre (association for
companies using call center)

Comparable Performance Indicators (comparison of quality of service between
U.K. telecommunications companies)

DACOM (Korean operator)

Distance Education (World Wide Web Virtual Library)

E-TECH, Inc. (two-way cable modems, HDSL, ADSL, and transmission
products)

FCS: Federation of Communication Services (U.K. trade association for the
mobile telecommunications industry)

Federal Gateway (consolidates federal sites and provides pre-searched information)

Fujitsu Broadband Services (high-capacity phone systems, voice messaging
products, ATM equipment, SONET, IP telephony, call center products)

General Cable plc (U.K. cable company covering Yorkshire, West London, and Birmingham)

Global Telecom International (distributors of PhonClub: flat-rate long-distance service on IP)

Infozech (billing software services for telecommunications service providers)

Intelecon Research & Consultancy Ltd. (focus on rural telecommunications, emerging markets, and developing countries)

JSCSYS (Canadian consultancy and systems development for telecommunications service providers)

Kuyaké Technologies, Inc. (strategic telecommunications consulting services)

Loral Space & Communication

myfreeoffice.com (long-distance, calling cards, and Internet access)

NCI: Net Communications, Inc. (global IP-based voice and data services)

Phillips Tarifica (consultancy specializing in tariffs)

Phillips TelecomWeb (news, research, and analysis in the telecommunications industry)

Primal Systems (software solutions to aid customer acquisition and retention in the telecommunications industry)

Priority Communications, Inc. (national telephone and voice processing solutions)

Profundus Consulting Ltd. (telecommunications marketing consultancy with a European focus)

RomTelecom (Romanian operator)

Shiron Satellite Communications (satellite communications products including satellite L-band modems, VSATs, and InterSKY for two-way fast Internet via satellite, based in Israel)

Shunra Software Ltd. (developed The Cloud, a tool that simulates Internet and IP WAN intranet conditions)

Stormin Protection Products, Inc. (surge suppressor for communications equipment)

SureCom Solutions (digital business telephone and voice processing solutions)

TCIL: Telecommunications Consultants India Ltd. (telecommunications consultancy and engineering company)

Tele2 UK (high-speed data connection over advanced digital wireless network)

Telecom Today (telecommunications news and information)

Telekom-Control (Austrian Regulator)

TelePost Office 2000 (virtual office provider)

WebTelecom (Internet phone numbers that support calls from any H323 standard phone)

XCP, Inc. (phone card printer vending machines)

@Home Network (Internet access over cable modem)

1st Family ("free" Internet access bundled with flat-rate long-distance IP telephony)

ACTEL Enterprises Pty Ltd. (strategy consultants specializing in cellular, WLL, and VAS)

Alcatel (site for Alcatel's space activities)

Analysys Satellite Communications Database

Andersen Management International A/S (management consultants specialized in telecommunications)

Arcome SA

Atlantic Fibre Optics (manufacturer of fiber optic test equipment)

BABT: British Approvals Board for Telecommunications (independent European approval body for telecommunications equipment, as well as offering training and consultancy services)

Barry Reichelt Telecommunications Consulting Services Pty Ltd. (audits, tenders, Y2K, training)

BiblioNet (IT and telecommunications resources)

BiblioNet: Standards Direct (ordering service for standards documents)

BTI Communications Group Ltd. (equipment, local, long-distance, Internet, computer, telephony, and wireless)

Business Communications Ltd. (global consultancy specializing in telecommunications and media convergence)

Cable Community (representation group for Cable & Wireless customers)

Cadzow Communications Consulting Ltd. (consultants on TETRA, communications security, and technology management)

CD3 Multimedia (graphic designers for multimedia)

Cell Tech Ltd. (transmission, RF on fiber, GSM /DCS turnkey solutions)

Centre for Telecommunication Networks (University of Durham, United Kingdom)

CLEC News (on-line telecommunications magazine)

CLEMESSY (telecommunications test systems)

Click GSM (mobile phone network in Egypt)

Comisión de Regulación de Telecomunicaciones (Colombian Telecommunications Regulatory Commission)

Connogue Limited (radiocommunications mobile and broadcasting, spectrum management, regulatory policy, and strategy)

CUT: The Campaign for Unmetered Telecommunications (U.K. lobbying group)

Cutler & Company (specialists in convergence)

Datapoint (data, voice, and video integration products)

Demon Internet

Domosys Corp. (communications solutions for residential and commercial LANs)

DTS: Dynamic Technology Systems, Inc. (value-added service provider of tele-communications and information technology)

Economic TeleDevelopment Forum (forum on relationship between telecom-munications infrastructure investment and enhanced community eco-nomic development)

EHPT: Ericsson Hewlett-Packard Telecommunications AB (telecommunica-tions management solutions especially in fixed-mobile convergence)

E-Journal (The World Wide Web Virtual Library: Electronic Journals)

El.pub (telecommunications) (telecommunications section of EC-sponsored R&D news and resources site)

Elektrobit Group (engineering consultancy for wireless communications)

EQUIP'TRANS (French data transmission equipment manufacturer)

ESRI (GIS tools for geographical analysis in telecommunications)

European Electronics Limited (transmission and reception solutions: terres-trial, satellite, microwave, mobile, LAN, and WAN)

European Interconnect Atlas (country-by-country description of the status of interconnection in Europe, created by Analysys Ltd. for the European Commission)

Evans Associates (telecommunications network planning, design, and imple-mentation)

FreBon International Corporation (implements point-to-point and multipoint videoconferencing networks)

Genesys Telecommunications Laboratories, Inc. (provides open CTI frame-work and software for telecommunications including call routing, report-ing, and multimedia applications)

Global Internetworking (provider of local and long-distance high-speed band-width and leased line services)

Infrared Communication Systems, Inc. (infrared systems for data networking and telecommunications applications)

Insight Research Corporation (market research and competitive analysis exclu-sively for the telecommunications industry)

InterSatCom, Inc. (satellite Internet provider)

IPL Telecommunications (systems integrator and software engineering)

iSpring Interactive (multimedia consulting, design, and production)

ITU Regulatory Colloquium (news and discussion forum on telecommunica-tions regulation, free access to ITU Regulatory Colloquium reports)

LCPI

Linx Communications (one-number follow-me services and virtual PBX solu-tions)

Matra Marconi Space

Min-X (marketplace for IP telephony minutes)

Mitel (manufacturer of telecommunications products and semiconductors)

Monte R Lee & Co (telecommunications engineering company)

Mott MacDonald (telecommunications engineering arm of multidisciplinary consultancy)

MSUA: Mobile Satellite Users Association (forum for users, and good industry links)

National Communication Service, Inc. (full telecommunications provision for multisite companies)

O-m6 Telecommunications (manufactures of fiber optic cable assemblies)

ONCTek LLC (network consulting services)

Optimus (GSM operator in Portugal)

Pacific Access (technology services and solutions)

Phillips Business Information, Inc. (satellite, cable, and telecommunications information)

Phillips Omnicom Training (telecommunications and datacommunications training, including distance learning programs)

Phone Savings: ADM & Associates (full-service telecommunications brokers)

PROD-EL: Prodotti Electtronici SpA (radio simulcast networks analog and digital, radio terminals, and turnkey systems)

Route-2 (IP telephony rate broker)

Satellite Today (news service by Phillips TelecomWeb)

SomerData (digital communications data capture and recording systems)

Superintendencia de Telecomunicaciones (regulator in Guatemala)

Tartanchase Computer Services Ltd. (supplies datacommunications and computer networking equipment)

T-COM (T1, T3, and Sonet test equipment manufacturer)

Technical Communications Corporation (encryption systems for data, voice, and fax network communications security)

Teldat (internetworking, routers, transmission equipment [XDSL, baseband], multimedia, etc.)

Telecommunication and Informatics Unit (World Bank work program)

Telecommunications Industry at a Glance (ITU rankings of operators and country subscriber data)

Tele-M@rket (news and discussion forum for the call center business)

Teleplan Italy (Italian consultancy using billing and tariff analysis to minimize telephone expenses for clients)

Telsoft, Inc. (manufactures DTMF and telephone line analyzers)

The Cell Site (information resource on Digital Mobile Telephone Systems)

Transpower Technologies, Inc. (magnetic components for Ethernet, telecommunications, datacommunications, and power applications)

Ultra Active Technology Ltd. (videoconferencing supplier in the Far East)

UUNET

Vega Group (systems and software engineering consultancy working in the space and telecommunications industry)

Vital Connections (service provider also offering autodialers, voice-mail systems, message on hold systems)

Warren Publishing, Inc. (daily telecommunications news)

Appendix D:
WTO Basic Telecommunications Services Agreement: Summary of Country Commitments (73 Countries)†

All commitments are effective as of February 5, 1998, unless otherwise noted. Unlimited foreign investment permitted unless otherwise noted. The "reference paper" is a set of pro-competitive regulatory principles.

Antigua & Barbuda. Market access and national treatment for enhanced service suppliers, mobile, PCS, trunked radio, and closed user groups. Mobile and fixed satellite transport services may be provided through arrangements with exclusive operator. Market access and national treatment for all other services as of 2012. No commitments on regulatory principles. MFN exemption for treatment accorded to CARICOM members with regard to terrestrial-based mobile services.

Argentina. Market access and national treatment for domestic data and telex, domestic and international fax, paging, trunk radio, and leased circuits (with a preference given to existing supplier until 11/8/00). Duopoly for cellular. Access for PCS to be decided in "The light of present and future needs." Market access and national treatment for all other services as of 11/8/00, including those provided via nongeostationary, non-fixed satellite services. Adopted the reference paper. MFN exemption for access to geostationary fixed satellite systems.

---

†The information contained in Appendix D has been taken from the U.S. Department of Commerce (March 2000).

Australia. Market access and national treatment for all services. Majority Australian ownership required in Vodafone. Foreign ownership in Telstra of up to 35% of initial sale (11.7% of total equity) allowed, with a limit of 5% per individual or associated group of foreign investors. Also, limits on the amount of individual foreign ownership in Optus. Adopted the reference paper.

Austria. Market access and national treatment for all services. Adopted the reference paper.

Bangladesh. Two private operators provide local and long-distance services in competition with government operator. Four licenses issued for cellular. All service suppliers must use facilities of network operator, including VSAT and gateway earth station services. No regulatory commitments. MFN exemption on accounting rates. Market access and national treatment for telecommunications terminal equipment rental, maintenance, connection, repair, and consulting services.

Barbados. Market access and national treatment for enhanced services (including Internet), VSAT services, and mobile terrestrial and satellite services. All other services reserved to exclusive suppliers until 1/1/12. Adopted the reference paper.

Belgium. Market access and national treatment for all services. Adopted the reference paper.

Belize. Market access (but not national treatment) for trunked radio services and teleconferencing as of 1/1/03. Trunked radio services must be provided through joint venture with Belizean national. Market access (but not national treatment) for enhanced services and paging as of 1/1/08 over the facilities provided by the incumbent operator. Paging must be provided through joint venture with Belizean national. Adopted the reference paper.

Bolivia. Market access and national treatment for all services as of 11/28/01. Market access and national treatment as of 1/1/98 for local data, telex, telegraph, fax, closed user groups, and mobile services (cellular, paging, PCS, mobile satellite). Local services limited to existing companies. Adopted limited set of regulatory principles.

Brazil. Market access and national treatment for enhanced services, paging and nonpublic domestic and international services for closed user groups. Duopoly for analog/digital cellular mobile service. One hundred percent foreign ownership of nonpublic service providers; 49% limit on cellular and satellite until 7//99, thereafter none. Commitment to bind outcome of future reform legislation, which is expected to cover public and nonpublic services within 1 year of enactment. Use of foreign-licensed GSO space segment facilities allowed whenever they offer better technical, operational, and commercial conditions. MFN exemption for telecommunications services supplied for distribution of radio or television programming for direct reception by service consumers. *Has not yet ratified the Agreement.*

Brunei. Local service reserved for up to 10 years after privatization of government operator. Two operators have exclusivity in international services until 2010. New licenses in cellular may be issued in 2010. Adopted the reference paper.

Bulgaria. Market access and national treatment for closed user groups and VSAT (not connected to the public switched network), data, paging, mobile data. Market access and national treatment on 1/1/03 for resale of public voice, telegraph/telex, and cellular voice. Market access and national treatment for all other services and facilities on 1/1/05. Adopted the reference paper.

Canada. Market access and national treatment (with a limit of 46.7% foreign ownership) for all services except fixed satellite services and submarine cables. Market access and national treatment for fixed satellite services in 2000, with 100% foreign ownership, and submarine cable landings as of 10/1/98, with 100% foreign ownership. One hundred percent foreign ownership also permitted for resellers. No routing requirements on mobile satellite services between points in Canada and between Canada and points in the United States. Routing requirements on international mobile satellite services ends on 10/1/98. Routing requirements on international services ends on 1/1/00, except for fixed satellite services between Canada and points in the United States. Routing requirements on all satellites ends as of 3/1/00. No date fixed for end to routing restrictions on Canada-to-Canada points (other than by satellite). Licenses to operate earth stations for provision of Canada-United States fixed satellite service may be limited until 3/1/00. Adopted the reference paper. [Note: Through domestic regulatory action, Canada has abolished routing requirements on international services, including those between Canada and the United States, and restrictions on Canada-to-Canada points (other than by satellite).]

Chile. Market access and national treatment for long-distance and international wireline and wireless (including satellites). No commitment on local service, one-way transmission by satellite of direct-to-home, direct broadcast satellite, and digital audio services. Adopted the reference paper.

Colombia. Market access and national treatment for private networks for voice and data, paging, trunked radio, geostationary satellite, and local public voice services. Market access and national treatment for long-distance and international services but subject to an economic needs test. Seventy percent foreign ownership permitted for all services. Market access and national treatment for cellular services on 9/1/99 and PCS on 1/1/00 but subject to an economic needs test. Adopted the reference paper.

Côte D'Ivoire. Market access and national treatment for data, fax, telegraph, private circuits, mobile (voice and data), and PCS. Open for all satellite services except domestic and international public voice service. Market access and national treatment for local, long-distance, and international voice and telex on 1/1/05. Adopted the reference paper.

Cyprus. No commitments made (except for "consumption abroad") for domestic and international services pending review for regulatory study and decision of Council of Ministers on specific steps to gradually liberalize the market.

Czech Republic. Market access and national treatment for domestic and international data, telex, telegraph, fax, leased circuits, closed user groups, domestic mobile services, and video and audio transport services. Market access and national treatment for domestic and international voice services on 1/1/01. Adopted the reference paper.

Denmark. Market access and national treatment for all services. Adopted the reference paper.

Dominica. Market access and national treatment for nonpublic data, fixed and mobile satellite systems, enhanced services (including Internet access), teleconferencing and telecommunications equipment sales, rental, maintenance, connection, repair, and consulting services. Adopted the reference paper. *Has not yet ratified the Agreement.*

Dominican Republic. Market access (but no national treatment) for all services (but must establish a commercial presence). Adopted the reference paper.

Ecuador. Market access and national treatment for domestic cellular services. No commitments on regulatory principles.

El Salvador. Market access and national treatment for all services. Adopted the reference paper.

Finland. Market access and national treatment for all services. Adopted the reference paper.

France. Market access and national treatment for all services. Indirect investment is unlimited, but there is 20% direct investment limit for radio-based networks and limit on investment in France Telecom. Adopted the reference paper.

Germany. Market access and national treatment for all services. Adopted the reference paper.

Ghana. Duopoly for domestic and international voice for 5 years, after which additional suppliers may be licensed. Market access and national treatment for data, telex/telegraph, fax, closed user groups (not connected to public switched network), Internet access services (excluding voice), teleconferencing, trunked radio, domestic fixed satellite, and global mobile satellite services but must provide through joint ventures with Ghanaian nationals. Market access and national treatment for telecommunications equipment rental, maintenance, connection, repair, and consulting services through joint venture with Ghanian nationals. Adopted reference paper.

Greece. Market access and national treatment for all services other than public voice telephony and facilities-based services, which will be open on 1/1/03. Adopted the reference paper.

Grenada. Market access and national treatment for enhanced services and mobile and fixed satellite services through arrangements with the incumbent operator. Market access and national treatment for all other services as of 1/1/06. Adopted reference paper.

Guatemala. Market access and national treatment for all services. Adopted the reference paper. *Has not yet ratified the Agreement.*

Hong Kong. Local wireline and wireless network services limited to current four providers. Market access and national treatment for switched resale, international data and fax by resale, call back, resale-based virtual private network service (not connected to the public switched network), mobile satellite services, and self-provision of external satellite circuits by a company or closed user group permitted. Hong Kong now allows external service based competition (such as international simple resale) and will allow international facilities based competition on 1/1/00. Adopted the reference paper.

Hungary. Market access and national treatment for data, telex/telegraph, fax, leased circuits, and satellite services (other than public voice). Market access and national treatment for international and long-distance public voice as of 1/1/03 and local public voice as of 1/1/04. Public land mobile services limited to one operator for NMT 450 MHz; two operators for GSM until 2003. Paging limited to one analog operator and two for pan-European service. One hundred percent foreign ownership permitted except 75% limit for Matav and Antenna Hungaria Rt. Adopted the reference paper.

Iceland. Market access and national treatment for all services. Adopted the reference paper.

India. Duopoly for local and long-distance wireline, duopoly for cellular, and 25% foreign investment limit. No resale allowed. Additional licenses may be issued based on economic needs test. Will review market access for national long-distance in 1999 and international in 2004. GSM technology mandated for cellular services. MFN exemption for accounting rates. Adopted some of the reference paper.

Indonesia. Local services (public switched voice, circuit switched data, and teleconferencing) limited to provision by PT Telkom and five regional joint operation scheme operators. Long-distance exclusive to PT Telkom. Packet switched public data provided by three operators on a nonexclusive basis. Domestic telex/telegraph services must be provided through joint venture or joint operation. Selection of new mobile cellular and PCS operators is subject to economic needs test and frequency availability and must be through joint venture (and, in case of PCS, with state-owned company). New paging operators subject to economic needs test, frequency availability, and public interest test and a joint venture or joint operation is required. Duopoly for international public switched voice, circuit switched data, teleconferencing services, and Internet access services (until 2005). Duopoly for domestic and international satellite services and international telex/telegraph. Call back prohibited. Thirty-five percent foreign ownership limit. Adopted the reference paper. Government will review number of suppliers permitted to provide international services in 2005, long-distance services in 2006 and local services in 2011.

Ireland. Market access and national treatment for all services other than public voice telephony and facilities-based services which, will be open as of 1/1/00. Mobile operators may interconnect with networks other than the state-controlled company to supply international services from 1/1/99. Adopted the reference paper. (Note: The Irish Prime Minister has announced that Ireland's market would be fully liberalized on 12/1/98.)

Israel. Market access and national treatment for private networks, cellular, paging, satellite voice and data, and international fax, with limit of 80% foreign investment for cellular. Market access and national treatment for all other services as of 1/1/02 with limitation of 74% foreign ownership. Permits 80% foreign investment for wireless service providers. Adopted the reference paper.

Italy. Market access and national treatment for all services, with limitation on investment in Stet. Adopted the reference paper.

Jamaica. Market access and national treatment for data transmission and enhanced services, including Internet access, mobile data, PCS and paging services, satellite-based video transmission services (excluding video telephony), and telecom equipment, sales, rental, maintenance, connection, repair, and consulting services. Market access and national treatment for trunked radio, teleconferencing and international voice, data and video transmission services supplied to firms involved information processing located within free zones but restrictions on use of facilities and interconnection apply. Call back services and refile are not permitted. Voice telephony reserved to exclusive supply until September 2013. Adopted the reference paper.

Japan. Market access and national treatment for all services, with a 20% limit on foreign investment in KDD and NTT. Adopted the reference paper.

Korea. Market access and national treatment for all services. Foreign investment limitations. Thirty-three percent investment limit on facilities until

1/1/01 then 49% (with individual shareholdings limited to 33% for wireless and 10% for wire-line); 20% for Korea Telecom until 1/1/01 then 33% with individual shareholdings limited to 3%; 49% for resellers until 1/1/01, 100% thereafter. Adopted the reference paper.

Luxembourg. Market access and national treatment for all services. Adopted the reference paper.

Malaysia. Market access and national treatment for all services only through acquisition of up to 30% of the shares of existing licensed public telecom operators. No resale permitted. Adopted some of the reference paper.

Mauritius. Market access and national treatment for fax, paging, private mobile radio, GMPCS and telecommunications equipment rental, sales, and maintenance services. Market access and national treatment for other services market by 2004 with no foreign investment restrictions other than company registration requirement. Committed to adopt reference paper in future.

Mexico. Market access and national treatment for all services except for requirement to use Mexican satellites for the provision of domestic services until 2002. One hundred percent foreign ownership for cellular services; 49% for all other services. Adopted the reference paper.

Morocco. Market access (but no national treatment) for domestic packet switched data, frame relay, mobile, paging, and PCS. Market access to "point-to-point voice telephone service" as of 1/1/02. Adopted own set of regulatory commitments on interconnection and regulatory authority.

Netherlands. Market access and national treatment for all services. Adopted the reference paper.

New Zealand. Market access and national treatment for all services, except limit of 49.9% foreign investment in NZ Telecom. Adopted the reference paper.

Norway. Market access and national treatment for all services. Adopted the reference paper.

Pakistan. Market access and national treatment for domestic data, VSAT, telex, fax, video conferencing, tele-medicine, tele-education terminal end services. Market access (but no national treatment) for other services on 1/1/04. Adopted own regulatory principles. MFN exception on accounting rates.

Papua New Guinea. Will review exclusivity for all telecom services 2 years prior to expiration of license in 2002. Call back not allowed. Adopted the reference paper. *Has not yet ratified the Agreement.*

Peru. Market access and national treatment for local data, telex/telegraph, fax, private leased circuits, cellular, PCS, paging, trunking services, and mobile data. Market access and national treatment for all other services as of June 1999. Adopted the reference paper. (Note: Peru ended its voice monopoly on 8/1/1998.)

Philippines. Market access and national treatment for facilities-based services except satellites, subject to a limit of 40% foreign ownership. No commitment on resale or leased circuits/closed user groups. Partial adoption of the reference paper. *Has not yet ratified the Agreement.*

Poland. Market access and national treatment for local public voice (wireline), domestic and international data, international fax, cellular, pan-European paging systems, and international private leased circuits. Market access and national treatment for telex and telegraph as of 1/1/00. Market access and national treatment for all other services as of 1/1/03, subject to a limit of 49% foreign ownership for international and domestic long-distance services including cellular. Public digital cellular must be provided thru TPSA's international network until 1/1/03. Adopted the reference paper.

Portugal. Market access and national treatment for resale of all services. Market access and national treatment for alternative facilities services as of 1/1/99. Market access and national treatment for public voice telephony, telex and telegraph as of 1/1/00. Adopted the reference paper.

Romania. Market access and national treatment for nonpublic voice, data, telex/telegraph, fax, paging, and VSAT (not connected to PSN). Market access for analog cellular as of 4/1/02. Market access and national treatment for all other services as of 1/1/03. Adopted the reference paper.

Senegal. Market access and national treatment for all services (except cellular and mobile satellites) as of 2006, although may be earlier due to government commitment to consider additional access in 2003. Three cellular providers as of 1998. Adopted the reference paper.

Singapore. Market access and national treatment for local and international resale, resale of cellular mobile and paging services, paging, mobile data, and trunked radio. Market access and national treatment for all other services as of 4/1/00. Seventy-four percent (combined direct and indirect) foreign ownership permitted. Adopted the reference paper.

Slovak Republic. Market access and national treatment for data, telex/telegraph, fax, private leased circuits, paging, trunked radio, domestic mobile and PCS (excluding analog cellular voice), and closed user groups (not connected to the PSN). Market access and national treatment for all other services as of 1/1/03. Adopted the reference paper.

South Africa. Market access and national treatment for paging, PCS, and trunked radio services. Duopoly for local, long-distance, international, data, telex, fax, and private leased circuits services as of 1/1/04. Duopoly for mobile but will license one more supplier within 2 years. No commitment on satellite-based services but will schedule commitments within 1 year of adoption of legislation in this area. Limitation of 30% foreign ownership. Liberalization of resale services to take place between 2000 and 2003. Government will consider feasibility of additional suppliers of public switched services and satellites services by 12/31/03. Government will consider feasibility of additional cellular suppliers by 12/31/98. Adopted the reference paper.

Spain. Market access and national treatment for all services from 12/1/98. Foreign investment limit for the government-owned firms (Telefonica and Retevision). Adopted the reference paper.

Sri Lanka. Market access and national treatment for data, paging, and mobile cellular but number of operators limited. Duopoly for international services allowed as of 1/1/00. Foreign ownership in excess of 40% requires government approval. Adopted the reference paper. MFN exception for accounting rates.

Suriname. Market access and national treatment, using leased lines, for data, Internet and Internet access services, closed user groups, mobile data, paging, trunked radio services, telecom equipment sales and rental, maintenance, connection, repair, and consulting services. Foreign investment limited to 40%. Market access and national treatment for other services as of 1/1/03. Adopted the reference paper.

Sweden. Market access and national treatment for all services. Adopted the reference paper.

Switzerland. Market access and national treatment for all services. Adopted the reference paper.

Thailand. Market access and national treatment for all services as of 1/1/06. Until then, market access for local services provided only with case-by-case cabinet approval. Foreign investment limited to 20%. No regulatory commitments.

Trinidad & Tobago. Market access and national treatment for enhanced services, trunked radio, cellular, mobile data, paging, PCS, telecommunications equipment sales, rental, maintenance, connection, repair, and consulting services. Market access and national treatment for all other services as of 2010. Adopted the reference paper.

Tunisia. Market access and national treatment for telex, packet-switched data and teleconference services as of 1/1/99; mobile telephone, paging, teleconferencing and frame relay from 1/1/00 and local voice telephony as of 1/1/03. Foreign ownership limited to 49% (10% in Tunisia Telecom). No regulatory commitments.

Turkey. Market access and national treatment for mobile, paging, and private data networks, with a limitation of 49% foreign ownership. Market access and national treatment for all other services as of 1/1/06. Adopted some additional commitments on regulatory principles. MFN exception for accounting rates and fees for transit land connections and use of satellite ground stations.

Uganda. Market access and national treatment for closer user groups (resale of excess capacity is not allowed), Internet data, paging, private mobile and trunked radio, and global mobile personal satellite communications services. Voice duopoly subject to review after 2003 with no foreign investment restrictions except that neither of the exclusive providers can hold ownership in the other. Until 2003, two more cellular operators will be licensed in addition to the one existing operator. Adopted the reference paper.

United Kingdom. Market access and national treatment for all services. Adopted the reference paper.

United States. Market access and national treatment for all services, with a limit of 20% direct foreign investment. Adopted the reference paper. MFN exception for one-way satellite transmission of direct-to-home, direct broadcast satellite, and digital audio transmission services.

Venezuela. Market access and national treatment for all services as of 11/27/00. Adopted some of the reference paper.

Appendix E:
Information Economy Data

The New Economy

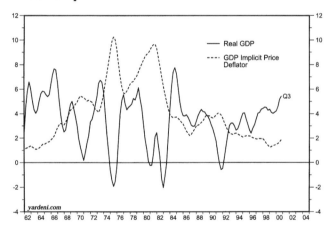

Figure E.1 GDP (yearly percent change in four-quarter average). In the Old Economy, booms led to higher inflation, causing the Fed to tighten, which eventually triggered a bust. In the New Competitive Economy (NCE), global competition keeps a lid on inflation and forces companies to cut costs and boost productivity. By offering consumers better goods and services at reasonable prices, they sell more units, especially as wages rise faster than prices. Low inflation leads to stronger growth!

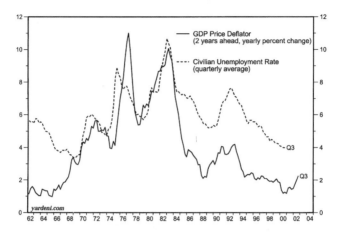

Figure E.2 Inflation and unemployment. In the NCE, increased competition has reduced inflation, which has led to lower unemployment. In the past, uncompetitive markets boosted inflation and led to rising unemployment.

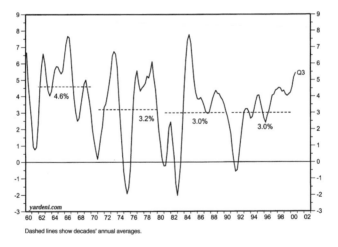

Dashed lines show decades' annual averages.

Figure E.3 Real GDP (yearly percent change in four-quarter average). The economy has the potential to grow faster than it did during the previous three decades.

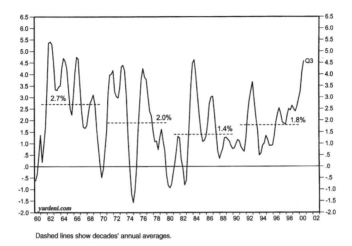

Dashed lines show decades' annual averages.

Figure E.4 Productivity: nonfarm business (yearly percent change in four-quarter average). During the current decade, productivity should grow by at least 2.5% per year, the best performance since the 1960s.

High-Tech Revolution

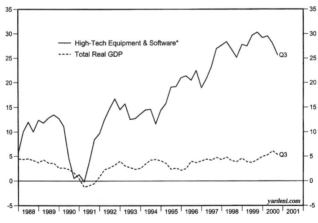

* High-tech is business purchases of high-tech equipment and software plus consumer spending on computers and software.

Figure E.5 Real GDP (yearly percent change). In the New Competitive Economy, high-tech spending now accounts for one-quarter of real GDP growth.

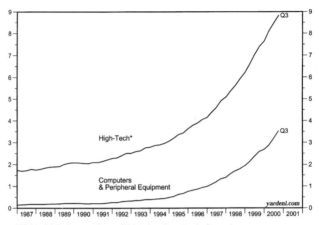

Figure E.6 Real GDP: high-tech equipment and software (as a percent of total real GDP). Spending on high-tech now accounts for almost 9.0% of real GDP, up from less than 2% in 1987. During Tech I over the last 10 years, the PC and the Internet were the New, New Things. Over the next 10 years, cheap broadband telecom will drive Tech II.

Age Wave

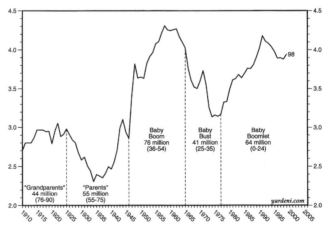

Figure E.7 Live births (in millions). In 2000, the Baby Boomers will be 36–54 years old. The numbers in parentheses are the youngest and oldest ages of group members during 2000.

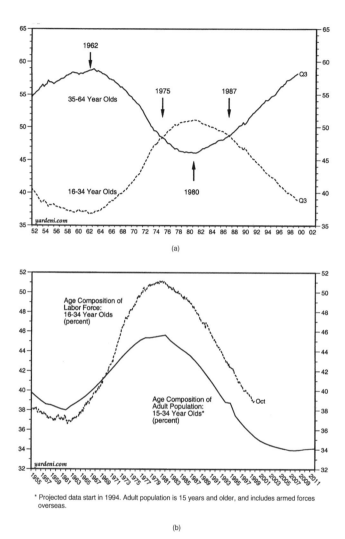

(a)

(b)

* Projected data start in 1994. Adult population is 15 years and older, and includes armed forces overseas.

Figure E.8 (a) Age composition of labor force (percent)—oldest and youngest Baby Boomers turned 16 in 1962 and 1980, respectively. (b) The age wave. In 1980, the Baby Boomers all turned 16 years or older. Maturity will increase through 2010. The labor force is the most mature since the early 1960s.

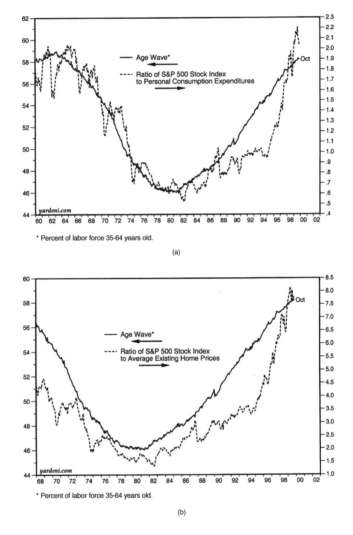

* Percent of labor force 35-64 years old.

(a)

* Percent of labor force 35-64 years old.

(b)

Figure E.9 (a) The age wave and stock prices. (b) The age wave and asset prices. Stock prices have risen faster than consumption and home prices as the Baby Boomers have aged.

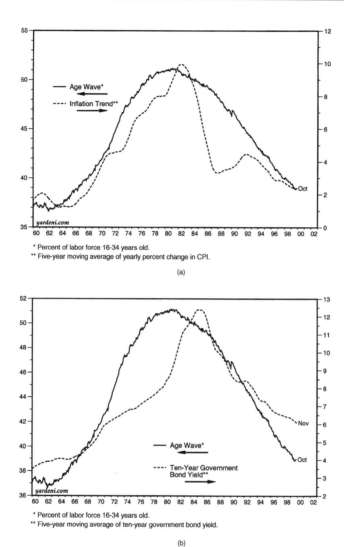

* Percent of labor force 16-34 years old.
** Five-year moving average of yearly percent change in CPI.

(a)

* Percent of labor force 16-34 years old.
** Five-year moving average of ten-year government bond yield.

(b)

Figure E.10 (a) The age wave and inflation. (b) The age wave and bond yield. Age wave is a big influence on inflation trend and bond yield and suggests both will remain low at least through the end of the coming decade.

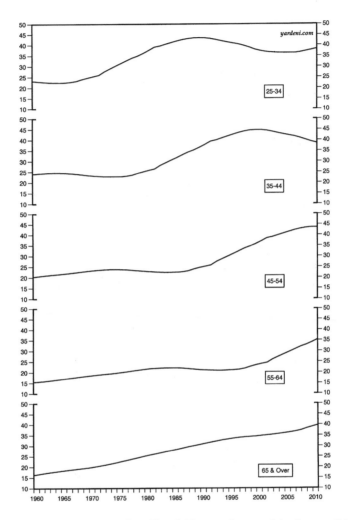

Figure E.11 U.S. population by age (in millions). (*Source:* Bureau of the Census, U.S. Department of Commerce, January 2000.)

Income and Productivity

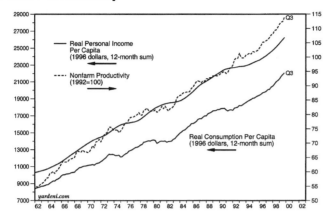

Figure E.12 Real person income and consumption per capita versus productivity. Per capita real personal income and consumption are at record highs. Productivity is the main determinant of these two measures of the standard of living and it is also at a record high.

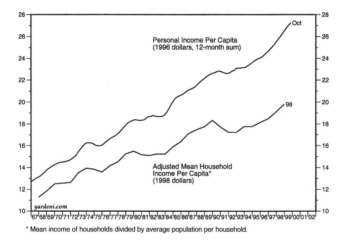

* Mean income of households divided by average population per household.

Figure E.13 Income per capita (in thousands of dollars). Both median and mean family household income data show virtually no growth over the past 25 years. Both measures need to be divided by the average size of the household unit, which has been falling. Adjusted this way, mean household income per capita has been growing in line with personal income per capita!

Income Growth

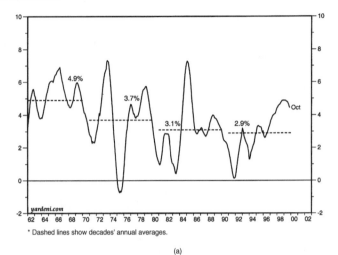

(a)

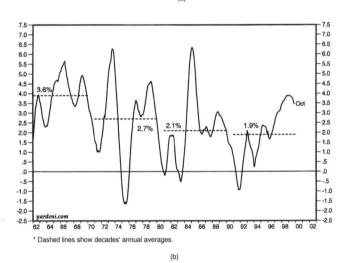

(b)

Figure E.14 (a) Real personal income (yearly percent change in 12-month average). (b) Real per capita personal income (yearly percent change in 12-month average). Real personal income growth was on a downtrend over the past three decades as growth in productivity stagnated. This long-term trend started to be reversed in the 1990s as productivity staged a secular rebound.

Consumption Growth

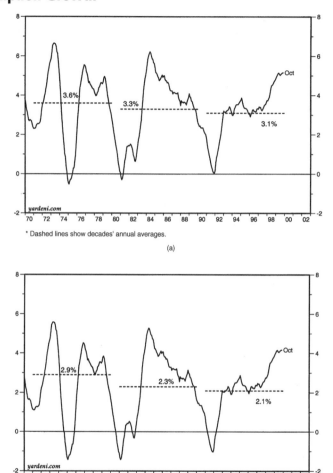

(a)

(b)

Figure E.15 (a) Real consumption expenditures (yearly percent change in 12-month average). (b) Real consumption per capita (yearly percent change in 12-month average). Consumer spending growth has been on a slight downtrend since the 1970s, but is likely to grow faster in the decade ahead.

Consumption

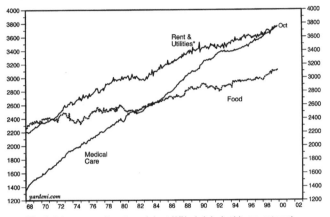

* Rent includes owner- and tenant-occupied rent. Utilities include electricity, gas, water and other sanitary services, fuel oil and coal.

(a)

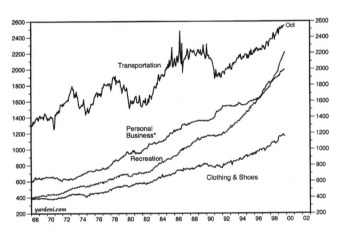

* Includes brokerage charges and investment counseling, bank service charges, services furnished without payment by financial intermediaries except life insurance carriers and private noninsured pension plans, expense of handling life insurance legal services, funeral and burial expenses and other.

(b)

Figure E.16 (a) Real consumption per capita (1996 dollars, saar). (b) Real consumption per capita (1996 dollars, saar). On a per capita basis, the fastest growing category of real consumption is recreation.

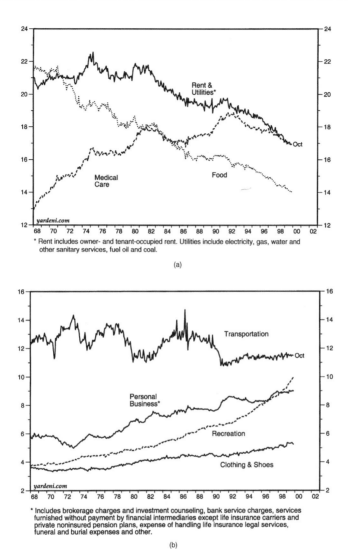

* Rent includes owner- and tenant-occupied rent. Utilities include electricity, gas, water and
other sanitary services, fuel oil and coal.

(a)

* Includes brokerage charges and investment counseling, bank service charges, services
furnished without payment by financial intermediaries except life insurance carriers and
private noninsured pension plans, expense of handling life insurance legal services,
funeral and burial expenses and other.

(b)

Figure E.17 (a) Real consumption by category (as a percent of total real consumption). (b) Real consumption by category (as a percent of total real consumption). In the decade ahead, consumers are likely to spend relatively more on medical care and continue to spend more on recreation.

Income Shares

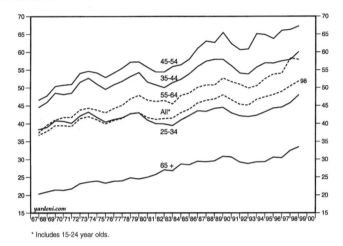

Figure E.18 Mean income of households by age of head (in thousands of 1998 dollars). Older workers tend to earn more than younger ones, presumably because they are more experienced, productive, and committed to their jobs. In 2005, the Baby Boomers will be 41–59 years old, so there will be more households earning more money than ever before. (*Source:* Bureau of the Census, Income Statistics Branch, unpublished data.)

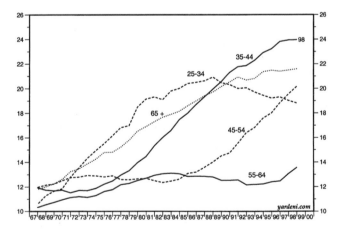

Figure E.19 Number of households by age of head (in millions). By 2010, the Baby Boomers will be 46–64 years old and have the biggest share of income. (*Source:* Bureau of the Census, Income Statistics Branch, unpublished data.)

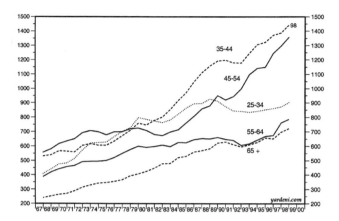

Figure E.20 Aggregate money income by age of household (mean income times number of households) (in billions of 1998 dollars). In the decade ahead, the biggest gain in total group income will be for 55–64-year-olds. (*Source:* Bureau of the Census, Income Statistics Branch, unpublished data.)

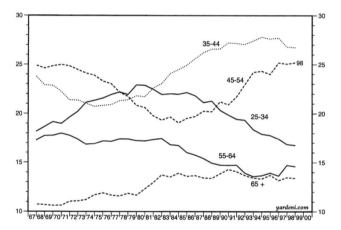

Figure E.21 Aggregate money income by age of household (mean income in 1998 dollars times number of households) (as a percent of total money income). By 2005, 35–54-year-olds will probably account for 65% of personal income, up from over 52% currently. (*Source:* Bureau of the Census, Income Statistics Branch, unpublished data.)

Personal Savings

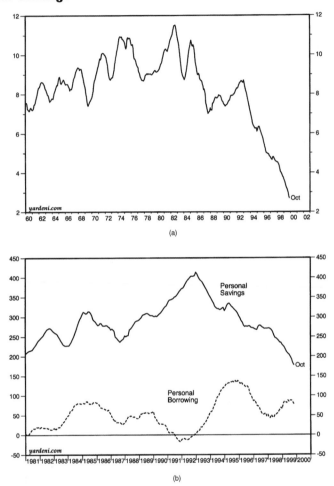

(a)

(b)

Figure E.22 (a) Personal savings rate (12-month average, percent). (b) Personal savings versus consumer borrowing (12-month sum, billion dollars). As the Baby Boomers have aged, the savings rate has plunged. They poured money into stocks. Soaring stock prices boosted their net worth and reduced their need to save.

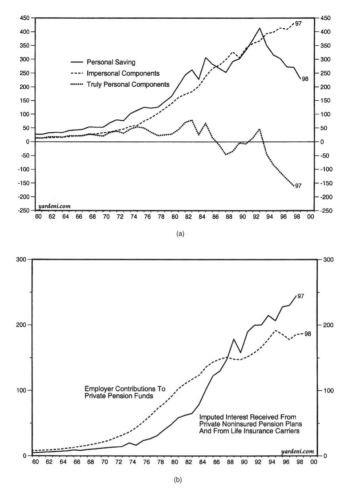

(a)

(b)

Figure E.23 (a) Personal savings and its components (billion dollars). (b) "Impersonal" components of personal income (billion dollars). Personal savings include two big "impersonal" sources of income that are not taxed and can only be saved. They exceed $400 billion and currently exceed total personal savings. Indeed, excluding these two, "truly" personal savings has been mostly negative since 1987.

Labor Force

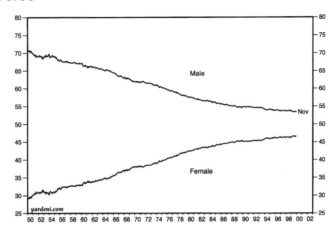

Figure E.24 Civilian labor force (as a percent of total labor force). Males now account for less than 55% of the labor force, down from 70% in the early 1950s. Females account for 46%, up from 29% in the early 1950s.

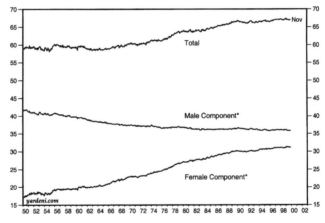

* Labor force divided by noninstitutional working-age population. Male and female components are persons that are 16 years and older.

Figure E.25 Civilian labor force participation rate (labor force divided by noninstitutional working-age population) (percent). The labor force participation rate is hovering around 67%. Since the 1950s, the decline in the male components of the labor force participation rate has been more than offset by females.

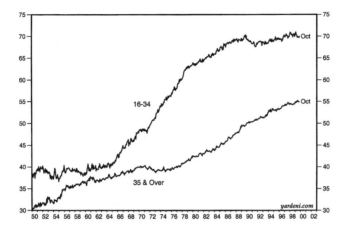

Figure E.26 Female civilian labor force participation rate by age (female labor force divided by female noninstitutional working-age population) (percent). A record 55% of all women who are age 35 or older are in the labor force. The labor force participation rate of younger women has stabilized around 70% in recent years.

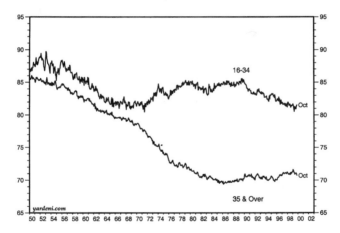

Figure E.27 Male civilian labor force participation rate by age (male labor force divided by male noninstitutional working-age population) (percent). The percent of older males participating in the labor force is hovering around 70%, while the younger ones' participation rate is heading down toward 80%.

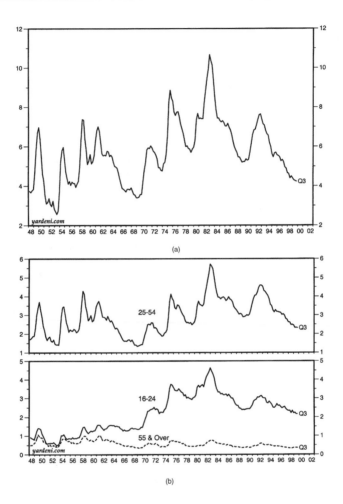

Figure E.28 (a) Civilian unemployment rate (percent). (b) Total unemployed by age (as a percent of total labor force). The unemployment rate shifted upward during the 1970s and early 1980s relative to inflation because there was a huge influx of young workers who tend to have higher unemployment rates than older ones. Since the early 1980s, the unemployment rate has been shifting back down. It is currently at a 29-year low.

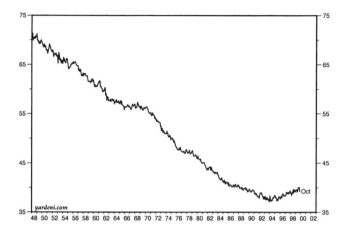

Figure E.29 Labor force: men age 55 and over (as a percent of male civilian noninstitutional population age 55 and older). There was a dramatic long-term downward trend in the labor force participation rate of older men since the late 1940s until flattening out in the mid-1990s. Now there is a slight uptrend.

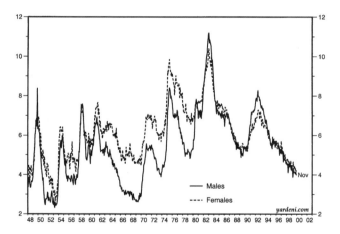

Figure E.30 Unemployment rates by sex (percent). During the 1950s, 1960s, and 1970s, the female unemployment rate usually exceeded the male rate. Since the 1980s, the two rates have been nearly identical. Both are down to about 4.0%.

Population: Numbers

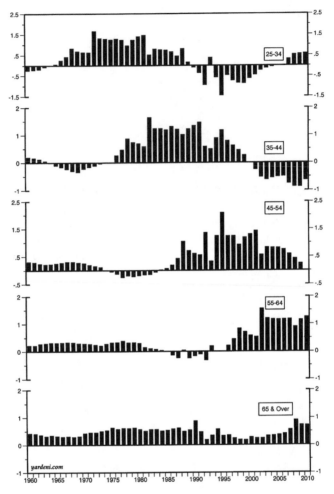

Figure E.31 U.S. population by age (annual change, millions). (*Source:* Bureau of the Census, U.S. Department of Commerce, January 2000.)

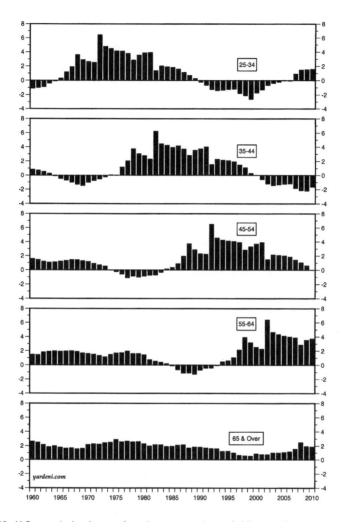

Figure E.32 U.S. population by age (yearly percent change). (*Source:* Bureau of the Census, U.S. Department of Commerce, January 2000.)

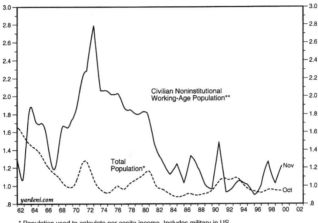

* Population used to calculate per capita income. Includes military in US.

Figure E.33 Population (yearly percent change in 12-month average). The working-age population grew much faster than the total population during the late 1960s and 1970s as the Baby Boomers flooded into the labor markets. This growth rate has been much closer to that of the population through most of the 1990s.

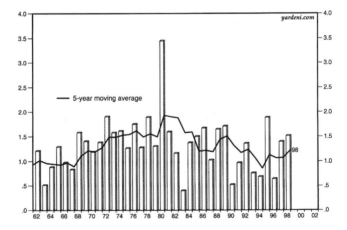

Figure E.34 Household formation (millions). Household formation is likely to rise in the decades ahead as the children of Baby Boomers become independent adults. (*Source:* Bureau of the Census, U.S. Department of Commerce, Series P-20, January 2000.)

Population: Characteristics

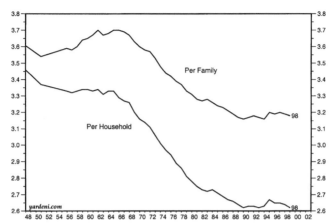

Figure E.35 Average population. Since the mid-1960s, there has been a significant downtrend in the average number of people in both family and household living units. However, both have flattened in recent years. Households are smaller than families, on average, because there are many single-person households: People are marrying later, getting divorced, and living longer. (*Source:* Current Population Reports, January 2000.)

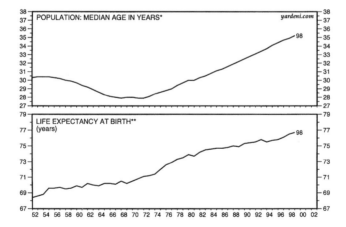

Figure E.36 (a) Population median age in years.* (b) Life expectancy at birth.** The median age of the population is 35, up from 28 twenty years ago, and it will continue to rise over the rest of the decade because the Baby Boomers are aging and senior citizens are living longer. Life expectancy is more than 76 years, compared to about 70 years during the late 1960s. *(*Source:* Bureau of the Census, Population Division, U.S. Department of Commerce, release PPL-57, United States Population Estimates by Age, Sex, Race, and Hispanic Origin, January 2000.) **(*Source:* U.S. National Center for Health Statistics, Vital Health Statistics of the United States, January 2000.)

Consumer

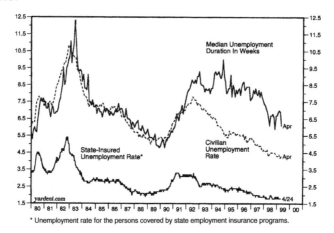

* Unemployment rate for the persons covered by state employment insurance programs.

Figure E.37 Unemployment rates. The rate of insured unemployment is at a record low of 1.8%. Civilian jobless rate edged up in April, but remains near a 29-year low!

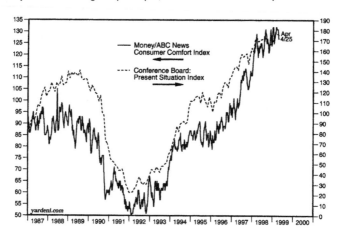

Figure E.38 Consumer confidence. Consumer spending grew at its fastest pace this decade during Q1. All three confidence measures remain at high levels, predicting that the consumer spending spree is likely to continue.

Productivity

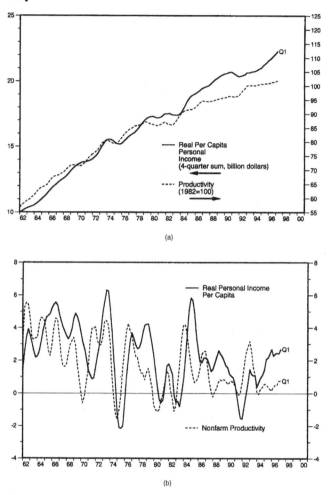

(a)

(b)

Figure E.39 (a) Real per capita personal income versus productivity (four-quarter average). (b) Productivity and real personal income per capita (yearly percent change in four-quarter moving average). Historically, real per capita personal income correlated well with productivity. Recently, the sales data suggest that productivity is growing much faster than official statistics show.

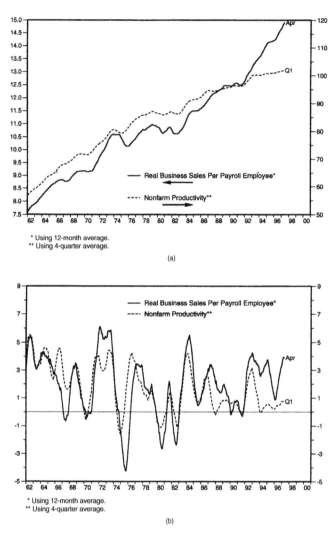

Figure E.40 (a) Productivity. (b) Productivity (yearly percent change). Historically, real business sales per employee correlated well with productivity. Recently, the sales data suggest that productivity is growing much faster than official statistics show.

Appendix F:
Domestic Telecommunications Production and Distribution

High-Tech Orders

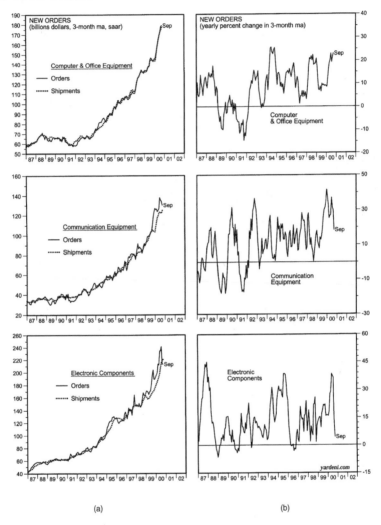

(a)

(b)

Figure F.1 New high-tech orders: (a) in billions of dollars, 3-month m.a., saar; (b) yearly percentage change in 3-month m.a.

High-Tech Production

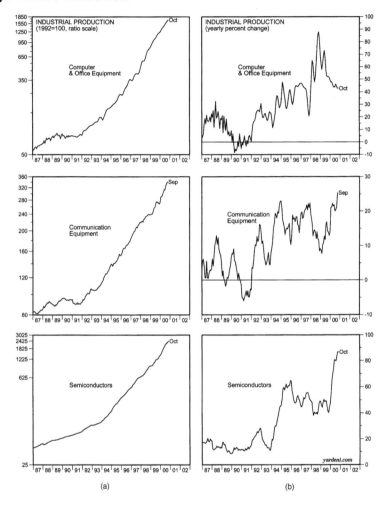

Figure F.2 High-tech industrial production: (a) 1992 = 100, ratio scale; (b) yearly percent change.

High-Tech GDP

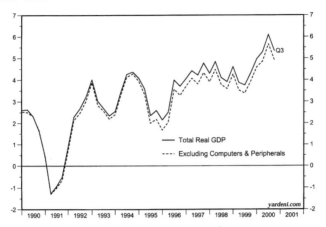

Figure F.3 Real GDP (yearly percent change). Over the past several years computer spending and investment have added half a percentage point on annual GDP growth.

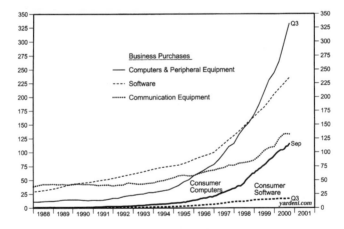

Figure F.4 Purchases of high-tech equipment and software (in billions of 1996 dollars, saar). Except for telecommunications, high-tech spending remains very strong.

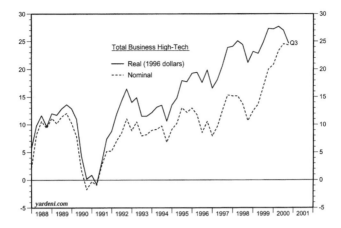

Figure F.5 Business purchases of high-tech equipment and software (information processing equipment and software spending, which includes office, computing, and accounting machinery, communications equipment, photographic equipment, and scientific and engineering instruments) (yearly percent change). During the third quarter, the growth rate of business high-tech spending was 24% in nominal terms, and 25% in real terms.

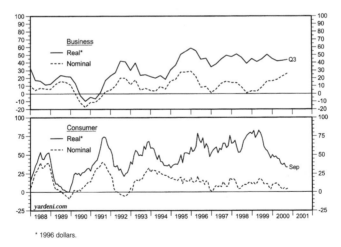

* 1996 dollars.

Figure F.6 Purchases of high-tech equipment: computers (yearly percent change). Business purchases of computers were very strong through Q3. Growth in computer sales to consumers has slowed sharply since mid-1999.

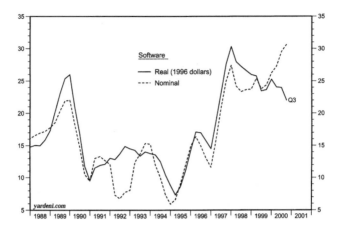

Figure F.7 Business purchases of high-tech equipment and software (yearly percent change). Real business software outlays are up 22%. Nominal spending is up over 30%, a record.

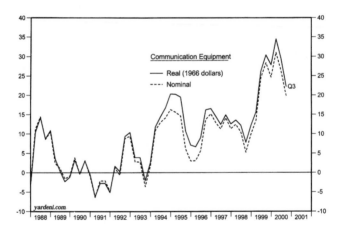

Figure F.8 Business purchases of producers of durable equipment (yearly percent change). Growth rate of business spending on communications equipment is falling.

Communications Equipment

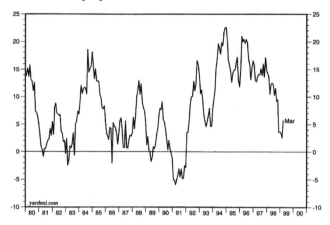

Figure F.9 Communications equipment: industrial production (yearly percent change). (*Source:* Board of Governors of the Federal Reserve System, January 2000.)

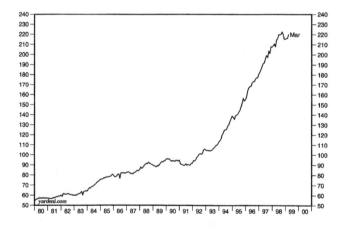

Figure F.10 Communications equipment: industrial production (1992 = 100). (*Source:* Board of Governors of the Federal Reserve System, January 2000.)

Figure F.11 Communications equipment: inventories-to-shipments ratio. (*Source:* Bureau of the Census, U.S. Department of Commerce, January 2000.)

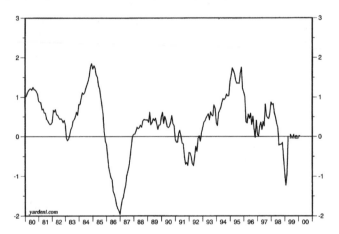

Figure F.12 Communications equipment: orders and shipments (in billions of dollars, 3-month m.a., saar). (*Source:* Bureau of the Census, U.S. Department of Commerce, January 2000.)

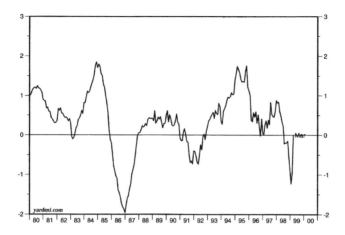

Figure F.13 Communications equipment: inventories (yearly change, in billions). (*Source:* Bureau of the Census, U.S. Department of Commerce, January 2000.)

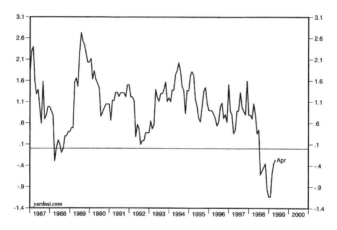

Figure F.14 Communications equipment: producer price index (yearly percent change). (*Source:* Bureau of Labor Statistics, U.S. Department of Labor, January 2000.)

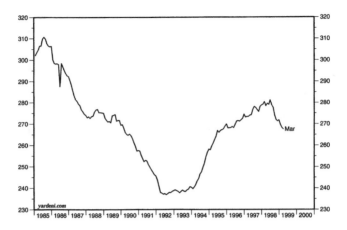

Figure F.15 Communications equipment: employment (in thousands, nsa). (*Source:* Bureau of Labor Statistics, U.S. Department of Labor, June 2000.)

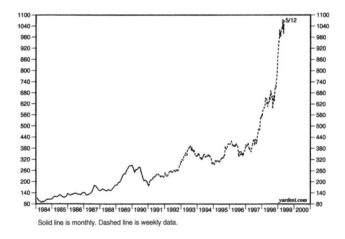

Solid line is monthly. Dashed line is weekly data.

Figure F.16 S&P telecommunications, long-distance (currently includes AT&T, MCI, and Sprint) (1984 = 100).

Computer Prices

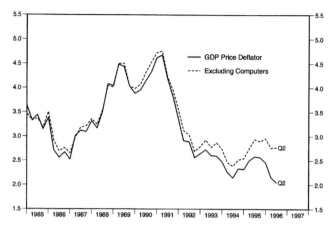

Figure F.17 Implicit price deflators (yearly percent change). Computers are a major source of disinflation for GDP deflator.

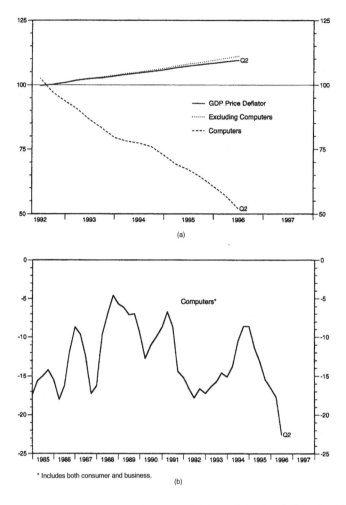

Figure F.18 (a) Implicit price deflators (1992 = 100). (b) Implicit price deflator (yearly percent change). Computer prices are plunging in the GDP accounts. They are down 23% over the past year and 49% over the past four years.

Computers and GDP

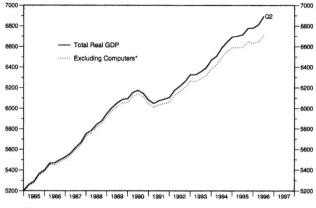

* Consumer and business expenditures on computers.

(a)

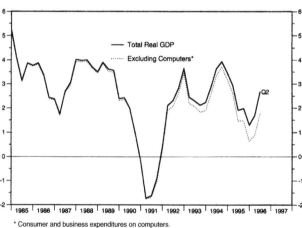

* Consumer and business expenditures on computers.

(b)

Figure F.19 (a) Real GDP (in billions of 1992 dollars, saar). (b) Real GDP (yearly percent change). Computer spending has added a full percentage point to real GDP growth over the past year. Over the past year, real GDP is up 2.7% and 1.8%, including and excluding computers, respectively: It accounts for a whopping one-third of the increase in real GDP.

Computers and Production

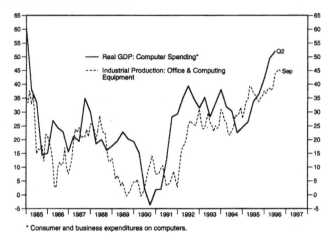

Figure F.20 Real GDP and industrial production (yearly percent change). Spending on computers is a major source of recent GDP boom, soaring 51% over the past year, with business up 42% and consumer up 73%. Consumers now account for 32% of total computer spending, up from 16% in 1990.

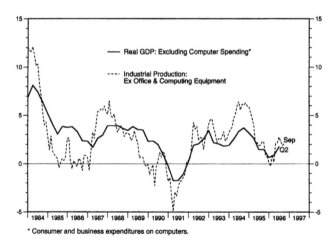

Figure F.21 Real GDP and industrial production (yearly percent changes). Low-tech economy is growing slowly. Industrial production excluding computers is up only 2.1% over the past 12 months.

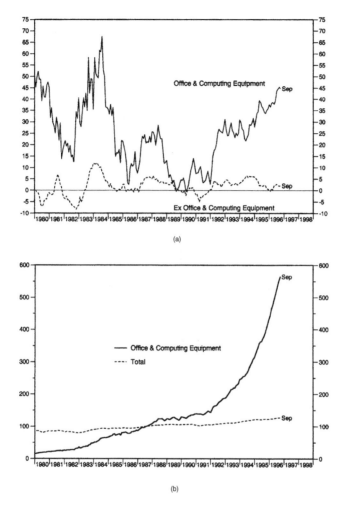

(a)

(b)

Figure F.22 (a) Industrial production (yearly percent change). (b) Industrial production (1987 = 100). High-tech economy is growing much faster than low-tech economy. Over the past 12 months, industrial production is up 3.5% and 2.1%, including and excluding computers, respectively.

Computer Industry

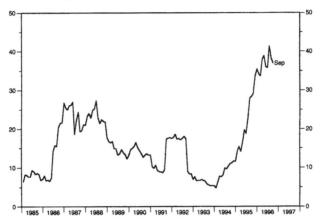

Figure F.23 Technology sector merger and acquisitions volume (12-month sum, in billions of dollars). There has been a big jump in M&A activity in technology. (*Source:* Securities Data Corp.)

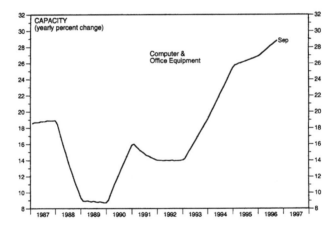

Figure F.24 Capacity (yearly percent change). There has been extraordinary expansion in the computer industry. Capacity is up 29% over the past 12 months and up 243% since the start of 1990.

Appendix G: Telecommunications Industrial Production

G7 Industrial Production

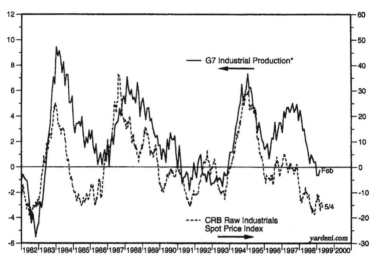

(a)

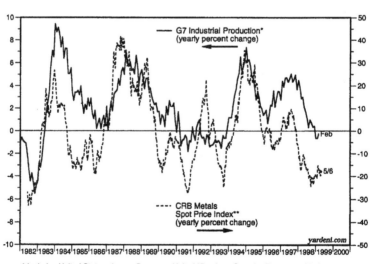

(b)

Figure G.1 G7 production and CRB (yearly percent change). Growth rate in G7 industrial production is still declining, but at a slower pace. Falling industrial commodity prices suggest growth in both G7 and non-G7 industrial output remains suppressed.

U.S. Industrial Production

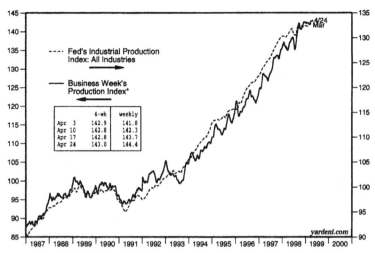

* Four-week average. Includes auto, truck, coal, lumber, electric power, rail-freight traffic, crude oil refining and steel production.

(a)

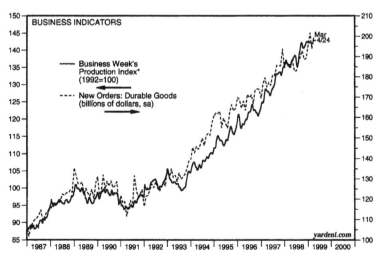

* Four-week average. Includes auto, truck, coal, lumber, electric power, rail-freight traffic, crude oil refining and steel production.

(b)

Figure G.2 (a) U.S. industrial production indexes (1992 = 100). (b) Business indicators. *Business Week's* production index continues to climb to new highs, but pace is slowing. U.S. industrial activity remains subdued as strong domestic demand is partially offset by weak global demand for U.S. goods.

Capacity Utilization: National Figures

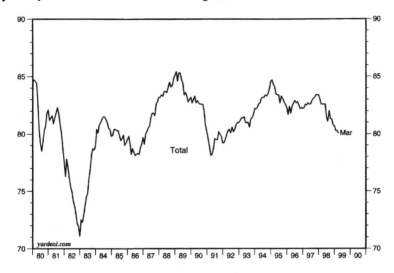

Figure G.3 United States: capacity utilization (percent).

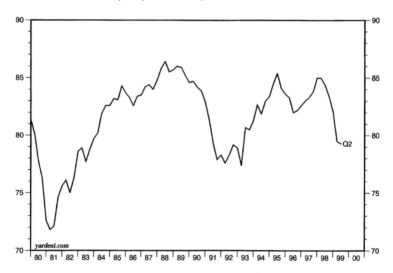

Figure G.4 United Kingdom: capacity utilization (percent).

Figure G.5 Japan: capacity utilization (1995 = 100).

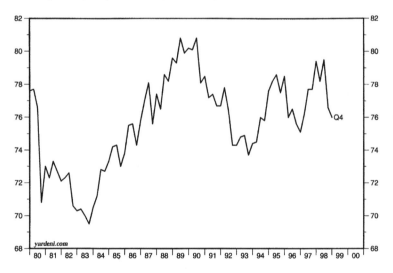

Figure G.6 Italy: capacity utilization (percent).

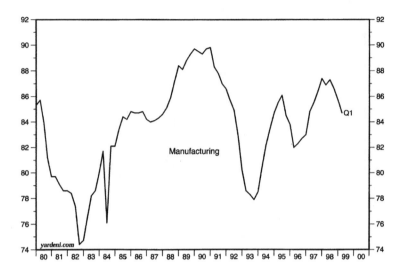

Figure G.7 Western Germany: capacity utilization (percent).

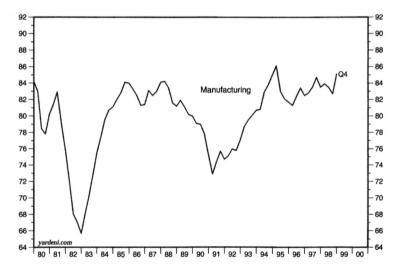

Figure G.8 Canada: capacity utilization (percent).

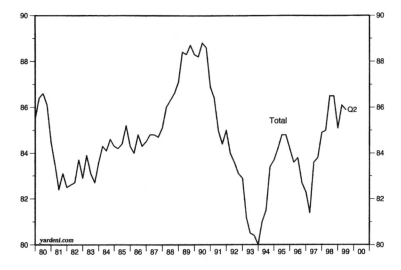

Figure G.9 France: capacity utilization (percent).

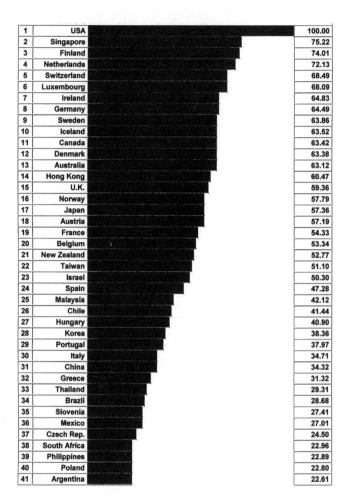

1	USA	100.00
2	Singapore	75.22
3	Finland	74.01
4	Netherlands	72.13
5	Switzerland	68.49
6	Luxembourg	68.09
7	Ireland	64.83
8	Germany	64.49
9	Sweden	63.86
10	Iceland	63.52
11	Canada	63.42
12	Denmark	63.38
13	Australia	63.12
14	Hong Kong	60.47
15	U.K.	59.36
16	Norway	57.79
17	Japan	57.36
18	Austria	57.19
19	France	54.33
20	Belgium	53.34
21	New Zealand	52.77
22	Taiwan	51.10
23	Israel	50.30
24	Spain	47.28
25	Malaysia	42.12
26	Chile	41.44
27	Hungary	40.90
28	Korea	38.36
29	Portugal	37.97
30	Italy	34.71
31	China	34.32
32	Greece	31.32
33	Thailand	29.31
34	Brazil	28.68
35	Slovenia	27.41
36	Mexico	27.01
37	Czech Rep.	24.50
38	South Africa	22.96
39	Philippines	22.89
40	Poland	22.80
41	Argentina	22.61

Figure G.10 The world competitiveness scoreboard (ranking as of April 19, 2000). (*Source:* International Institute for Management Development, April 2000.)

About the Author

James K. Shaw, Ph.D., is an associate professor in the College of Professional Studies at the University of San Francisco. Dr. Shaw is also a member of the adjunct faculties at Saint Mary's College Graduate School of Business and the University of California, Berkeley's Worldwide Programs Division. Dr. Shaw teaches graduate coursework in electronic commerce and information economics, and he has acted as a corporate and government consultant in telecommunications policy for the past decade. Dr. Shaw has also coauthored, with William Wolff, *The Business-Government Relationship* (University of Maryland Press, 1983), and authored *Strategic Management in Telecommunications* (Artech House, 2000).

Index

Winning Telco Customers Using Marketing Databases,
  Rob Mattison

World-Class Telecommunications Service Development,
  Ellen P. Ward

For further information on these and other Artech House titles, including previously considered out-of-print books now available through our In-Print-Forever® (IPF®) program, contact:

Artech House
685 Canton Street
Norwood, MA 02062
Phone: 781-769-9750
Fax: 781-769-6334
e-mail: artech@artechhouse.com

Artech House
46 Gillingham Street
London SW1V 1AH UK
Phone: +44 (0)20 7596-8750
Fax: +44 (0)20 7630-0166
e-mail: artech-uk@artechhouse.com

Find us on the World Wide Web at:
www.artechhouse.com